ANGLES AND ARCS (1.2, 2.2)

θ in Degrees

$$\frac{\theta°}{360°} = \frac{s}{C}$$

C (circumference)

θ in Radians

$$\theta = \frac{s}{R}$$
$$s = R\theta$$

DEGREES AND RADIANS (1.2, 2.2)

$$\frac{\theta°}{180°} = \frac{\theta}{\pi \text{ rad}}$$

$\frac{1}{360}$ circumference

1°

1 radian

SIGNIFICANT DIGITS (1.4, 6.1)

ANGLE TO NEAREST	SIGNIFICANT DIGITS FOR SIDE MEASURE
1°	2
10′ or 0.1°	3
1′ or 0.01°	4
10″ or 0.001°	5

TRIGONOMETRIC FUNCTIONS (2.3, 2.6)

$$\sin x = \frac{b}{R} \qquad \csc x = \frac{R}{b}$$
$$\cos x = \frac{a}{R} \qquad \sec x = \frac{R}{a}$$
$$\tan x = \frac{b}{a} \qquad \cot x = \frac{a}{b}$$

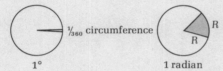

$$R = \sqrt{a^2 + b^2} > 0$$

(x in degrees or radians)

For x any real number and T any trigonometric function,
$$T(x) = T(x \text{ rad})$$
For a unit circle:

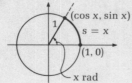

(cos x, sin x)

s = x

(1, 0)

x rad

PYTHAGOREAN THEOREM

$$a^2 + b^2 = c^2$$

SPECIAL TRIANGLES (2.4)

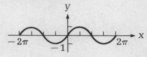

30°–60° triangle   45° triangle

GRAPHING TRIGONOMETRIC FUNCTIONS (3.2–3.5)

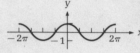

y = sin x   y = cos x

$$y = A \sin(Bx + C) \qquad y = A \cos(Bx + C)$$

$$\text{Amplitude} = |A| \qquad \text{Period} = \frac{2\pi}{B}$$

$$\text{Frequency} = \frac{B}{2\pi}$$

$$\text{Phase shift} = \left|\frac{C}{B}\right| \begin{cases} \text{left if } C/B > 0 \\ \text{right if } C/B < 0 \end{cases}$$

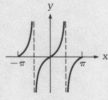

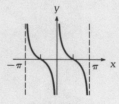

y = tan x   y = cot x

$$y = A \tan(Bx + C) \qquad y = A \cot(Bx + C)$$

$$\text{Period} = \frac{\pi}{B}$$

$$\text{Phase shift} = \left|\frac{C}{B}\right| \begin{cases} \text{left if } C/B > 0 \\ \text{right if } C/B < 0 \end{cases}$$

(Continued inside back cover)

THIRD EDITION

ANALYTIC TRIGONOMETRY
WITH APPLICATIONS

THIRD EDITION

ANALYTIC TRIGONOMETRY
WITH APPLICATIONS

RAYMOND A. BARNETT
Merritt College

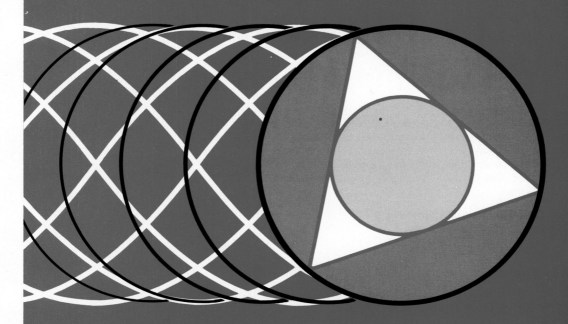

WADSWORTH PUBLISHING COMPANY
Belmont, California
A Division of Wadsworth, Inc.

Mathematics Editor: Richard Jones
Production Editor: Phyllis Niklas
Designer: John Edeen/Janet Bollow
Copy Editor: Betty Berenson
Technical Illustrator: Carl Brown/AYXA Art
Cover: John Edeen

Printed in the United States of America
1 2 3 4 5 6 7 8 9 10—88 87 86 85 84

LIBRARY OF CONGRESS CATALOGING IN PUBLICATION DATA

Barnett, Raymond A.
   Analytic trigonometry with applications.

   Includes index.
   1.   Trigonometry, Plane.   I.   Title.
QA533.B32   1984   516.3'4   83-14647

ISBN 0-534-02858-6

CONTENTS

Test #1   Ch 1,2
#2   3,4,5
#3   6,7,8,9

CHAPTER 4

CHAPTER 5

CHAPTER 6

CHAPTER 7

PREFACE

Extensive use and feedback are indispensable to the evolution of a really effective text for both students and instructors. In its third edition, *Analytic Trigonometry with Applications* has evolved out of such a process.

Prerequisites for the course are $1\frac{1}{2}$–2 years of high school algebra and 1 year of high school geometry or their equivalent.

KEY CHANGES FROM THE SECOND EDITION

1. More applications have been added to the earlier chapters and to Chapter 8.
2. Scientific calculators are used extensively (see "Comments on Hand Calculators" on page xii). However, this is not a book on calculators (or logs). It is a book on trigonometric functions, their properties and use.
3. Chapter 2 on trigonometric functions has been rewritten in part and has been reorganized. Calculator evaluation and exact values for special angles and real numbers are emphasized. Table evaluation is included, but is optional. Trigonometric functions are still first introduced with angle domains. But now there is a whole section on circular functions that are developed in such a way that a student should be able to shift back and forth between the two approaches as the need dictates.
4. In Chapter 3 on graphing, the formal use of translation of axes has been replaced with an informal approach. The graphing of the tangent, cotangent, secant, and cosecant functions has been rewritten, and it proceeds in a more natural way.
5. Chapter 5 on identities has been expanded, both in the development and in the exercise sets. A separate section on product and factor identities is now included.
6. The whole text, including Chapter 8, has been gone over very carefully by several people independently, to ensure the correct choice of physical units and the consistency of use of significant digits.

KEY FEATURES FROM THE SECOND EDITION RETAINED

1. The book is written for students. An open, easy-to-read, informal format is used. Most concepts are illustrated by examples, and almost every exam-

ple is followed by a matching problem to encourage an active rather than passive involvement. (Answers to matching problems are placed at the end of each section just before the exercise set.)

2. To gain reader interest quickly, the text moves directly into trigonometric concepts. Review material from prerequisite courses is found in the appendixes and can be reviewed as needed or treated in a systematic way.

3. Problem sets, except in Chapter 8 and the appendixes, are divided into A (routine, easy mechanics), B (more difficult mechanics), and C (difficult mechanics and theoretical) levels. Answers to most of the odd-numbered exercises and almost all the chapter review exercises are included at the end of the book.

4. Chapter review exercises are included at the end of each chapter except Chapter 8, which is a resource chapter on applications.

5. The content satisfies the requirements for many technical courses, including physics, analytic geometry, and the calculus sequence.

6. A functional use of a second color is employed to increase clarity of exposition and figures.

7. The trigonometric functions are defined first in terms of angle domains, using degree and radian measure side-by-side, and then in terms of real number domains. All of this is done early in the text and is reinforced throughout. By the end of the course, students should be relatively comfortable with all three modes, and should be able to shift from one to the other without difficulty.

8. Periodic properties are emphasized, and many applications are included to illustrate the importance of these properties.

9. Historical remarks are included where appropriate to provide perspective.

10. An instructor's manual is available to instructors without charge. Included in the manual are tests (and easy-to-use answer keys) and answers to even-numbered text problems. The format is $8\frac{1}{2}$ by 11 inches to facilitate reproduction.

ERROR CHECK

Because of the careful checking and proofing by a number of very competent people (acting independently), the author believes this book to be substantially error-free. For any errors remaining, the author would be grateful if they were sent to: Mathematics Editor, Wadsworth Publishing Company, 10 Davis Drive, Belmont, CA 94002.

ACKNOWLEDGMENTS

The publication of a book requires the effort and skills of many people in addition to an author. I wish to thank the many users and reviewers for their many helpful suggestions and comments. In particular, I wish to thank Steven Heath, Southern Utah State College; Stephanie Jorgensen, Oregon State University; David Race, North Texas State University; Gladys Rockind, Oakland Community College—Auburn Hills; Thomas Spradley, American River College; and Philip Stoddard, Hartnell College.

A special thanks is due to Richard St. Andre of Central Michigan University. He provided the author with many new and interesting applications, checked all answers in the text, and proofed all the material in the production process.

COMMENTS ON HAND CALCULATORS

Hand calculators are of two basic types relative to their internal logic (the way they compute): algebraic and Reverse Polish Notation (RPN). Throughout this book we will identify algebraic calculator steps with A and Reverse Polish Notation calculator steps with P. For example, below we show how each type of calculator would compute

$$\frac{(5)(3)(2) - (7)(6)}{2(11)}$$

	Press	Display
A:	$\boxed{5}$ $\boxed{\times}$ $\boxed{3}$ $\boxed{\times}$ $\boxed{2}$ $\boxed{-}$ $\boxed{7}$ $\boxed{\times}$ $\boxed{6}$ $\boxed{=}$ $\boxed{\div}$ $\boxed{2}$ $\boxed{\div}$ $\boxed{11}$ $\boxed{=}$	-0.54545455
P:	$\boxed{5}$ $\boxed{\text{ENTER}}$ $\boxed{3}$ $\boxed{\times}$ $\boxed{2}$ $\boxed{\times}$ $\boxed{7}$ $\boxed{\text{ENTER}}$ $\boxed{6}$ $\boxed{\times}$ $\boxed{-}$ $\boxed{2}$ $\boxed{\div}$ $\boxed{11}$ $\boxed{\div}$	-0.54545455

Some people prefer the algebraic logic and others prefer the Polish. Which is better is still being debated. The answer seems to rest with the type of problems one encounters and individual preferences. The author owns both types and uses the one with Polish logic most frequently. However, many people prefer the algebraic type.

In any case, irrespective of the type of calculator you own, it is essential that you read the user's manual for your calculator. A large variety of calculators are on the market, and each is slightly different from the others.

Therefore, take the time to read the manual. Don't try to read and understand everything the calculator can do. That will only tend to confuse you. Read only those sections pertaining to the operations you are or will be using; then return to the manual as necessary when you encounter new operations.

In many places in this text, calculator steps for new types of calculations will be shown (similar to the steps shown above). These are only aids. Try the calculation first, and then use the aid only if you get stuck.

It is important to remember that a calculator is not a substitute for thinking. It can save you a great deal of time in certain types of problems, but you still must know how and when to use it.

RIGHT TRIANGLE RATIOS 1

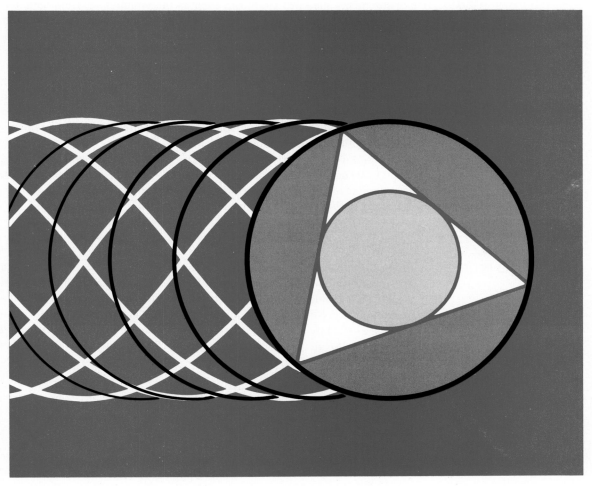

†At the beginning of each chapter, review material that appears in the appendixes is listed close to its first point of use.

1.1   HISTORICAL BEGINNINGS

If you were asked to find your height, you would no doubt take a ruler or tape measure and measure it directly. But if you were asked to find the area of your bedroom floor in square feet, you would not be likely to measure the area directly by taking squares 1 ft long on each side, laying them out over the entire floor, and counting them. Instead, you would probably find the area **indirectly** by using the formula $A = ab$ from plane geometry, where A represents the area of the room and a and b are the lengths of its sides, as indicated in the figure.

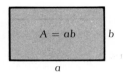

In general, **indirect measurement** is a process of determining unknown measurements from known measurements by a reasoning process.

How do we measure quantities such as the volumes of containers, the distance to the center of the earth, the area of the surface of the earth, and the distances to the sun and the stars? All these measurements are accomplished indirectly by the use of special formulas and deductive reasoning.

The Greeks in Alexandria, during the period 300 BC–200 AD, contributed substantially to the art of indirect measurement by developing formulas for finding areas, volumes, and lengths. Using these formulas, they were able to determine the circumference of the earth with an error of only about 2% and to estimate the distance to the moon. We will examine these measurements as well as others in the sections that follow.

It was during the early part of this Greek period that trigonometry, the study of triangles, was born. Hipparchus (160–127 BC), one of the greatest astronomers of the ancient world, is credited with making the first systematic study of the indirect measurement of triangles.

1.2   ANGLES, DEGREES, AND ARCS

ANGLES AND DEGREES

Central to the study of trigonometry is the concept of angle. A point P on a line divides the line into two parts. Either part together with the end point is called a **half-line** or **ray.** A geometric figure consisting of two rays with a common end point is called an **angle** (Figure 1). The two rays that form an angle are called the **sides** of the angle, and the common end point is called the **vertex.** We may refer to the angle in Figure 1 in any of the ways indicated:

Angle θ	$\angle \theta$
Angle P	$\angle P$
Angle QPR	$\angle QPR$

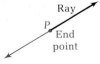

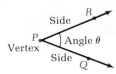

FIGURE 1
Angle θ

Tell why
(Egyptians) ✓

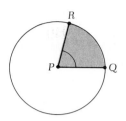

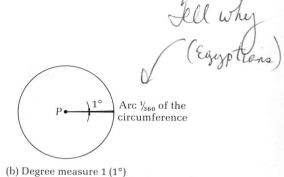

Arc $\frac{1}{360}$ of the circumference

FIGURE 2

(a) Central angle RPQ subtended by arc $\widehat{RQ}$

(b) Degree measure 1 (1°)

The symbol "$\angle$" denotes "angle." In addition to the Greek letter theta (θ), the letters alpha (α), beta (β), and gamma (γ) are often used to denote angles.

How can we compare angles of different sizes? By defining a standard unit of measure. Just as a line segment can be measured in inches, meters, or miles, an angle can be measured in degrees or radians. We will postpone our discussion of radian measure of angles until Section 2.2 and concentrate here on the degree measure of angles.

Given an arc $\widehat{RQ}$ of a circle with center P, the angle RPQ is said to be the **central angle** that is **subtended** by the arc $\widehat{RQ}$ (Figure 2a). A central angle of a circle subtended by an arc equal to $\frac{1}{360}$ of the length of the circumference of the circle is said to have **degree measure 1,** written **1°** (Figure 2b). This definition of degree measure is independent of the size of the circle.

A degree can be divided further using either decimals or minutes and seconds. For the latter we divide a degree into minutes and seconds just as an hour is divided into minutes and seconds; that is, each degree is divided into 60 equal parts called **minutes,** and each minute is divided into 60 equal parts called **seconds.** Symbolically, minutes are represented by ′ and seconds are represented by ″. Thus,

$$5°12'32''$$

is a concise way of writing 5 degrees, 12 minutes, and 32 seconds.

EXAMPLE 1   How many minutes are in 5°45′?

Solution $$5°45' = (5 \cdot 60 + 45)' = 345'$$

Calculator operations (read the user's manual for your calculator):†

A: ⑤ ✕ 60 ➕ 45 ═   P: ⑤ ENTER 60 ✕ 45 ➕

†A: Algebraic operating system (also may be symbolized by AOS); P: Reverse Polish Notation (also may be symbolized by RPN). [*Note:* After some practice with your own scientific calculator, you should be able to carry out the calculator operations directly in terms of the original problem without having to refer to the detailed calculator steps shown.]

PROBLEM 1[†]   How many seconds are in $2°20'13''$?

CONVERSION TO
DECIMAL DEGREES

It is often necessary to convert degree-minute-second forms into decimal forms. In fact, most scientific calculators compute with degrees in decimal form.

EXAMPLE 2   Convert $12°35'13''$ to decimal degrees.

Since

$$35' = \frac{35°}{60} \quad \text{and} \quad 13'' = \frac{13°}{3,600}$$

then

$$12°35'13'' = \left(12 + \frac{35}{60} + \frac{13}{3,600}\right)°$$

$$\approx 12.587° \qquad \text{To three decimal places}$$

A: 12 + ((35 ÷ 60)) + ((13 ÷ 3600)) =

P: 12 ENTER 35 ENTER 60 ÷ + 13 ENTER 3600 ÷ +

CONVERSION ACCURACY

If an angle is measured to the nearest second, then the converted decimal form should not go beyond three decimal places.

PROBLEM 2   Convert $73°13'43''$ to decimal degrees.

ANGLES AND ARCS

It follows from the definition of degree that a central angle subtended by an arc ¼ the circumference of a circle, called a **right angle**, has degree measure 90; ½ the circumference of a circle, called a **straight angle**, has degree measure 180; and the whole circumference of a circle has degree measure 360. In general, for a circle with circumference C units, the degree measure θ of an angle subtended by an arc of s units is given by

$$\theta° = \frac{s}{C} 360° \qquad \theta \text{ in decimal degrees; } s \text{ and } C \text{ in same units}$$

This is equivalent to the proportion indicated in Figure 3.

[†]Answers to matched problems are located at the end of each section just before the exercise.

FIGURE 3

$$\frac{\theta°}{360°} = \frac{s}{C}$$

θ in decimal degrees; s and C in same units

If we know any two of the three quantities s, C, or θ, we can find the third by using simple algebra. For example, in a circle with circumference 25 in., the degree measure of a central angle θ subtended by an arc of length 5 in. is given by

$$\theta° = \frac{5}{25}360° = 72°$$

A: 5 ÷ 25 × 360 =

P: 5 ENTER 25 ÷ 360 ×

(It should be noted that "an angle of 6°" means "an angle of degree measure 6," and "$\theta = 72°$" means "the degree measure of θ is 72.")

APPROXIMATION OF
EARTH'S
CIRCUMFERENCE

The early Greeks were aware of the relationship shown in Figure 3, and Eratosthenes (240 BC) used it in his famous calculation of the circumference of the earth. He reasoned as follows: It was well-known that at Syene (now Aswan), during the summer solstice, the noon sun was reflected on the water in a deep well (this meant the sun shone straight down the well and must be directly overhead). Eratosthenes reasoned that if the sun rays entering the well were continued down into the earth, they would pass through its center (see Figure 4). On the same day at the same time, 5,000 stadia (approx. 500 mi) due north, in Alexandria, sun rays crossed a vertical pole at an angle of 7.5° as indicated in Figure 4. Since sun rays are very nearly parallel when they reach the earth, Eratosthenes concluded that $\angle ACS$ was also 7.5° (Why?).[†]

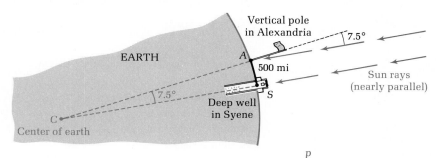

Vertical pole
in Alexandria

7.5°

EARTH

A

500 mi

Sun rays
(nearly parallel)

7.5°

S

Deep well
in Syene

C

FIGURE 4   Center of earth

[†]If line p crosses parallel lines m and n, then angles α and β have the same measure.

Even though Eratosthenes' reasoning was profound, his final calculation of the circumference of the earth requires only elementary algebra:

$$\frac{s}{C} = \frac{\theta^\circ}{360^\circ}$$

$$\frac{500 \text{ mi}}{C} \approx \frac{7.5^\circ}{360^\circ}$$

A: [360] [÷] [7.5] [×] [500] [=]

$$C \approx \frac{360}{7.5}(500 \text{ mi}) = 24{,}000 \text{ mi}$$

P: [360] [ENTER] [7.5] [÷] [500] [×]

The value calculated today is 24,875 mi.

The diameter of the earth (D) and the radius (R) can be found from this value using the formulas $C = 2\pi R$ and $D = 2R$ from plane geometry.[†]

EXAMPLE 3   How large an arc is intercepted by a central angle of 6.23° on a circle with radius 10 cm? (Use $\pi \approx 3.14$.)

Solution   Since

$$\frac{s}{C} = \frac{\theta^\circ}{360^\circ} \qquad \text{and} \qquad C = 2\pi R$$

then

$$\frac{s}{2\pi R} = \frac{\theta^\circ}{360^\circ} \qquad \text{Replace } C \text{ with } 2\pi R$$

$$\frac{s}{2(3.14)(10 \text{ cm})} \approx \frac{6.23^\circ}{360^\circ}$$

$$s \approx \frac{2(3.14)(10 \text{ cm})(6.23)}{360} \approx 1.09 \text{ cm}$$

A: [2] [×] [3.14] [×] [10] [×] [6.23] [÷] [360] [=]

P: [2] [ENTER] [3.14] [×] [10] [×] [6.23] [×] [360] [÷]

[†]The constant π has a long and interesting history; a few important dates are listed below:

1650 BC	Rhind Papyrus	$\pi \approx {}^{256}\!/_{81} = 3.16049 \ldots$
240 BC	Archimedes	$3^{10}\!/_{71} < \pi < 3\frac{1}{7}$ $(3.1408 \ldots < \pi < 3.1428 \ldots)$
470 AD	Tsu Ch'ung-chih	$\pi \approx {}^{355}\!/_{113} = 3.1415929 \ldots$
1674 AD	Leibniz	$\pi = 4(1 - \frac{1}{3} + \frac{1}{5} - \frac{1}{7} + \frac{1}{9} - \frac{1}{11} + \cdots)$
		$\approx 3.1415926535897932384626$
		(This and other series can be used to compute π to any decimal accuracy desired.)
1761 AD	Johann Lambert	Showed π to be irrational

PROBLEM 3   How large an arc is intercepted by a central angle of 50.73° on a circle with radius 5 m? (Use $\pi \approx 3.14$.)

APPROXIMATION
OF SUN'S DIAMETER

EXAMPLE 4   If the distance from the earth to the sun is about 93,000,000 mi, find the approximate diameter of the sun if it subtends an angle of $\frac{1}{2}$° on the surface of the earth. (Use $\pi \approx 3.14$.)

Mention →

Solution   For small central angles in circles with very large radii, the **intercepted arc** (arc opposite the central angle) and its **chord** (the straight line joining the end points of the arc) are approximately the same length. We thus use the intercepted arc to approximate its chord in many practical problems, particularly when the length of the intercepted arc is easier to compute. We apply these ideas to finding the diameter of the sun as follows (see Figure 5):

$$\frac{s}{2\pi R} = \frac{\theta°}{360°}$$

$$s = \frac{2\pi R\theta}{360} \approx \frac{2(3.14)(9.3 \times 10^7 \text{ mi})(0.5)}{3.6 \times 10^2} \approx 810{,}000 \text{ mi}$$

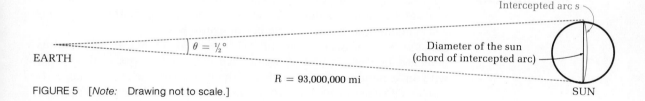

Intercepted arc s

EARTH $\theta = \frac{1}{2}$°   Diameter of the sun
                                                (chord of intercepted arc)

$R = 93{,}000{,}000$ mi   SUN

FIGURE 5   [*Note:* Drawing not to scale.]

PROBLEM 4   If the moon subtends an angle of about $\frac{1}{2}$° on the surface of the earth when it is 239,000 mi from the earth, estimate its diameter. (Use $\pi \approx 3.14$.)

ANSWERS TO   **1.** 8,413″   **2.** 73.229°   **3.** 4.42 m   **4.** 2,080 mi
MATCHED PROBLEMS

EXERCISE 1.2   *Indicate the number of degrees in a central angle subtended by the given arc on a circle.*

A   **1.** $\frac{1}{2}$ a circumference   **2.** $\frac{1}{4}$ a circumference
        3. $\frac{1}{8}$ a circumference   **4.** $\frac{1}{3}$ a circumference
        5. $\frac{2}{3}$ a circumference   **6.** $\frac{1}{12}$ a circumference

Convert to the indicated quantities using a calculator where desired.

7. $10°33' = ?'$ 8. $40°27' = ?'$ 9. $1°10'12'' = ?''$

10. $3°2'25'' = ?''$ 11. $72' = ?°?'$ 12. $84'' = ?'?''$

B   *Use a calculator to change to decimal degrees.*

13. $43°21'4''$ 14. $61°52'11''$ 15. $2°12'47''$

16. $23°5'21''$ 17. $103°17'41''$ 18. $228°40'51''$

Find s, C, θ, or R as indicated. (Use $\pi \approx 3.14$. See Figure 3.)

19. $C = 1,000$ cm, $\theta = 36°$, $s = ?$

20. $s = 12$ m, $C = 108$ m, $\theta = ?$

21. $s = 25$ km, $\theta = 20°$, $C = ?$

22. $C = 720$ mi, $\theta = 72°$, $s = ?$

C   23. $R = 5.4 \times 10^6$ mi, $\theta = 2.6°$, $s = ?$

24. $s = 3.8 \times 10^4$ cm, $\theta = 45.3°$, $R = ?$

25. $\theta = 12°31'4''$, $s = 50.2$ cm, $C = ?$

26. $\theta = 24°16'34''$, $s = 14.23$ m, $C = ?$

APPLICATIONS[†]   27. *Diameter of a distant object*   A round object 1,000 m away subtends an angle of 2°. Approximately what is its diameter (in meters)? (Use $\pi \approx 3.14$ and compute answer to three significant figures.)

*28. *Satellite telescopes*   In the Orbiting Astronomical Observatory launched in 1968, ground personnel can direct telescopes in the vehicle with an accuracy of 1' of arc. They claim this corresponds to hitting a 25¢ coin at a distance of 100 yd (see the figure). Show that this claim is approximately correct. (Diameter of quarter $\approx$ 0.94 in.)

25¢

1'

100 yd

29. *Sun diameter*   Approximate the diameter of the sun by using the average sun–earth distance of 92,956,000 mi (measured indirectly in 1958 by radar reflection off Venus) and the fact that the diameter of the sun subtends an angle of 32' on the surface of the earth. (Use $\pi \approx 3.142$ and compute answer to three significant figures.)

[†]The most difficult problems are double-starred (**); moderately difficult problems are single-starred (*); easier problems are not marked.

**30. *Moon distance and diameter* Hipparchus calculated the distance from the earth to the moon to be about 33.5 earth diameters. Using Eratosthenes' calculation for the circumference of the earth (24,000 mi), calculate the distance to the moon in miles. (Use $\pi \approx 3.14$.) Estimate the diameter of the moon if it subtends an angle of $\frac{1}{2}°$ on the surface of the earth. Calculate answers to two significant figures.

*31. *Nautical miles* A nautical mile is the length of 1′ of arc on the equator (or on any other circle on the surface of the earth having the same center as the equator; see the figure below). If the length of the equator is 24,860 statute mi (a statute mile is an ordinary mile 5,280 ft long), how many statute miles are in 1 nautical mi to three decimal places?

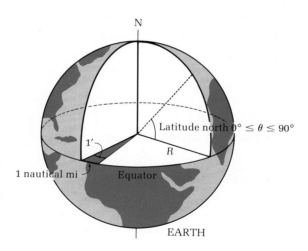

32. *Distance on earth's surface* If Washington, D.C., has a latitude of 38°44′N, what is its approximate distance in nautical miles from the equator (see the figure above)?

1.3 SIMILAR TRIANGLES

In Section 1.2, problems were included that required knowledge of the distances from the earth to the moon and the sun. How can inaccessible distances of this type be determined? Surprisingly, the ancient Greeks made fairly accurate calculations of these distances as well as many others. The basis for their methods is the following elementary theorem of Euclid:

THEOREM 1

EUCLID'S THEOREM

If two triangles are similar, their corresponding sides are proportional.

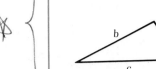
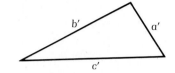

$$\frac{a}{a'} = \frac{b}{b'} = \frac{c}{c'}$$

SIMILAR TRIANGLES

(Recall from plane geometry that two triangles are similar if two angles of one triangle have the same measure as two angles of the other. If the two triangles happen to be right triangles—each contains a 90° angle—then they are similar if an **acute angle** (an angle between 0° and 90°) in one has the same measure as the acute angle in the other—Why?)

Let us now solve a couple of elementary problems using Euclid's theorem.

EXAMPLE 5   A tree casts a shadow of 32 ft at the same time a vertical yardstick casts a shadow of 2 ft. How tall is the tree?

Solution   Since the two triangles in Figure 6 are similar (Why?), we can write

$$\frac{x}{3 \text{ ft}} = \frac{32 \text{ ft}}{2 \text{ ft}}$$

$$x = \frac{3}{2}(32 \text{ ft}) = 48 \text{ ft}$$

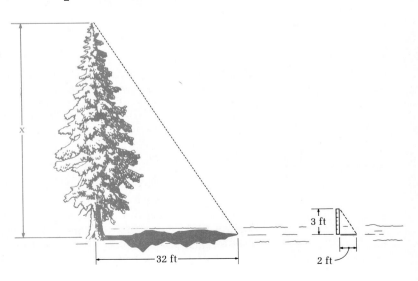

FIGURE 6

PROBLEM 5   Repeat Example 5 for the case when a tree casts a shadow of 31 ft and a 5 ft vertical pole casts a shadow of 0.5 ft.

EXAMPLE 6   Find the length of the proposed air vent indicated in Figure 7.

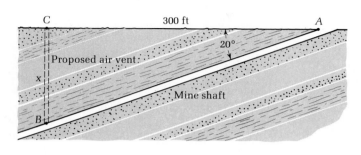

Discuss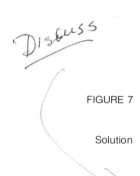

FIGURE 7

Solution   Let us make a careful scale drawing of the mine shaft relative to the proposed air vent as follows: Pick any convenient length, say 2 in., for $A'C'$; copy the 20° angle CAB and the 90° angle ACB using a protractor (see Figure 8). Now measure $B'C'$ (approx. 0.74 in.) and set up a proportion. Thus,

$$\frac{x}{0.74 \text{ in.}} = \frac{300 \text{ ft}}{2 \text{ in.}}$$

$$x \approx \frac{0.74}{2}(300 \text{ ft}) \approx 110 \text{ ft}$$

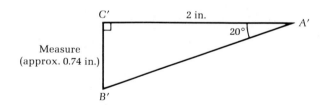

FIGURE 8
Scale drawing

 [*Note:*   The use of scale drawings for finding indirect measurements is included here only to help fix basic ideas. A more efficient method based on these ideas will be developed in Section 1.4.]

PROBLEM 6   Suppose in Example 6 $AC = 500$ ft and $\angle A = 30°$. If in a scale drawing $A'C'$ is chosen to be 3 in. and $B'C'$ is measured as 1.76 in., find BC in Figure 7.

Mention DISTANCE TO   Now let us calculate the distance to the moon using Euclid's elementary
THE MOON   theorem and a refinement of Eratosthenes' estimate of the circumference of the earth. Suppose stations A and B are established 4,150 mi apart at con-

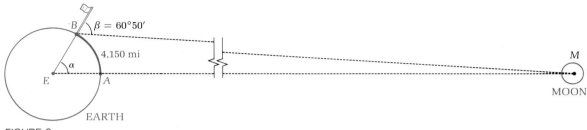

FIGURE 9

venient locations on earth (see Figure 9). At a certain time on a particular night the moon is directly overhead at A, and the observed angle between a vertical pole at B and the moon is $60°50'$. We now have enough information to find the distance from the earth to the moon!

We first calculate the radius of the earth EB using $C = 2\pi R$ with $C = 24{,}900$ mi and $\pi \approx 3.14$:

$$R = \frac{C}{2\pi} = \frac{24{,}900 \text{ mi}}{2(3.14)} \approx 3{,}960 \text{ mi}$$

Next we calculate the angle α:

$$\alpha° = \frac{s}{C}(360°) = \frac{4{,}150 \text{ mi}}{24{,}900 \text{ mi}}(360°) = 60°$$

Since $\angle EBM$ is the supplement of β ($\angle EBM + \beta = 180°$), we have

$$\angle EBM = 180° - \beta = 180° - 60°50' = 119°10'$$

Triangle EBM is determined!

We now carefully draw a similar triangle $E'B'M'$ to scale, choosing a convenient value for $E'B'$ (see Figure 10). We then measure $E'M'$ (approx. 30 in.) and solve for EM using an appropriate proportion:

$$\frac{EM}{E'M'} = \frac{EB}{E'B'}$$

$$EM = \frac{E'M'}{E'B'}(EB) \approx \frac{30 \text{ in.}}{0.5 \text{ in.}}(3{,}960 \text{ mi}) \approx 240{,}000 \text{ mi}$$

FIGURE 10
Scale drawing

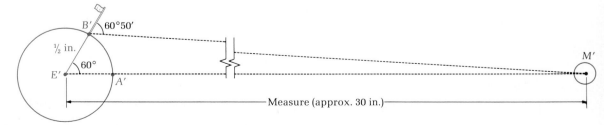

ANSWERS TO   **5.** 310 ft   **6.** 290 ft
MATCHED PROBLEMS

EXERCISE 1.3

A   Given two similar triangles as in the figure below, find the unknown
    length indicated.

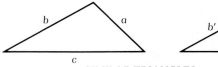

SIMILAR TRIANGLES

1. $a = 5$, $b = 15$, $a' = 2$, $b' = ?$
2. $b = 3$, $c = 21$, $b' = 1$, $c' = ?$
3. $c = ?$, $a = 12$, $c' = 18$, $a' = 2$
4. $a = 51$, $b = ?$, $a' = 24$, $b' = 8$
5. $b = 52{,}000$, $c = 1{,}300$, $b' = 8$, $c' = ?$
6. $a = 630{,}000$, $b = ?$, $a' = 15$, $b' = 0.5$

B   Given the similar triangles in the figure below, find the indicated quantities
    to two significant figures. (A calculator may be useful for some of these
    problems.)

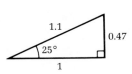
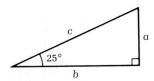

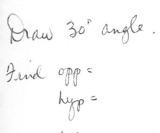
Draw 30° angle.

Find opp =

hyp =

opp / hyp =

7. $a = 10$ m, $b = ?$, $c = ?$
8. $a = 5$ mi, $b = ?$, $c = ?$
9. $b = 50$ in., $a = ?$, $c = ?$
10. $b = 30$ cm, $a = ?$, $c = ?$
11. $c = 2 \times 10^5$ km, $a = ?$, $b = ?$
12. $c = 5 \times 10^{-7}$ m, $a = ?$, $b = ?$
13. $a = 23.4$ km, $b = ?$, $c = ?$
14. $a = 63.19$ cm, $b, = ?$, $c = ?$
15. $b = 2.489 \times 10^9$ yd, $a = ?$, $c = ?$
16. $b = 1.037 \times 10^{13}$ m, $a = ?$, $c = ?$
17. $c = 8.39 \times 10^{-5}$ mm, $a = ?$, $b = ?$
18. $c = 2.86 \times 10^{-8}$ cm, $a = ?$, $b = ?$

C   Find the unknown quantities. (If you have a protractor, actually make a
    scale drawing and complete the problem using your own measurements

and calculations. If you don't have a protractor, use the quantities supplied by the author.)

19.   Suppose in the figure below, $\angle A = 70°$, $\angle C = 90°$, and $a = 100$ ft. If a scale drawing is made of the triangle by choosing a' to be 2 in. and c' is then measured to be 2.13 in., estimate c in the original triangle.

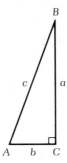

20.   Repeat Problem 19 choosing $a' = 5$ in. and c', a measured quantity, to be 5.28 in.

APPLICATIONS   **21.**   Find the height of the tree in the figure below given that $AC = 12$ ft, $CD = 1.1$ ft, and $DE = 5.5$ ft.

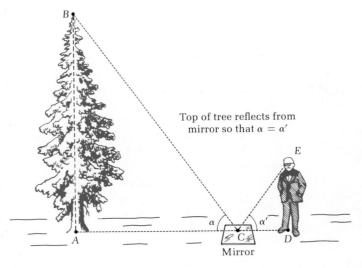

Top of tree reflects from mirror so that $\alpha = \alpha'$

22.   Find the height of the tree in the figure above given that $AC = 25$ ft, $CD = 2$ ft 3 in., and $DE = 5$ ft 9 in.

Problems 23 and 24 are optional for those who have protractors and can make scale drawings.

23. The figure below illustrates a method that is used to determine depths of moon craters from observatories on earth. If sun rays strike the surface of the moon so that $\angle BAC = 15°$ and AC is measured to be 4.0 km, how high is the rim of the crater above its floor?

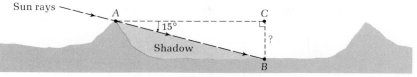

DEPTH OF MOON CRATER

24. The figure below illustrates how the height of a mountain on the moon can be determined from earth. If sun rays strike the surface of the moon so that $\angle BAC = 30°$ and AC is measured to be 4,000 m, how high is the mountain?

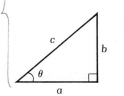

HEIGHT OF MOON MOUNTAIN

1.4   TRIGONOMETRIC RATIOS AND RIGHT TRIANGLES

If in a right triangle we are given the measure of two sides, or the measure of an acute angle and one side, then the triangle is determined. Our problem is then to find the measures of the remaining sides and acute angles. This is called **solving a triangle.** In this section we will show how this can be done without the use of scale drawings. The concepts introduced here will be generalized extensively as we progress through the book.

TRIGONOMETRIC RATIOS

We see in Figure 11 that there are six possible ratios of the sides of a right triangle that can be computed for each angle θ. These ratios are referred to as

$$\frac{b}{c} \quad \frac{c}{b}$$

$$\frac{a}{c} \quad \frac{c}{a}$$

$$\frac{b}{a} \quad \frac{a}{b}$$

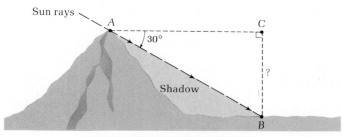

FIGURE 11

trigonometric ratios, and because of their importance, each is given a name: sine (sin), cosine (cos), tangent (tan), cosecant (csc), secant (sec), and cotangent (cot). And each is written in abbreviated form as follows:

DEFINITION 1

Do w/
x, y, r

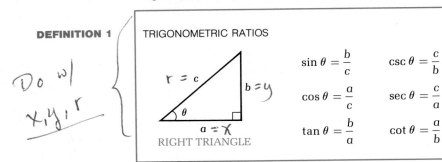

TRIGONOMETRIC RATIOS

RIGHT TRIANGLE

$$\sin \theta = \frac{b}{c} \qquad \csc \theta = \frac{c}{b}$$

$$\cos \theta = \frac{a}{c} \qquad \sec \theta = \frac{c}{a}$$

$$\tan \theta = \frac{b}{a} \qquad \cot \theta = \frac{a}{b}$$

Side b is often referred to as the **side opposite** angle θ, a as the **side adjacent** to angle θ, and c as the **hypotenuse.** Using these designations, the ratios above become

**DEFINITION 1
(alternate form)**

SOHCAHTOA

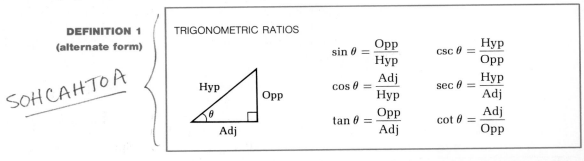

TRIGONOMETRIC RATIOS

$$\sin \theta = \frac{\text{Opp}}{\text{Hyp}} \qquad \csc \theta = \frac{\text{Hyp}}{\text{Opp}}$$

$$\cos \theta = \frac{\text{Adj}}{\text{Hyp}} \qquad \sec \theta = \frac{\text{Hyp}}{\text{Adj}}$$

$$\tan \theta = \frac{\text{Opp}}{\text{Adj}} \qquad \cot \theta = \frac{\text{Adj}}{\text{Opp}}$$

These trigonometric ratios should be memorized because they will be used extensively in work that follows. It is also important to note that the right angle in a right triangle can be oriented in any position and that the names of the angles are arbitrary, but that the hypotenuse is always opposite the right angle.

For all right triangles, no matter how large or small, it follows from Euclid's theorem concerning similar triangles (page 10) that for a given acute angle θ, $\sin \theta$ will always be the same. This is easy to see, since if the measure of the acute angle θ is the same in two right triangles, the triangles are similar. And if they are similar, their corresponding sides are proportional (see Figure 12). If we can compute $\sin \theta$ accurately for one triangle, then we have

eg. sin 30° = $\frac{1}{2}$

FIGURE 12
Similar triangles:
$$\sin \theta = \frac{b}{c} = \frac{b'}{c'}$$

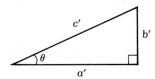

sin θ for all triangles similar to the original! The same reasoning applies to the other five trigonometric ratios.

For the trigonometric ratios to be useful in solving right triangle problems, we must be able to find each for any acute angle. Scientific calculators can approximate (almost instantly) these ratios to eight or ten significant digits. Trigonometric tables can also be used, but are not nearly as efficient.

Evaluation using scientific calculators will be emphasized throughout this book. Table evaluation is included, but is optional.

CALCULATOR
EVALUATION

Most scientific calculators have a choice of three trigonometric modes: degree (decimal), radian, or grad. Our interest now is in **degree mode.** Later we will discuss radian mode in detail. The grad mode (of special interest in certain areas of engineering) will not be considered in this book.

Refer to your user's manual accompanying your calculator to determine how it is to be set in degree mode, and set it that way. *This is an important step and should not be overlooked.* Many errors in using calculators for finding trigonometric ratios can be traced to calculators being set in the wrong mode.

If you look at the function keys on your scientific calculator, you will find three buttons labeled

$$\boxed{\text{sin}} \qquad \boxed{\text{cos}} \qquad \boxed{\text{tan}}$$

These buttons are used to find sine, cosine, and tangent ratios, respectively, for angles in decimal degrees. The calculator can also be used to compute cosecant, secant, and cotangent ratios using the reciprocal[†] relationships below, which follow directly from Definition 1.

RECIPROCAL RELATIONSHIPS

$$\csc \theta \sin \theta = \frac{c}{b} \cdot \frac{b}{c} = 1 \qquad \text{thus} \qquad \csc \theta = \frac{1}{\sin \theta}$$

$$\sec \theta \cos \theta = \frac{c}{a} \cdot \frac{a}{c} = 1 \qquad \text{thus} \qquad \sec \theta = \frac{1}{\cos \theta}$$

$$\cot \theta \tan \theta = \frac{a}{b} \cdot \frac{b}{a} = 1 \qquad \text{thus} \qquad \cot \theta = \frac{1}{\tan \theta}$$

[†]Recall that two numbers a and b are **reciprocals** of each other if $ab = 1$; then we may write $a = 1/b$ and $b = 1/a$.

EXAMPLE 7   Evaluate to four significant digits using a scientific calculator.
(A)   sin 23.72°   (B)   tan 54°37′
(C)   sec 49.31°   (D)   cot 12°52′

Solutions   (A)   Set calculator in degree mode, enter 23.72, and push sin button:
sin 23.72° = 0.4023 $\boxed{23.72}$ $\boxed{\text{sin}}$

(B)   Convert to decimal degrees and proceed as in (A).
tan 54°37′ = tan(54.61666 . . .)° $54°37′ = (54 + \frac{37}{60})° = (54.61666 . . .)°$
      = 1.408 $\boxed{54.61666 \dots}$ $\boxed{\text{tan}}$

(C)   Use the reciprocal relationship sec θ = 1/cos θ.
sec 49.31° = 1.534 $\boxed{49.31}$ $\boxed{\text{cos}}$ $\boxed{1/x}$

(D)   Convert to decimal degrees and proceed as in (C).
cot 12°52′ = cot(12.8666 . . .)°   Convert to decimal degrees
      = 4.378 $\boxed{12.8666 \dots}$ $\boxed{\text{tan}}$ $\boxed{1/x}$

PROBLEM 7   Evaluate to four significant digits using a scientific calculator.
(A)   cos 38.27°   (B)   sin 37°44′
(C)   cot 49.82°   (D)   csc 77°53′

TABLE EVALUATION   Tables I and II in Appendix C include trigonometric ratios for angles from
(OPTIONAL)   0° to 90°. Table I is for angles in degrees and minutes and Table II is for
angles in decimal degrees.

USING TABLES I AND II TO FIND TRIGONOMETRIC RATIOS

Angles between 0° and 45° are listed in the left-hand column, and we
read the tables down from the top for these angles. Angles between 45°
and 90° are listed in the right-hand column, and we read the tables up
from the bottom.

You might wonder why a table from 0° to 45° can be used to find all six
trigonometric ratios corresponding to angles from 0° to 90°. Referring to
Definition 1 (the definition of the six trigonometric ratios), we see that:

Show w/
30° - 60°

COMPLEMENTARY RELATIONSHIPS

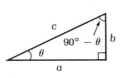

$$\sin \theta = \frac{b}{c} = \cos(90° - \theta)$$

$$\tan \theta = \frac{b}{a} = \cot(90° - \theta)$$

$$\sec \theta = \frac{c}{a} = \csc(90° - \theta)$$

Thus, the sine of θ is the same as the cosine of the complement of θ (which is $90° - \theta$ in the triangle shown), the tangent of $90° - \theta$ is the same as the cotangent of the complement of θ, and so on. (Recall, two positive angles are **complementary** if their sum is $90°$.) We duplicate a small portion of Table I here, and note several values:

	sin	cos	tan	cot	sec	csc	
20°00′	0.3420	0.9397	0.3640	2.747	1.064	2.924	70°00′
10′	0.3448	0.9387	0.3673	2.723	1.065	2.901	69°50′
20′	0.3475	0.9377	0.3706	2.699	1.066	2.878	40′
30′	0.3502	0.9367	0.3739	2.675	1.068	2.855	30′
40′	0.3529	0.9356	0.3772	2.651	1.069	2.833	20′
20°50′	0.3557	0.9346	0.3805	2.628	1.070	2.812	10′
21°00′	0.3584	0.9336	0.3839	2.605	1.071	2.790	69°00′
	cos	sin	cot	tan	csc	sec	

Complementary angles

$\sin 20°00'' = 0.3420 = \cos 70°00'$ $\tan 69°20' = 2.651 = \cot 20°40'$
$\sec 20°50' = 1.070 = \csc 69°10'$

SOLVING TRIANGLES

Solving triangles using a calculator or Tables I and II is best illustrated through examples. We note at the outset that accuracy of the computations involved is governed by the following table (which is also reproduced inside the front cover for easy reference):

ANGLE TO NEAREST	SIGNIFICANT DIGITS FOR SIDE MEASURE
1°	2
10′ or 0.1°	3
1′ or 0.01°	4
10″ or 0.001°	5

When we use the equal sign $(=)$ in the following computations it should be understood that equality holds only to the number of significant digits justified by the above table. The approximation symbol $(\approx)$ is often used when we want to emphasize the approximation.

EXAMPLE 8   Solve the right triangle.

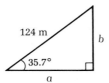

124 m

b

35.7°

a

Solution   *Solve for the complementary angle*

$$90° - \theta = 90° - 35.7° = 54.3°$$

Solve for b
Since $\theta = 35.7°$ and $c = 124$ m, we look for a trigonometric ratio that involves θ and c (the known quantities) and b (an unknown quantity). Referring to Definition 1 (which should be memorized), we see that both sine and cosecant involve all three quantities. We choose sine, and proceed as follows:

$$\sin \theta = \frac{b}{c}$$

$$b = c \sin \theta$$

$$= (124 \text{ m})(\sin 35.7°)$$

$$= 72.4 \text{ m}$$

A: $\boxed{35.7}$ $\boxed{\sin}$ $\boxed{\times}$ $\boxed{124}$ $\boxed{=}$

P: $\boxed{124}$ $\boxed{\text{ENTER}}$ $\boxed{35.7}$ $\boxed{\sin}$ $\boxed{\times}$

Solve for a
Now that we have b, we can use the tangent, cotangent, cosine, or secant (or even the Pythagorean theorem) to find a. We choose the cosine. Thus,

$$\cos \theta = \frac{a}{c}$$

mention why not

$$a = c \cos \theta$$

$$= (124 \text{ m})(\cos 35.7°)$$

$$= 101 \text{ m}$$

PROBLEM 8   Solve the triangle in Example 7 with $\theta = 28.3°$ and $c = 62.4$ cm.

EXAMPLE 9   Solve the triangle for θ and $90° - \theta$ to the nearest 10′ (using a calculator or Table I) and for a to three significant figures.

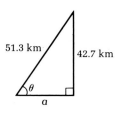

Solution   *Solve for θ*

$$\sin \theta = \frac{b}{c} = \frac{42.7 \text{ km}}{51.3 \text{ km}}$$

$$\sin \theta = 0.832$$

Given the sine of θ, how do we find θ? We can find θ directly as follows (a detailed discussion of what is behind the process will be given in Chapter 5):

Calculator solution

The button $\boxed{\sin^{-1}}$ or the combination $\boxed{\text{inv}}$ $\boxed{\sin}$ takes us from a trigonometric ratio back to the corresponding angle in decimal degrees (if the calculator is in degree mode).

$$\begin{aligned}\theta &= \text{inv sin } 0.8320 \\ &= 56.30° \\ &= 56°20'\end{aligned}$$

$\boxed{0.8320}$ $\boxed{\text{inv}}$ $\boxed{\sin}$
$(0.3)(60) = 18' \approx 20'$
To nearest 10'

Or (for some calculators):

$$\begin{aligned}\theta &= \sin^{-1} 0.8320 \\ &= 56.30° \\ &= 56°20'\end{aligned}$$

$\boxed{0.8320}$ $\boxed{\sin^{-1}}$

Table solution
(optional)

We use Table I, reading up from the bottom to the nearest 10' to obtain

$$\theta = 56°20'$$

Solve for the complementary angle

$$\begin{aligned}90° - \theta &= 90° - 56°20' \\ &= 33°40'\end{aligned}$$

Solve for a

Use cosine, secant, cotangent, or tangent. We will use tangent:

$$\tan \theta = \frac{b}{a}$$

$$\begin{aligned}a &= \frac{b}{\tan \theta} \\ &= \frac{42.7 \text{ km}}{\tan 56°20'} \\ &= 28.4 \text{ km}\end{aligned}$$

A: $\boxed{42.7}$ $\boxed{\div}$ $\boxed{56.333\ldots}$ $\boxed{\tan}$ $\boxed{=}$

P: $\boxed{42.7}$ $\boxed{\text{ENTER}}$ $\boxed{56.333\ldots}$ $\boxed{\tan}$ $\boxed{\div}$

Check

We check by using the Pythagorean theorem.[†]

$$28.4^2 + 42.7^2 \overset{?}{=} 51.3^2$$
$$2{,}630 \overset{\checkmark}{=} 2{,}630$$

Compute both sides to three significant digits

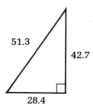

51.3   42.7

28.4

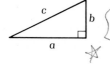

[†]Pythagorean theorem: A triangle is a right triangle if and only if the sum of the squares of the two shorter sides is equal to the square of the longest side:

$$a^2 + b^2 = c^2$$

PROBLEM 9   Repeat Example 9 with $b = 23.2$ km and $c = 30.4$ km.

In the next section we will consider a large variety of applications involving the techniques discussed in this section.

ANSWERS TO
MATCHED PROBLEMS

7. (A) 0.7851   (B) 0.6120   (C) 0.8445   (D) 1.023
8. $90° - \theta = 61.7°$, $b = 29.6$ cm, $a = 54.9$ cm
9. $\theta = 49°45'$, $90° - \theta = 40°15'$, $a = 19.6$ km

EXERCISE 1.4

A   *For the triangle below, identify each of the given ratios without looking at Definition 1. For example,* $\csc \theta = c/b$.

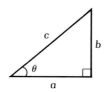

1. $\cos \theta$	2. $\sin \theta$	3. $\tan \theta$
4. $\cot \theta$	5. $\sec \theta$	6. $\csc \theta$

For the triangle above, identify each of the ratios by name without looking at Definition 1. For example, c/a is sec θ.

7. b/c	8. a/c	9. b/a
10. a/b	11. c/b	12. c/a

Find to three significant digits.

13. $\sin 25.6°$	14. $\cos 36.4°$	15. $\tan 35°20'$
16. $\cot 12°40'$	17. $\sec 44.8°$	18. $\csc 18.3°$
19. $\cos 72.9°$	20. $\sin 63.1°$	21. $\cot 54.9°$
22. $\tan 48.3°$	23. $\csc 67°30'$	24. $\sec 51°40'$

B   *Solve the right triangle (labeled as in the figure at the beginning of the exercise) given the information in each problem.*

25. $\theta = 58°40'$, $c = 15.0$ mm	26. $\theta = 62°10'$, $c = 33.0$ cm
27. $\theta = 83.7°$, $b = 3.21$ km	28. $\theta = 32.4°$, $a = 42.3$ m
29. $\theta = 71.5°$, $b = 12.8$ in.	30. $\theta = 44.5°$, $a = 2.30 \times 10^6$ m
31. $a = 22.4$ cm, $90° - \theta = 33°40'$	32. $c = 3.45$ in., $90° - \theta = 17°50'$
33. $b = 63.8$ ft, $c = 134$ ft (angles to nearest 10′)	34. $b = 22.0$ km, $a = 46.2$ km (angles to nearest 10′)

35. $b = 132$ mi, $a = 108$ mi
(angles to nearest 0.1°)

36. $a = 134$ m, $c = 182$ m
(angles to nearest 0.1°)

C   In Problems 37–44 solve the right triangles (labeled as in the figure at the beginning of the exercise):

37. $\theta = 37.46°$, $b = 5.317$ cm

38. $\theta = 29.83°$, $c = 4.032$ m

39. $a = 23.82$ mi, $\theta = 83°12'$

40. $a = 6.482$ m, $\theta = 35°44'$

41. $b = 42.39$ cm, $a = 56.04$ cm
(angles to nearest 1′)

42. $a = 123.4$ ft, $c = 163.8$ ft
(angles to nearest 1′)

43. $b = 35.06$ cm,
$c = 50.37$ cm
(angles to nearest 0.01°)

44. $b = 5.207$ mm,
$a = 8.030$ mm
(angles to nearest 0.01°)

45. Show that $(\sin \alpha)^2 + (\cos \alpha)^2 = 1$, using Definition 1 and the Pythagorean theorem.

1.5   RIGHT TRIANGLE APPLICATIONS

Now that you know how to solve right triangles using a calculator or Tables I and II, we are in a position to consider a variety of interesting and significant applications.

EXAMPLE 10   Solve the mine shaft problem in Example 6 (page 11).

Solution

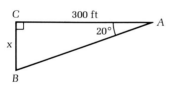

$$\tan \theta = \frac{\text{Opp}}{\text{Adj}}$$

$$\tan 20° = \frac{x}{300 \text{ ft}}$$

$$x = (300 \text{ ft})(\tan 20°) = 110 \text{ ft}$$

To two significant figures

PROBLEM 10   Solve the mine shaft problem in Example 10 if $AC = 500$ ft and $\angle A = 30°$.

Before proceeding with the next two examples, we introduce two new terms: **angle of elevation** and **angle of depression.** An angle measured from the horizontal upward is called an angle of elevation; one measured from the horizontal downward is called an angle of depression (see Figure 13).

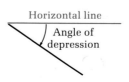

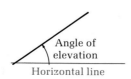

FIGURE 13

EXAMPLE 11   From an aerial photograph, a portion of the Grand Canyon is found to be 8.72 km wide. From one rim of the canyon the angle of depression to the bottom of the canyon at the other rim was measured to be 6.6°. How deep is the canyon? (Assume both rims are the same altitude above sea level.)

Solution   We first sketch a picture and label the known parts (change kilometers to meters).

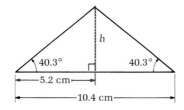

$$\tan 6.6° = \frac{\text{Opp}}{\text{Adj}}$$

$$= \frac{d}{8{,}720 \text{ m}}$$

$$d = (8{,}720 \text{ m})(\tan 6.6°)$$

$$= 1{,}010 \text{ m}$$

PROBLEM 11   The horizontal shadow of a vertical tree is 23.4 m long when the angle of elevation of the sun is 56.3°. How tall is the tree?

EXAMPLE 12   Find the altitude of an isosceles triangle with two equal angles of 40.3° and a base of 10.4 cm.

Solution

$$\tan 40.3° = \frac{h}{5.2 \text{ cm}}$$

$$h = (5.2 \text{ cm})(\tan 40.3°)$$

$$= 4.41 \text{ cm}$$

PROBLEM 12   Find the altitude of an isosceles triangle if the two equal sides are 5.3 m and the angle between them is 52°.

ANSWERS TO MATCHED PROBLEMS   **10.** 290 ft   **11.** 35.1 m   **12.** 4.8 m

_____ **EXERCISE 1.5**   *Use Tables I and II or a calculator to solve the problems in this exercise.*

A   ① An 8.0 m ladder is placed against a building as indicated in the figure at the top of the next page. How high will the top of the ladder reach up the building?

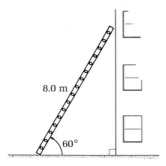

8.0 m

60°

2. In Problem 1, how far is the foot of the ladder from the wall of the building?

3. Use the information in the figure below to find the distance x from the boat to the base of the cliff.

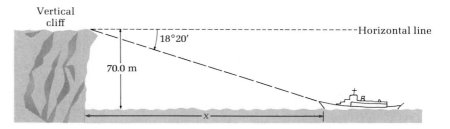

Vertical cliff

18°20′

70.0 m

Horizontal line

x

4. In Problem 3, how far is the boat from the top of the cliff?

5. Find the depth of the moon crater in Problem 23, Exercise 1.3 (page 15).

6. Find the height of the mountain on the moon in Problem 24, Exercise 1.3 (page 15).

7. A glider is flying at an altitude of 8,240 m. The angle of depression from the glider to the control tower at an airport is 15°40′. What is the horizontal distance in kilometers from the glider to a point directly over the tower?

8. The height of a cloud or fog cover over an airport can be measured as indicated in the figure below. Find h in meters if $b = 1.00$ km and $\alpha = 23.4°$.

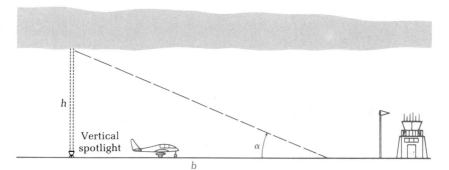

h

Vertical spotlight

α

b

B   **9.** What is the altitude of an equilateral triangle with side 4.0 m?
10. The altitude of an equilateral triangle is 5.0 cm. What is the length of a side? [An **equilateral triangle** has all sides (and all angles) equal.]
11. Find the length of one side of a nine-sided regular polygon inscribed in a circle with radius 8.32 cm (see the figure below).

8.32 cm

12. What is the radius of a circle inscribed in the polygon in Problem 11? (The circle will be tangent to each side of the polygon and the radius will be perpendicular to the tangent line at the point of tangency.)
13. Most airlines approach San Francisco International Airport on a straight 3° glide path starting at the San Mateo Bridge 5.5 mi out from the field. For the purpose of ground noise reduction, California Bear Airlines approaches the field on a 6° glide path at a distance of 5.5 mi, and then switches to a 3° glide path at a distance of 1.5 mi (see the figure below). What is the height x in feet of the California Bear plane when it switches to a 3° glide path? (Use 1 mi = 5,280 ft.)[†]

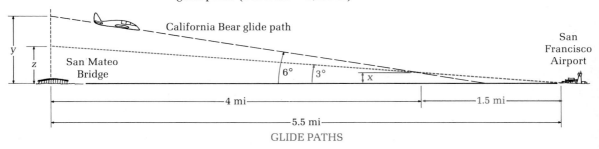

California Bear glide path

San Mateo Bridge

San Francisco Airport

y

z

6°   3°   ↑ x

—4 mi—

—1.5 mi—

—5.5 mi—

GLIDE PATHS

14. Using the information in Problem 13, what is the height difference $(y - z)$ between a California Bear plane and other planes at the San Mateo Bridge? Give the answer in feet.[†]

C   **15.** Use the information in the figure at the top of the next page to find the length x of the island.

†The author is indebted to Dennis Zill of Loyola-Marymount University, Los Angeles, Ca., for providing the material for Problems 13 and 14.

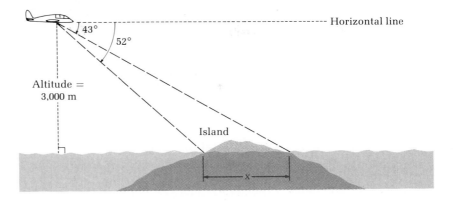

16. Use the information in the figure below to find the height y of the mountain.

$$\tan 40° = \frac{y}{x}$$

$$\tan 25° = \frac{y}{1.0 + x}$$

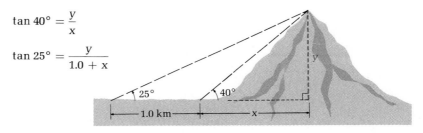

17. Show that

$$h = \frac{d}{\cot \alpha - \cot \beta}$$

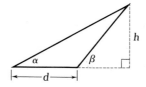

18. Use the results in Problem 17 to find the height of the mountain in Problem 16.

19. In physics one can show that the velocity (v) of a ball rolling down an inclined plane (neglecting air resistance and friction) is given by

$$v = g(\sin \theta)t$$

where g is the gravitational constant and t is time (see the figure). Galileo

GALILEO'S EXPERIMENT

(1564–1642) used this equation in the form

$$g = \frac{v}{(\sin \theta)t}$$

so he could determine g after measuring v experimentally. (There were no timing devices available then that were accurate enough to measure the velocity of a free-falling body. He had to use an inclined plane to slow the motion down, and then he was able to calculate an approximation for g.) Find g if at the end of 2 sec a ball is traveling at 11.1 ft/sec down a plane inclined at 10.0°.

20.   Find R in the figure. The circle is tangent to all three sides of the isosceles triangle. [*Hint:* The radius of a circle and a tangent line are perpendicular at the point of tangency. Also, the altitude of the isosceles triangle will pass through the center of the circle and will divide the original triangle into two congruent triangles.]

EXERCISE 1.6   CHAPTER REVIEW

A   **1.** $2°1'20'' = ?''$

2.   An arc of $\frac{1}{6}$ the circumference of a circle subtends a central angle of how many degrees?

3.   Given two similar triangles as in part A of Exercise 1.3 (page 13), find a if $c = 20,000$, $a' = 2$, and $c' = 5$.

4.   Change $36°23'$ to decimal degrees (to two decimal places).

5.   If an office building casts a shadow of 31 ft at the same time a vertical yardstick casts a shadow of 2.0 in., how tall is the building?

6.   For the triangle shown here identify each ratio:

(A) $\sin \theta$ (B) $\sec \theta$ (C) $\tan \theta$
(D) $\csc \theta$ (E) $\cos \theta$ (F) $\cot \theta$

7.   Solve the right triangle in Problem 6, given $c = 20.2$ cm and $\theta = 35.2°$.

B   **8.**   Find the degree measure of a central angle subtended by an arc of 8 cm in a circle with circumference 20 cm.

9.   If the minute hand of a clock is 2 in. long, how far does the tip of the hand travel in 20 min? (Use $\pi \approx 3.14$.)

10.   For the triangle in Problem 6, identify each of the following ratios relative to angle θ by name:

(A) a/c (B) b/a (C) b/c (D) c/a (E) c/b (F) a/b

11.   Solve the right triangle in Problem 6, given $\theta = 62°20'$ and $a = 4.00 \times 10^{-8}$ m.

12.   Solve the right triangle in Problem 6, given $b = 13.3$ mm and $a = 15.7$ mm. (Find angles to the nearest 0.1°.)

13.   Find angles in Problem 12 to the nearest 10′.

14.   If an equilateral triangle has a side of 10 ft, what is its altitude to two significant figures?

C   **15.**   A curve of a railroad track follows an arc of a circle of radius 1,500 ft. If the arc subtends a central angle of 36°, how far will a train travel on this arc? (Use $\pi \approx 3.14$.)

16.   Solve the triangle in Problem 6, given $90° - \theta = 23°43'$ and $c = 232.6$ km.

17.   Solve the triangle in Problem 6, given $a = 2,421$ m and $c = 4,883$ m. (Find angles to the nearest 0.01°.)

18.   Use a calculator to find csc 72.3142° to four decimal places.

TRIGONOMETRIC FUNCTIONS

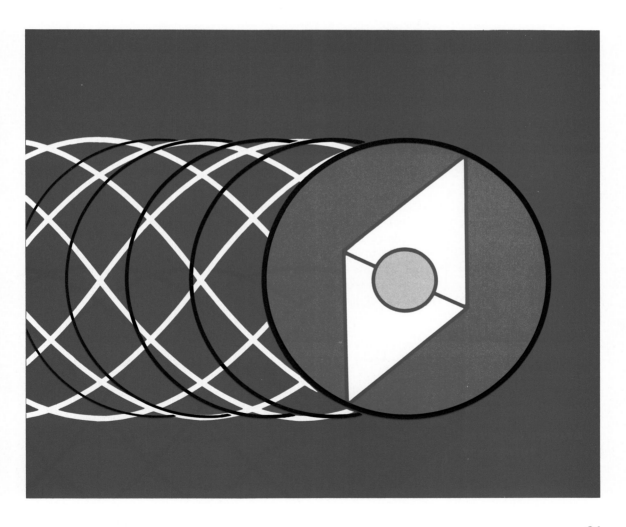

2.1 INTRODUCTORY REMARKS

The trigonometric ratios studied in Chapter 1 provide a powerful tool for indirect measurement. It was for this purpose only that trigonometry was used for nearly 2,000 years. Surveying, map-making, navigation, construction, military uses, and astronomy created the need that resulted in the extensive development of trigonometry as a tool for indirect measurement.

A turning point in thinking about trigonometry occurred after the development of the rectangular coordinate system (René Descartes, 1596–1650). The trigonometric ratios, through the use of this system, were generalized into trigonometric functions. This generalization increased their usefulness far beyond the dreams of those originally responsible for this development. The Swiss mathematician Leonhard Euler (1707–1783) made substantial contributions in this area.

Through the demands of modern science, the periodic nature of these new functions soon became apparent, and they were quickly put to use in the study of various types of periodic phenomena. The trigonometric functions began to be used on problems that had nothing whatsoever to do with angles and triangles.

In this chapter we will generalize the concept of trigonometric ratios along the lines suggested above. Before we undertake this task, however, we will introduce a new measure of angle called the **radian.**

2.2 DEGREES AND RADIANS

DEGREE AND RADIAN
MEASURE OF ANGLES

In Chapter 1 we defined an angle and its degree measure. Recall that a central angle in a circle has angle measure 1° if it subtends an arc $\frac{1}{360}$ of the circumference of the circle. Another angle measure that will be of considerable use to us is radian measure. A central angle subtended by an arc of length equal to the radius of the circle is defined to be an angle of **radian measure 1** (see Figure 1). These definitions of angle measure are independent of the size of the defining circle (Why?).

FIGURE 1
Degree and radian measure

Thus, when we write $\theta = 2°$, we are referring to an angle of degree measure 2; when we write $\theta = 2$ rad, we are referring to an angle of radian measure 2.

It follows from the above definition that the radian measure of a central angle θ subtended by an arc of length s is found by determining how many times the length of the radius R, used as a unit length, is contained in the arc length s. In terms of a formula, we have the following:

RADIAN MEASURE OF CENTRAL ANGLES

$$\theta = \frac{s}{R} \text{ radians (rad)}$$

Also, $s = R\theta$

Because of their importance, the formulas in the box should be understood before proceeding any further.

What is the radian measure of a central angle subtended by an arc of 32 cm in a circle of radius 8 cm?

$$\theta = \frac{32 \text{ cm}}{8 \text{ cm}} = 4 \text{ rad}$$

Note: The units in which the arc length and radius are measured cancel; hence, we are left with a "unitless" or pure number. For this reason, the word *radian* is often omitted when we are dealing with the radian measure of angles unless a special emphasis is desired.

What is the radian measure of an angle of 180°? A central angle of 180° is subtended by an arc $\frac{1}{2}$ of the circumference of the circle. Thus, if C is the circumference of a circle, then $\frac{1}{2}$ of the circumference is given by

$$s = \frac{C}{2} = \frac{2\pi R}{2} = \pi R$$

and

$$\theta = \frac{s}{R} = \frac{\pi R}{R} = \pi \text{ rad}$$

Hence, 180° corresponds to π rad. Remember this! The radian measures of many special angles can be obtained based on this correspondence. For example, 90° is 180°/2; therefore, 90° corresponds to $\pi/2$ rad. Since 360° is twice 180°, 360° corresponds to 2π rad. Similarly, 60° corresponds to $\pi/3$ rad, 45° to $\pi/4$ rad, and 30° to $\pi/6$ rad. These special angles and their degree and radian measures will be referred to frequently throughout this book. We summarize these special correspondences here for ease of reference:

RADIANS	$\frac{\pi}{6}$	$\frac{\pi}{4}$	$\frac{\pi}{3}$	$\frac{\pi}{2}$	π	2π
DEGREES	30	45	60	90	180	360

In general, we can use the following proportion to convert degree measure to radian measure and vice versa:

RADIAN–DEGREE CONVERSION FORMULAS

$$\frac{\theta°}{180°} = \frac{\theta}{\pi \text{ rad}}$$ or $$\theta° = \frac{180}{\pi}\theta \quad \text{Radians to degrees}$$

$$\theta = \frac{\pi}{180}\theta° \quad \text{Degrees to radians}$$

EXAMPLE 1 (A) Find the radian measure of 50°.
(B) Find the degree measure of 1.5 rad.

Handwritten margin note: Dr

Handwritten: $\pi^R = 180°$

Solutions (A) $$\frac{\theta°}{180} = \frac{\theta}{\pi \text{ rad}}$$

Handwritten margin notes:
Dv
$similar\ problem$
$40° = \underline{\quad}^R$
$\frac{\pi}{12}{}^R = \underline{\quad}^°$

$$\theta = \frac{\pi}{180}\theta°$$

$$= \frac{\pi}{180}(50)$$

$$= \frac{5}{18}\pi \text{ rad} \qquad \text{Exact form}$$

$$\approx 0.87 \text{ rad} \qquad \text{Approximation}$$

Handwritten right margin:
$\pi^R = 180°$
$1^R = \frac{180°}{\pi}$
OR
$1° = \frac{\pi^R}{180}$

(B) $$\frac{\theta°}{180} = \frac{\theta}{\pi \text{ rad}}$$

$$\theta° = \frac{180}{\pi}\theta$$

$$= \frac{180}{\pi}(1.5)$$

$$= \frac{270°}{\pi} \qquad \text{Exact form}$$

$$\approx 85.9° \qquad \text{Approximation}$$

PROBLEM 1 (A) Find the radian measure of an angle of 20°.
(B) Find the degree measure of an angle of 1 rad.

GENERALIZED ANGLE

Before generalizing the concept of trigonometric ratios we must generalize the concept of angle. To this end, we start with a rectangular coordinate system and two rays coinciding with the nonnegative x axis. One ray, called the **initial side** of the angle, remains fixed; the other ray, called the **terminal side** of the angle, is rotated until it reaches its terminal position. When the terminal side is rotated counterclockwise, the angle formed is considered positive (see Figures 2a and 2c); when it is rotated clockwise, the angle formed is considered negative (see Figure 2b). Angles located in a coordinate system in this manner are said to be in their **standard position.** We can thus consider angles of any size, positive or negative.

FIGURE 2
Generalized angles

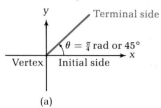

(a)

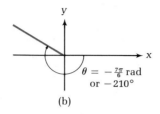

(b)

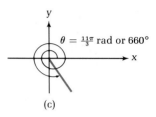

(c)

EXAMPLE 2   Sketch the following angles in their standard positions:
(A) $-60°$ (B) $3\pi/2$ rad   (C) -3π rad   (D) $405°$

Solutions   (A)

(B)

(C)

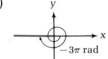

(D)

PROBLEM 2   Sketch the following angles in their standard positions:
(A) $120°$ (B) $-\pi/6$ rad   (C) $7\pi/2$ rad   (D) $-495°$

ANSWERS TO MATCHED PROBLEMS
1.   (A) $\pi/9$ rad ≈ 0.35 rad   (B) $(180/\pi)° \approx 57.3°$

2.   (A)

(B)

(C)

(D)

EXERCISE 2.2

A   *Remembering that 180° corresponds to π rad, mentally convert each degree measure to radian measure in terms of π.*

1. 90°   **2.** 45°   **3.** 60°
4. 30°   **5.** 120°   **6.** 150°

Remembering that π rad corresponds to 180°, mentally convert each radian measure to degree measure.

7. π/4 rad   **8.** π/2 rad   **9.** π/6 rad
10. π/3 rad   **11.** 5π/6 rad   **12.** 2π/3 rad

13. Convert to radian measure: 30°, 60°, 90°, 120°, 150°, 180°. (Start with 30° and take multiples.)
14. Convert to radian measure: 45°, 90°, 135°, 180°. (Start with 45° and take multiples.)

Sketch each angle in its standard position.

15. 60°   **16.** 45°   **17.** −30°
18. −45°   **19.** 240°   **20.** −135°

B   **21.** If the radius of a circle is 3 cm, find the radian measure of an angle subtended by an arc of length: (A)   6 cm   (B)   4.5 cm   (Use $\theta = s/R$.)
22. If the radius of a circle is 4 m, find the radian measure of an angle subtended by an arc of length: (A)   12 m   (B)   10 m
23. If the radius of a circle is 5 mm, find the length of the arc subtended by an angle of: (A)   2 rad   (B)   0.3 rad
24. If the radius of a circle is 4 cm, find the length of the arc subtended by an angle of: (A)   1 rad   (B)   0.25 rad

Find the radian measure for each angle. Express the answer in exact form. (See Example 1.)

25. 18°   **26.** 9°   **27.** 27°
28. 36°   **29.** 130°   **30.** 140°

Find the degree measure of each angle. Express the answer in exact form. (See Example 1.)

31. 1.6 rad   **32.** 0.5 rad   **33.** π/12 rad
34. π/36 rad   **35.** π/60 rad   **36.** π/180 rad

Sketch each angle in its standard position.

37. −π/6 rad   **38.** −π/3 rad   **39.** 300°
40. 390°   **41.** −7π/3 rad   **42.** −11π/4 rad

C   *In which quadrant† does the terminal side of each angle lie?*

43. 432°   **44.** 821°   **45.** $-14\pi/3$ rad

46. $-17\pi/4$ rad   **47.** 1,243°   **48.** $-942°$

Use a calculator to find each value to four decimal places. Convert radians to decimal degrees.

49. 57.3421° = ? rad   **50.** 103.2187° = ? rad

51. 0.3184 rad = ?°   **52.** 1.0394 rad = ?°

53. 26°23′14″ = ? rad   **54.** 179°3′43″ = ? rad

APPLICATIONS   **55.** *Radian measure* What is the radian measure of the smaller angle made by the hands of a clock at 2:30? Express the answer in terms of π and as a decimal fraction to two decimal places.

56. *Sun's diameter* The sun is about 1.5×10^8 km from the earth. If the angle subtended by the diameter of the sun on the surface of the earth is 9.3×10^{-3} rad, approximately what is the diameter of the sun? [*Hint:* Use the intercepted arc to approximate the diameter.]

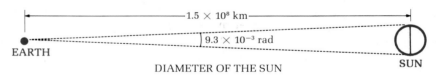

DIAMETER OF THE SUN

*57. *Earth's orbit* Assume that the earth's orbit is circular. A line from the earth to the sun sweeps out an angle of how many radians in 1 week? Express the answer in terms of π and as a decimal fraction to two decimal places.

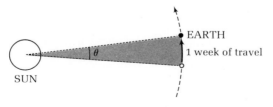

†Recall that a cartesian coordinate system divides a plane into four parts called **quadrants.** These are numbered in a counterclockwise direction starting in the upper right-hand corner.

58. *Revolutions and radians*   How many radians are in 5 revolutions? In 3.6 revolutions? In *n* revolutions? (Give answers in exact form in terms of π.)

*****59.** *Revolutions and radians*   Through how many radians does a pulley with 10 cm diameter turn when 5 m of rope has been pulled through it without slippage? How many revolutions result?

******60.** *Radians and arc length*   A bicycle has a front wheel with a diameter of 24 cm and a back wheel with a diameter of 60 cm. Through what angle (in radians) does the front wheel turn if the back wheel turns through 12 rad?

2.3   TRIGONOMETRIC FUNCTIONS

In Chapter 1 we introduced the concept of trigonometric ratios and tied this idea to right triangles. We were able to use these ratios to define six trigonometric functions with angle domains restricted to 0°–90°, or 0–$\pi/2$ rad. In this section we will introduce more general definitions that will apply to angle domains of arbitrary size, positive, negative, or zero, in degree or radian measure. We will then move one (giant) step further and define these functions for arbitrary real numbers. With these new functions we will be able to do everything we did with the trigonometric ratios in the first chapter, plus a great deal more. In Section 2.5 we will approach the subject from a more modern point of view where angles are not a necessary part of the definition. Each approach has its advantage for certain applications and uses.

TRIGONOMETRIC FUNCTIONS WITH ANGLE DOMAINS

We start with an arbitrary generalized angle θ (see Section 2.2) located in a rectangular coordinate system in a standard position. We then choose an arbitrary point $P(a, b)$ on the terminal side of θ, but away from the origin. If R is the distance of $P(a, b)$ from the origin, we can form six ratios involving R and the coordinates of P. We will use these six ratios to define six trigonometric functions as given in Definition 1 at the top of page 39.

REMARKS

1.   The ratios in the definition may be negative as well as positive, depending on the quadrant in which the terminal side of θ lies.

2.   The right triangle formed by dropping a perpendicular from $P(a, b)$ to the horizontal axis is called the **reference triangle** associated with the angle θ. We will often refer to this triangle.

3.   It is important to observe that these definitions include as a special case, the trigonometric ratios for right triangles studied in Section 1.4 when θ is restricted to positive acute angles. As we will soon see, the definitions also include much more.

These six definitions of the trigonometric functions must be learned— a great deal depends on them. It is important to note that because of the properties of similar triangles, these definitions are independent of the choice of $P(a, b)$ on the terminal side of θ. (See Problems 1–4 in Exercise 2.3.)

DEFINITION 1

TRIGONOMETRIC FUNCTIONS WITH ANGLE DOMAINS

For an arbitrary angle θ:

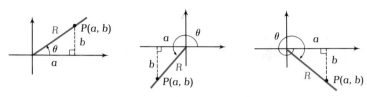

$$\sin \theta = \frac{b}{R} \qquad \csc \theta = \frac{R}{b} \quad b \neq 0 \qquad R = \sqrt{a^2 + b^2} > 0$$

$$\cos \theta = \frac{a}{R} \qquad \sec \theta = \frac{R}{a} \quad a \neq 0$$

$P(a, b)$ is an arbitrary point on the terminal side of θ, $(a, b) \neq (0,0)$

$$\tan \theta = \frac{b}{a} \quad a \neq 0 \qquad \cot \theta = \frac{a}{b} \quad b \neq 0$$

Domains: Sets of all possible angles for which the ratios are defined.
Ranges: Subsets of the set of real numbers

again,
use x, y
for a, b.

EXAMPLE 3   Find the value of each of the six trigonometric functions for the illustrated angle θ with the terminal side that contains $P(-4, -3)$.

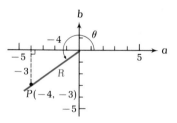

Solution

$(a, b) = (-4, -3)$

$R = \sqrt{a^2 + b^2} = \sqrt{(-4)^2 + (-3)^2} = \sqrt{25} = 5$

$$\sin \theta = \frac{b}{R} = \frac{-3}{5} = -\frac{3}{5} \qquad \csc \theta = \frac{R}{b} = \frac{5}{-3} = -\frac{5}{3}$$

$$\cos \theta = \frac{a}{R} = \frac{-4}{5} = -\frac{4}{5} \qquad \sec \theta = \frac{R}{a} = \frac{5}{-4} = -\frac{5}{4}$$

$$\tan \theta = \frac{b}{a} = \frac{-3}{-4} = \frac{3}{4} \qquad \cot \theta = \frac{a}{b} = \frac{-4}{-3} = \frac{4}{3}$$

PROBLEM 3   Find the value of each of the six trigonometric functions if the terminal side of θ contains the point $(-8, -6)$.   [*Note:* This point lies on the terminal side of the same angle as in Example 3.]

EXAMPLE 4

$\mathcal{D}o$

Find the value of each of the other five trigonometric functions for the given angle θ — without finding θ — given that the terminal side of θ is in quadrant III and

$$\cos \theta = \frac{-3}{5}$$

Solution

The information given is sufficient for us to locate a reference triangle in quadrant III for θ, even though we do not know θ. We sketch the reference triangle, label what we know, then complete the problem as indicated.

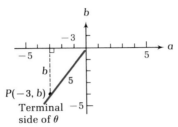

Since $\cos \theta = \dfrac{a}{R} = \dfrac{-3}{5}$, we know that $a = -3$ and $R = 5$ (R is never negative). If we can find b, we can determine the values of the other five functions using Definition 1.

Use the Pythagorean theorem to find b:

$$(-3)^2 + b^2 = 5^2$$
$$b^2 = 25 - 9 = 16 \qquad \text{b is negative since $P(a, b)$ is in}$$
$$b = -4 \qquad \qquad \text{quadrant III}$$

Thus,

$$(a, b) = (-3, -4) \qquad \text{and} \qquad R = 5$$

We can now find the other five functions using Definition 1.

$$\sin \theta = \frac{b}{R} = \frac{-4}{5} = -\frac{4}{5} \qquad\qquad \csc \theta = \frac{R}{b} = \frac{5}{-4} = -\frac{5}{4}$$

$$\tan \theta = \frac{b}{a} = \frac{-4}{-3} = \frac{4}{3} \qquad\qquad \cot \theta = \frac{a}{b} = \frac{-3}{-4} = \frac{3}{4}$$

$$\sec \theta = \frac{R}{a} = \frac{5}{-3} = -\frac{5}{3}$$

PROBLEM 4   Repeat Example 4 for $\tan \theta = -\frac{3}{4}$ and the terminal side of θ in quadrant II.

TRIGONOMETRIC
FUNCTIONS WITH
REAL NUMBER
DOMAINS

Let us now turn to the problem of defining trigonometric functions for real number domains. We first note that to each real number x there corresponds an angle of x radians, and to each angle of x radians there corresponds the real number x. We define trigonometric functions with real number domains in terms of trigonometric functions with angle domains.

DEFINITION 2

> TRIGONOMETRIC FUNCTIONS WITH REAL NUMBER DOMAINS
>
> For x any real number, we define
>
> $$\sin x = \sin(x \text{ rad}) \qquad \csc x = \csc(x \text{ rad})$$
> $$\cos x = \cos(x \text{ rad}) \qquad \sec x = \sec(x \text{ rad})$$
> $$\tan x = \tan(x \text{ rad}) \qquad \cot x = \cot(x \text{ rad})$$
>
> Domains: Subsets of the set of real numbers
> Ranges: Subsets of the set of real numbers

Thus, $\sin 3 = \sin(3 \text{ rad})$, $\cos 1.23 = \cos(1.23 \text{ rad})$, $\tan(-9) = \tan(-9 \text{ rad})$, and so on.

Because of the above definition, we will often omit "rad" after x and interpret x as a real number or an angle with radian measure x, whichever fits the context in which x appears.

At first glance Definition 2 appears artificial, but we will see that it frees the trigonometric functions from angles and opens them up to a large variety of significant applications not directly connected to angles.

CALCULATOR
EVALUATION

We used scientific calculators in Section 1.4 to approximate trigonometric ratios for acute angles in degree measure. These same calculators are internally programmed to approximate (to eight or ten significant digits) trigonometric functions for *any* angle (however large or small, positive or negative) in degree or in radian measure, or for *any* real number. Table use (for those still desiring that approach) is discussed in the optional Section 2.5. In Section 2.4 we will show how to obtain exact values for certain special angles (integer multiples of 30° and 45° or integer multiples of $\pi/6$ and $\pi/4$) without the use of calculators or tables.

> IMPORTANT
>
> 1. Set the calculator in **degree mode** when evaluating trigonometric functions of angles in degree measure (degree measure must be in decimal degrees).
> 2. Set the calculator in **radian mode** when evaluating trigonometric functions of angles in radian measure or trigonometric functions of real numbers.

We will also use the reciprocal relationships stated in Section 1.4 (and generalized using Definitions 1 and 2) to evaluate secant, cosecant, and cotangent.

RECIPROCAL RELATIONSHIPS

For x any real number or angle in degree or radian measure:

$$\csc x = \frac{1}{\sin x} \qquad \sin x \neq 0$$

$$\sec x = \frac{1}{\cos x} \qquad \cos x \neq 0$$

$$\cot x = \frac{1}{\tan x} \qquad \tan x \neq 0$$

EXAMPLE 5 Evaluate to four significant digits using a calculator.

(A) $\sin 286.38°$ (B) $\tan(3.472 \text{ rad})$
(C) $\cot 5.063$ (D) $\cos(-107°35')$
(E) $\sec(-4.799)$ (F) $\csc 192°47'22''$

Solutions

(A) $\sin 286.38° = -0.9594$

Degree mode
| 286.38 | | sin |

(B) $\tan(3.472 \text{ rad}) = 0.3430$

Radian mode
| 3.472 | | tan |

(C) $\cot 5.063 = -0.3657$

Radian mode
| 5.063 | | tan | | 1/x |

(D) $\cos(-107°35') = \cos(-107.5833\ldots)$ Change to decimal degrees

$\qquad\qquad = -0.3021$

Degree mode
| 107.5833... | | +/− | | cos |

Radian mode
| 4.799 | | +/− | | cos | | 1/x |

(E) $\sec(-4.799) = 11.56$

(F) $\csc 192°47'22'' = \csc(192.7894\ldots)$ Change to decimal degrees

$\qquad\qquad = -4.517$

Degree mode
| 192.7894 | | sin | | 1/x |

PROBLEM 5 Evaluate to four significant digits using a calculator.

(A) $\cos 303.73°$ (B) $\sec(-2.805)$
(C) $\tan(-83°29')$ (D) $\sin(12 \text{ rad})$
(E) $\csc 100°52'43''$ (F) $\cot 9$

SUMMARY
OF SIGN PROPERTIES

In closing this important section, let us summarize the sign properties of the six trigonometric functions in tabular form. The table need not be memorized, since particular entries are readily determined from the definitions of the functions involved. In the table at the top of the next page, x is associated with an angle that terminates in the respective quadrant, $P(a, b)$ is a point on the terminal side of the angle, and $R = \sqrt{a^2 + b^2} > 0$.

S-tudents| All

Take | Calculus

	Quadrant I			Quadrant II			Quadrant III			Quadrant IV		
	a	b	R	a	b	R	a	b	R	a	b	R
	+	+	+	−	+	+	−	−	+	+	−	+
$\sin x = b/R$, $\csc x = R/b$		+			+			−			−	
$\cos x = a/R$, $\sec x = R/a$		+			−			−			+	
$\tan x = b/a$, $\cot x = a/b$		+			−			+			−	

ANSWERS TO MATCHED PROBLEMS

3. $\sin \theta = -\frac{3}{5}$, $\cos \theta = -\frac{4}{5}$, $\tan \theta = \frac{3}{4}$, $\csc \theta = -\frac{5}{3}$, $\sec \theta = -\frac{5}{4}$, $\cot \theta = \frac{4}{3}$

[What do you think the values of the six trigonometric functions would be if we picked another point, say $(-12, -9)$, on the terminal side of the same angle θ as in Example 3?   *Answer:*   The same. (Why?)]

4. $\sin \theta = \frac{3}{5}$, $\csc \theta = \frac{5}{3}$, $\cot \theta = -\frac{4}{3}$, $\cos \theta = -\frac{4}{5}$, $\sec \theta = -\frac{5}{4}$

5. (A) 0.5553   (B) −1.059   (C) −8.754
    (D) −0.5366   (E) 1.018   (F) −2.211

EXERCISE 2.3

A

Find the value of each of the six trigonometric functions if the terminal side of θ contains the point $P(a, b)$. Do the same for $Q(a, b)$.

1. $P(3, 4)$; $Q(6, 8)$ **2.** $P(-3, -4)$; $Q(-9, -12)$

3. $P(4, -3)$; $Q(12, -9)$ **4.** $P(-3, 4)$; $Q(-6, 8)$

Find the value of each of the other five trigonometric functions for the angle θ (without finding θ) given the indicated information. Sketching a reference triangle should prove helpful.

5. $\cos \theta = \dfrac{3}{5}$

θ is a quadrant I angle

6. $\sin \theta = \dfrac{3}{5}$

θ is a quadrant I angle

7. $\cos \theta = \dfrac{3}{5}$

θ is a quadrant IV angle

8. $\sin \theta = \dfrac{3}{5}$

θ is a quadrant II angle

9. $\csc \theta = -\dfrac{5}{4}$

θ is a quadrant III angle

10. $\tan \theta = -\dfrac{4}{3}$

θ is a quadrant II angle

11. $\csc \theta = -\dfrac{5}{4}$

θ is a quadrant IV angle

12. $\tan \theta = -\dfrac{4}{3}$

θ is a quadrant IV angle

Use a calculator to find Problems 13–24 to four significant digits. Make sure the calculator is in the correct mode (degree or radian) for each problem.

13. $\tan 89°$	**14.** $\sin 37°$	**15.** $\cos(3 \text{ rad})$
16. $\sin(4 \text{ rad})$	**17.** $\csc 162°$	**18.** $\sec 283°$
19. $\cot 341°$	**20.** $\csc 269°$	**21.** $\sin 13$
22. $\cos 7$	**23.** $\cot 2$	**24.** $\tan 1$

B   *Find the value of each of the six trigonometric functions for an angle θ having a terminal side containing the indicated point.*

25. $(\sqrt{3}, 1)$	**26.** $(1, 1)$	**27.** $(1, -\sqrt{3})$
28. $(-1, \sqrt{3})$	**29.** $(\sqrt{2}, -\sqrt{2})$	**30.** $(-2, -2)$

In which quadrants must the terminal side of an angle θ lie in order for:

31. $\cos \theta > 0$	**32.** $\sin \theta > 0$	**33.** $\tan \theta > 0$
34. $\cot \theta > 0$	**35.** $\sec \theta > 0$	**36.** $\csc \theta > 0$
37. $\sin \theta < 0$	**38.** $\cos \theta < 0$	**39.** $\cot \theta < 0$
40. $\tan \theta < 0$	**41.** $\csc \theta < 0$	**42.** $\sec \theta < 0$

Find the value of each of the other five trigonometric functions for an angle θ (without finding θ), given the indicated information. Sketching a reference triangle should prove helpful.

43. $\sin \theta = -\dfrac{2}{3}$

θ is a quadrant III angle

44. $\cos \theta = -\dfrac{3}{5}$

θ is a quadrant II angle

45. $\sin \theta = -\dfrac{2}{3}$

θ is a quadrant IV angle

46. $\cos \theta = -\dfrac{3}{5}$

θ is a quadrant III angle

47. $\sec \theta = \sqrt{3}$

θ is a quadrant IV angle

48. $\tan \theta = -2$

θ is a quadrant II angle

Use a calculator to find Problems 49–60 to four significant digits.

49. $\cos 308.25°$	**50.** $\sin 170.23°$	**51.** $\tan 1.371$
52. $\sin 3.519$	**53.** $\cot(-265.33°)$	**54.** $\csc(-45.27°)$
55. $\sec(-4.013)$	**56.** $\cot(-0.1578)$	**57.** $\cos 208°12'55''$
58. $192°45'13''$	**59.** $\csc 112°5'38''$	**60.** $\cot 321°18'5''$

C   **61.**   Which trigonometric functions are not defined when the terminal side of an angle lies along the vertical axis?

62.   Which trigonometric functions are not defined when the terminal side of an angle lies along the horizontal axis?

APPLICATIONS   ***63.**   *Piston motion*   The figure shows a piston connected to a wheel turning at 10 revolutions per second (rps). If P is at $(1, 0)$ when $t = 0$, then $\theta = 20\pi t$, where t is time in seconds. Show that

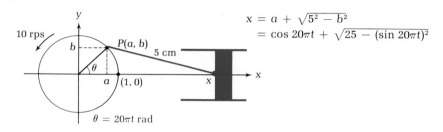

$$x = a + \sqrt{5^2 - b^2}$$
$$= \cos 20\pi t + \sqrt{25 - (\sin 20\pi t)^2}$$

$\theta = 20\pi t$ rad

*64.   In Problem 63, find the position of the piston (the value of x) for $t = 0$ and $t = 0.01$ sec.

65.   *Alternating current*   An alternating current generator produces an electric current (measured in amperes) that is described by the equation

   $I = 35 \sin(40\pi t - 10\pi)$

where t is time in seconds. (Section 8.3 has a detailed discussion of this subject.) What is the current I when $t = 0.13$ sec?

66.   *Alternating current*   What is the current I in Problem 65 when $t = 0.31$ sec?

67.   *Logistics*   A log of length L is being floated down a canal with a right angle turn as indicated in the figure below. Express the length of the log in terms of θ and the two widths of the canal (neglect the width of the log). The log is to touch the sides of the canal as shown. (In calculus this equation is used to find the largest log that will float around the corner.)

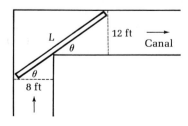

68.   *Logistics*   Refer to Problem 67. What are the lengths of the log associated with $\theta = 40°, 45°, 50°, 55°,$ and $60°$?

EXACT VALUE FOR SPECIAL ANGLES AND REAL NUMBERS

If an angle is an integer multiple of 30°, 45°, $\pi/6$ rad, or $\pi/4$ rad, and if a real number is an integer multiple of $\pi/6$ or $\pi/4$, then, for those values for which each trigonometric function is defined, the function can be evaluated exactly without the use of any calculator or table (which is different from finding approximate values using a calculator or table). With a little practice, you will be able to determine these exact values almost mentally. Working with exact values has advantages over working with approximate values in many situations.

Before we start evaluating trigonometric functions for special values, it is useful to visualize the definition of the trigonometric functions for both angle and real number domains in terms of a "function machine." Figure 3 illustrates a "cosine machine" with an angle domain and a "sine machine" with a real number domain.

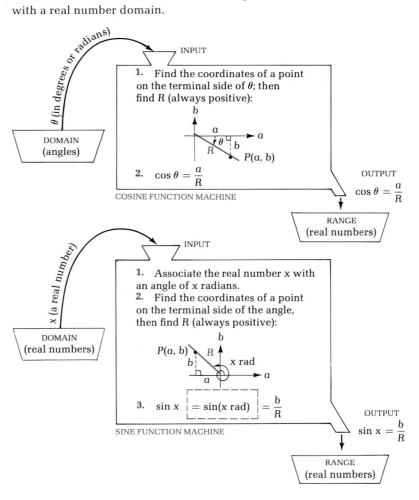

FIGURE 3
Cosine and sine
"function machines"

There are many significant applications of trigonometric functions, as we will see. Some require angle domains, and others require real number domains. Our definitions of the trigonometric functions enable us to shift from angle domains to real number domains, and vice versa, with relative ease.

We are now ready to evaluate trigonometric functions for special angles and special real numbers. For these angles and real numbers, it will be relatively easy to find exact coordinates of a point on the terminal side of the indicated angle. Once coordinates are found, R can be found, then values for all six trigonometric functions can be determined.

QUADRANTAL ANGLES

The easiest angles to deal with are **quadrantal angles;** that is, angles with their terminal side lying along a coordinate axis. These angles are integer multiples of 90° or $\pi/2$. It is easy to find coordinates of a point on a coordinate axis. Since any nonorigin point will do, we shall, for convenience, choose points 1 unit from the origin (Figure 4).

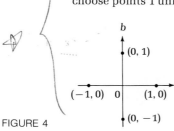

In each case
$R = \sqrt{a^2 + b^2} = 1$
a positive number

FIGURE 4

EXAMPLE 6   Find:

(A) $\sin 90°$ (B) $\cos \pi$ (C) $\tan(-2\pi)$ (D) $\cot(-180°)$

Solutions   For each, visualize the location of the terminal side of the angle relative to Figure 4. With a little practice, you should be able to do most of the following mentally.

find $\sin \theta$
$\cos \theta$
$\tan \theta$

where $\theta = 0°, 90°, 180°, 270°$

(A) $\sin 90° = \dfrac{b}{R} = \dfrac{1}{1} = 1$ $(a, b) = (0, 1), \quad R = 1$

(B) $\cos \pi = \dfrac{a}{R} = \dfrac{-1}{1} = -1$ $(a, b) = (-1, 0), \quad R = 1$

(C) $\tan(-2\pi) = \dfrac{b}{a} = \dfrac{0}{1} = 0$ $(a, b) = (1, 0), \quad R = 1$

(D) $\cot(-180°) = \dfrac{a}{b} = \dfrac{-1}{0}$ $(a, b) = (-1, 0), \quad R = 1$

Not defined

PROBLEM 6   Find

(A) $\sin(3\pi/2)$ (B) $\sec(-\pi)$ (C) $\tan 90°$ (D) $\cot(-270°)$

Notice that in Example 6(D), $\cot(-180°)$ is not defined. For what other values is the cotangent function not defined? Where are the other trigonometric functions not defined? These important questions are considered in Exercise 2.4.

SPECIAL 30°–60° AND 45° RIGHT TRIANGLES

If a reference triangle of a given angle is a 30°–60° right triangle or a 45° right triangle, then we will be able to find exact nonorigin coordinates on the terminal side of the given angle. Because the reference triangle is going to play a very important role in this process, we restate its definition as well as that of a **reference angle.**

REFERENCE TRIANGLE AND ANGLE

1.   To form a **reference triangle** for θ, drop a perpendicular from a point $P(a, b)$ on the terminal side of θ to the horizontal axis.
2.   The **reference angle** α is the angle (always taken positive) between the terminal side of θ and the horizontal axis.

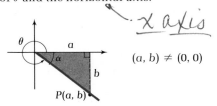

$(a, b) \neq (0, 0)$

If we take a 30°–60° right triangle, we note that it is one-half of an equilateral triangle, as indicated in Figure 5. Since all sides are equal in an equilateral triangle, we can apply the Pythagorean theorem to obtain a useful relationship among the three sides of the original triangle.

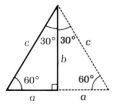

FIGURE 5

$$b = \sqrt{c^2 - a^2}$$
$$= \sqrt{(2a)^2 - a^2} \qquad \text{Since } c = 2a$$
$$= \sqrt{3a^2} = a\sqrt{3}$$

Similarly, using the Pythagorean theorem on a 45° right triangle, we obtain the following:

FIGURE 6

$$c = \sqrt{a^2 + a^2}$$
$$= \sqrt{2a^2}$$
$$= a\sqrt{2}$$

We summarize these results in the box, along with some frequently used special cases, for convenient reference. The ratios of the sides of these special triangles should be memorized, since they will be used often in this and subsequent sections. The two triangles in color in Figure 7 are the easiest to remember. The others can be obtained from these by multiplying or dividing the length of each side by the same nonzero quantity.

Make my chart for sin θ where θ = 0, 30, 45, 60, 90°

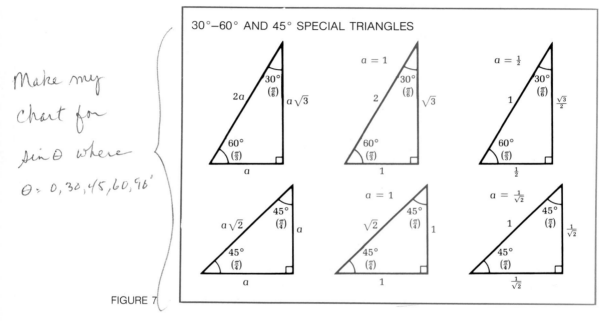

30°–60° AND 45° SPECIAL TRIANGLES

FIGURE 7

EVALUATION FOR
ANGLES OR REAL
NUMBERS WITH
30°–60° OR 45°
REFERENCE TRIANGLES

If an angle θ has a 30°–60° or 45° reference triangle, then it is an easy matter to find exact coordinates of a point P on the terminal side of θ and the exact distance of P from the origin. Then, using Definition 1 in Section 2.3, we can find the exact value of any of the six trigonometric functions for θ. Several examples will illustrate the process.

EXAMPLE 7 Evaluate exactly:
(A) $\sin 30°$, $\cos(\pi/6)$, $\cot(\pi/6)$ (B) $\cos 45°$, $\tan(\pi/4)$, $\csc(\pi/4)$

Solutions (A) Use the special 30°–60° triangle as the reference triangle for $\theta = 30°$ and $\theta = \pi/6$. Use the sides of the reference triangle to determine $P(a, b)$ and R; then use Definition 1.

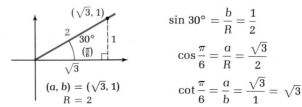

$$\sin 30° = \frac{b}{R} = \frac{1}{2}$$

$$\cos \frac{\pi}{6} = \frac{a}{R} = \frac{\sqrt{3}}{2}$$

$$\cot \frac{\pi}{6} = \frac{a}{b} = \frac{\sqrt{3}}{1} = \sqrt{3}$$

(B)   Use the special 45° triangle as the reference triangle for $\theta = 45°$ and $\theta = \pi/4$. Use the sides of the reference triangle to determine $P(a, b)$ and R; then use Definition 1.

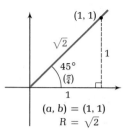

$$\cos 45° = \frac{a}{R} = \frac{1}{\sqrt{2}}$$

$$\tan \frac{\pi}{4} = \frac{b}{a} = \frac{1}{1} = 1$$

$$\csc \frac{\pi}{4} = \frac{R}{b} = \frac{\sqrt{2}}{1} = \sqrt{2}$$

PROBLEM 7   Evaluate exactly:

(A) $\cos 60°$, $\sin(\pi/3)$, $\tan(\pi/3)$ (B) $\sin 45°$, $\cot(\pi/4)$, $\sec(\pi/4)$

Before proceeding further, it is useful to observe (geometrically) multiples of $\pi/3$ (60°), $\pi/6$ (30°), and $\pi/4$ (45°). These are illustrated in Figure 8.

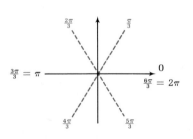

(a) Multiples of $\frac{\pi}{3}$ (60°)

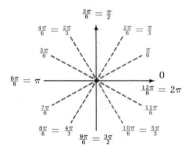

(b) Multiples of $\frac{\pi}{6}$ (30°)

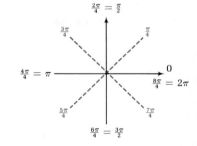

(c) Multiples of $\frac{\pi}{4}$ (45°)

FIGURE 8
Multiples of special angles

EXAMPLE 8   Evaluate exactly:

(A) $\sin 210°$ (B) $\cos(2\pi/3)$ (C) $\tan(7\pi/4)$
(D) $\sec 300°$ (E) $\cot(-5\pi/6)$ (F) $\csc(-240°)$

Solutions   Each angle has a 30°–60° or 45° reference triangle. Locate it, determine (a, b) and R, and then evaluate

(A) $\sin 210° = \dfrac{-1}{2} = -\dfrac{1}{2}$ (B) $\cos \dfrac{2\pi}{3} = \dfrac{-1}{2} = -\dfrac{1}{2}$

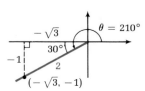

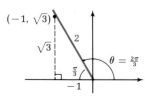

(C) $\tan \dfrac{7\pi}{4} = \dfrac{-1}{1} = -1$

(D) $\sec 300° = \dfrac{2}{1} = 2$

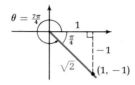

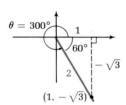

(E) $\cot\left(-\dfrac{5\pi}{6}\right) = \dfrac{-\sqrt{3}}{-1} = \sqrt{3}$

(F) $\csc(-240°) = \dfrac{2}{\sqrt{3}}$

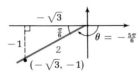

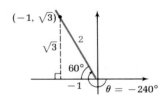

PROBLEM 8   Evaluate exactly:

(A) $\tan 210°$ (B) $\sin(2\pi/3)$ (C) $\cos(7\pi/4)$
(D) $\cot 300°$ (E) $\csc(-5\pi/6)$ (F) $\sec(-240°)$

Now let us reverse the problem. That is, suppose we are given the exact value of one of the six trigonometric functions that corresponds to one of the special reference triangles, then find θ.

EXAMPLE 9   Find the least positive θ in degree and radian measure for which each is true:

(A) $\sin \theta = \sqrt{3}/2$ (B) $\cos \theta = -1/\sqrt{2}$

Solutions   (A)   Draw a reference triangle in the first quadrant with side opposite reference angle $\sqrt{3}$ and hypotenuse 2. Observe that this is a special 30°–60° triangle.

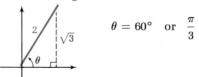

$\theta = 60°$ or $\dfrac{\pi}{3}$

(B)   Draw a reference triangle in the second quadrant with side adjacent reference angle -1 and hypotenuse $\sqrt{2}$. Observe that this is a special 45° triangle.

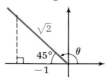

$\theta = 135°$ or $\dfrac{3\pi}{4}$

PROBLEM 9   Repeat Example 9 for:

(A) $\tan \theta = 1/\sqrt{3}$ (B) $\sec \theta = -\sqrt{2}$

We conclude this section with a summary of special values in Table 1. Some people memorize this table; others use Definition 1 in Section 2.3 and special triangles.

TABLE 1   SPECIAL VALUES

θ	$\sin \theta$	$\csc \theta$	$\cos \theta$	$\sec \theta$	$\tan \theta$	$\cot \theta$
0° or 0	0	N.D.	1	1	0	N.D.
30° or $\pi/6$	1/2	2	$\sqrt{3}/2$	$2/\sqrt{3}$	$1/\sqrt{3}$	$\sqrt{3}$
45° or $\pi/4$	$1/\sqrt{2}$	$\sqrt{2}$	$1/\sqrt{2}$	$\sqrt{2}$	1	1
60° or $\pi/3$	$\sqrt{3}/2$	$2/\sqrt{3}$	1/2	2	$\sqrt{3}$	$1/\sqrt{3}$
90° or $\pi/2$	1	1	0	N.D.	N.D.	0

N.D. = Not defined

ANSWERS TO
MATCHED PROBLEMS

6. (A) -1 (B) -1 (C)   Not defined   (D)   0
7. (A)   1/2, $\sqrt{3}/2$, $\sqrt{3}$ (B) $1/\sqrt{2}$,   1, $\sqrt{2}$
8. (A) $1/\sqrt{3}$ (B) $\sqrt{3}/2$ (C) $1/\sqrt{2}$ (D) $-1/\sqrt{3}$ (E) -2
        (F) -2
9. (A)   30°,   or $\pi/6$ (B)   135°,   or $3\pi/4$

EXERCISE 2.4

A   *Evaluate each of the following and leave answers in exact form.*

1. $\cos 0°$

2. $\sin 0°$

3. $\sin 30°$

4. $\cos 45°$

5. $\sin \dfrac{\pi}{2}$

6. $\cot \dfrac{\pi}{4}$

7. $\tan 45°$

8. $\tan 0$

9. $\tan \dfrac{\pi}{6}$

10. $\cot 0$

11. $\cos 90°$

12. $\cot \dfrac{\pi}{3}$

B   13. $\sin(-30°)$

14. $\cos(-60°)$

15. $\cos \dfrac{-\pi}{2}$

16. $\tan \pi$

17. $\tan 120°$

18. $\sin(-45°)$

19. $\cos \dfrac{-\pi}{6}$

20. $\tan \dfrac{-\pi}{4}$

21. $\cot 2\pi$

22. $\cos \dfrac{2\pi}{3}$

23. $\cot 150°$

24. $\cot(-60°)$

25. $\sin \dfrac{7\pi}{6}$

26. $\sin \dfrac{3\pi}{4}$

27. $\sin \dfrac{3\pi}{2}$

28. $\cos \dfrac{11\pi}{6}$ **29.** $\sin 225°$ **30.** $\cos 300°$

31. $\cot \dfrac{-5\pi}{4}$ **32.** $\tan \dfrac{-4\pi}{3}$ **33.** $\cos \dfrac{-5\pi}{6}$

34. $\sin \dfrac{-5\pi}{3}$ **35.** $\cot(-390°)$ **36.** $\tan(-405°)$

37. $\sin \dfrac{-15\pi}{4}$ **38.** $\sin \dfrac{-7\pi}{2}$ **39.** $\csc 420°$

40. $\sec 390°$ **41.** $\cos \dfrac{5\pi}{2}$ **42.** $\cos \dfrac{8\pi}{3}$

Find all angles between (A) 0° and 360° and (B) 0 and 2π for which the following functions are not defined:

43. tangent   **44.** cotangent   **45.** cosecant   **46.** secant

C   *Find the least positive θ in (A) degree measure and (B) radian measure for which each of the following is true:*

47. $\sin \theta = \dfrac{1}{2}$ **48.** $\cos \theta = \dfrac{1}{\sqrt{2}}$ **49.** $\cos \theta = \dfrac{-1}{2}$

50. $\sin \theta = \dfrac{-1}{2}$ **51.** $\tan \theta = -\sqrt{3}$ **52.** $\cot \theta = -1$

53. Find all the angles between 0° and 360° for which $\sin \theta = -\sqrt{3}/2$.
54. Find all the angles between 0 rad and 2π rad for which $\cos \theta = -\sqrt{3}/2$.

APPLICATIONS   *Later we will show that the area of an n-sided regular polygon inscribed in a circle of radius r (such as in the figure below) is given by*

$$A = \frac{nr^2}{2} \sin \frac{2\pi}{n}$$

Use this formula to find the areas of the polygons defined below.

INSCRIBED REGULAR POLYGON

55. $n = 3$, $r = 2$ cm   **56.** $n = 4$, $r = 5$ ft
57. $n = 6$, $r = 10$ in.   **58.** $n = 8$, $r = 4$ mm

2.5   TABLE EVALUATION (OPTIONAL)

For the terminal side of an angle θ not on a coordinate axis, let us see how Tables I–III, all involving angles restricted to intervals 0°–90° or 0–π/2 rad, can be used for angles outside these intervals. Reference triangles and corresponding reference angles play an important role in the process.

EXAMPLE 10   Find a reference triangle and a reference angle for:
(A) 212°23'   (B) 2.03 rad   (C) −6.83 rad

Solutions   Draw angles in standard position and form reference triangles; then compute reference angles.

(A) $\alpha = 212°23' − 180°$
      $= 32°23'$

(B) $\alpha = 3.14 − 2.03$
      $= 1.11$ rad

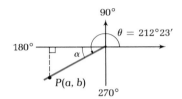

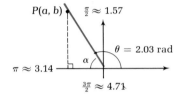

(C) $\alpha = 6.83 − 6.28$
      $= 0.55$ rad

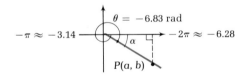

PROBLEM 10   Find reference angles for:
(A) 306°25'   (B) −2.34 rad   (C) 6.56 rad

Theorem 1 shows us how to use reference angles in evaluating trigonometric functions.

THEOREM 1

Let α be the reference angle corresponding to a given angle θ. Then:

$$\sin \theta = (\pm) \sin \alpha \qquad \csc \theta = (\pm) \csc \alpha$$
$$\cos \theta = (\pm) \cos \alpha \qquad \sec \theta = (\pm) \sec \alpha$$
$$\tan \theta = (\pm) \tan \alpha \qquad \cot \theta = (\pm) \cot \alpha$$

where the sign on the right of each equation is chosen the same as the sign of the particular function in the quadrant containing the terminal side of θ.

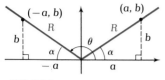

FIGURE 9

We will prove a small part of Theorem 1 to illustrate how it can be proved in general. Suppose the terminal side of θ is in quadrant II. We locate the reference triangle and the reference angle, and also place the reference triangle in the first quadrant with α in the standard position (Figure 9). Let (a, b) be a point on the terminal side of α in the first quadrant, excluding the origin. Then $(-a, b)$ is on the terminal side of θ. Thus, using Definition 1 in Section 2.3, we have

$$\sin \theta = \frac{b}{R} = \sin \alpha$$

$$\cos \theta = \frac{-a}{R} = -\frac{a}{R} = -\cos \alpha$$

$$\tan \theta = \frac{b}{-a} = -\frac{b}{a} = -\tan \alpha$$

and so on.

EXAMPLE 11   Use Table I, II, or III and Theorem 1 to find:
(A) $\cos 222°20'$ (B) $\tan(-4.32)$ (C) $\sin 425.13°$

Solutions   (A)   *Step 1*   Find reference angle α.

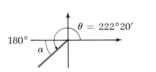

$$\alpha = 222°20' - 180°$$
$$= 42°20'$$

Step 2   Determine the sign of cosine in the third quadrant. It is negative, since a in $P(a, b)$ is negative in the third quadrant.
Step 3   Use Theorem 1 and Table I:

$$\cos 222°20' = -\cos 42°20'$$
$$= -0.7392$$

(B)

$$\alpha = 4.32 - 3.14$$
$$= 1.18$$

The terminal side of θ is in the second quadrant and the tangent function is negative in the second quadrant. Hence, using Theorem 1 and Table III, we obtain

$$\tan(-4.32) = -\tan 1.18$$
$$= -2.427$$

(C)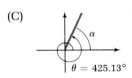

$$\alpha = 425.13° - 360°$$
$$= 65.13°$$

$\theta = 425.13°$

The terminal side of θ is in the first quadrant and the sine function is positive in the first quadrant. Hence, using Theorem 1 and Table II, we obtain

$$\sin 425.13° = \sin 65.13°$$
$$= 0.9073$$

PROBLEM 11   Use Table I, II, or III and Theorem 1 to find:
(A)   cot 280.34°   (B)   sin(−3.64)   (C)   cos(−53°40′)

ANSWERS TO   **10.**   (A)   53°35′   (B)   0.80 rad   (C)   0.28 rad
MATCHED PROBLEMS   **11.**   (A)   cot 280.34° = −cot 79.66° = −0.1825
(B)   sin(−3.64) = sin 0.50 = 0.4794
(C)   cos(−53°40′) = cos 53°40′ = 0.5925

EXERCISE 2.5

A   *Find the reference angle α for each angle (see Example 10).*

1.	170°	2.	150°
3.	195°	4.	230°
5.	340°	6.	280°
7.	−35°	8.	−68°
9.	−160°	10.	−210°
11.	−340°	12.	−260°
13.	165°40′	14.	108°50′
15.	195°10′	16.	342°20′
17.	345.37°	18.	203.72°
19.	143.14°	20.	189.53°
21.	3 rad	22.	2 rad
23.	−7 rad	24.	−6 rad
25.	4.08 rad	26.	2.93 rad
27.	−2.89 rad	28.	−1.37 rad

Select the appropriate sign on the right side of each equation to make the equation correct according to Theorem 1.

29. $\sin 195° = (\pm) \sin 15°$ 30. $\cos 170° = (\pm) \cos 10°$
31. $\tan 150° = (\pm) \tan 30°$ 32. $\cot 230° = (\pm) \cot 50°$

33. $\sec 108°50' = (\pm)\sec 71°10'$	34. $\csc 342°20' = (\pm)\csc 17°40'$
35. $\tan 189.53° = (\pm)\tan 9.53°$	36. $\sin 203.72° = (\pm)\sin 23.72°$
37. $\sec 2.93 = (\pm)\sec 0.21$	38. $\tan 4.08 = (\pm)\tan 0.94$
39. $\cos(-2.89) = (\pm)\cos 0.25$	40. $\sin(-1.37) = (\pm)\sin 1.37$

B   *Use Theorem 1 and Table I, II, or III to evaluate each of the following to three significant figures. Use $\pi \approx 3.14$ where necessary. Check the results against calculator values.*

41. $\sin 230°$	42. $\cos 150°$
43. $\cot 340°$	44. $\tan 280°$
45. $\cos(-35°)$	46. $\sin(-68°)$
47. $\sin 165°40'$	48. $\cos 108°50'$
49. $\tan 345.3°$	50. $\cot 203.7°$
51. $\sin 2$	52. $\sec 3$
53. $\tan(-6)$	54. $\cos(-7)$
55. $\csc 2.93$	56. $\cot 4.08$
57. $\cos 8$	58. $\tan(-13)$

C   59.   Prove Theorem 1 for the terminal side of θ in quadrant III.

60.   Prove Theorem 1 for the terminal side of θ in quadrant IV.

2.6   CIRCULAR FUNCTIONS

In this section we will give a more modern definition of the trigonometric functions, a definition given directly in terms of real numbers and not angles or triangles. We will then show how the angle approach and the more modern approach are related.

Since trigonometric functions with real number domains are used extensively in science and more advanced mathematics, it is useful to have this alternate definition of these functions. In addition, with this second approach we will be able to observe some very useful relationships that are not quite as apparent using the angle approach.

CIRCULAR FUNCTIONS

If we graph the equation $a^2 + b^2 = 1$ in a rectangular coordinate system, we obtain a circle with radius 1 called a **unit circle.** Using this circle, we can define the **circular functions** with real number domains, as stated in the box at the top of the next page (Definition 3).

We note that the definition of the circular functions does not involve any angles; it involves the coordinates of a point on the terminal end of an arc of length $|x|$ on a unit circle.

DEFINITION 3

CIRCULAR[†] FUNCTIONS

Let x be an arbitrary real number and U a unit circle with equation $a^2 + b^2 = 1$.

1. For x > 0:
Start at (1, 0) and proceed counterclockwise around the circumference of U until a distance (arc length) of x units has been covered. Let $P(a, b)$ be the point at the terminal end of the arc [(1, 0) is the initial end].

2. For x = 0:
$(a, b) = (1, 0)$

3. For x < 0:
Start at (1, 0) and proceed clockwise around the circumference of U until a distance (arc length) of |x| units has been covered. Let $P(a, b)$ be the point at the terminal end of the arc [(1, 0) is the initial end].

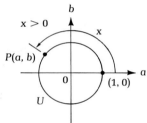

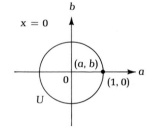

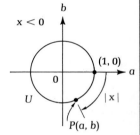

In all cases we define:

$$\sin x = b \qquad \cos x = a \qquad \tan x = \frac{b}{a} \quad a \neq 0$$

$$\csc x = \frac{1}{b} \quad b \neq 0 \qquad \sec x = \frac{1}{a} \quad a \neq 0 \qquad \cot x = \frac{a}{b} \quad b \neq 0$$

CIRCULAR FUNCTIONS
AND TRIGONOMETRIC
FUNCTIONS

We now show how the earlier definitions of the trigonometric functions (involving angle domains) can be related to the circular functions (involving real number domains).

To this end, let us look at the radian measure of an angle θ subtended by an arc of x units on a unit circle (Figure 10b). We see that for a unit circle the angle subtended by an arc of x units has a radian measure of x. Thus, every real number can be associated with an arc of x units on a unit circle

[†]It is common practice to also refer to the circular functions as trigonometric functions, and we will follow this practice.

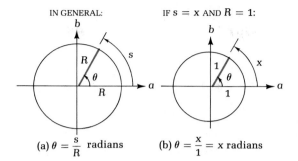

FIGURE 10   (a) $\theta = \dfrac{s}{R}$ radians   (b) $\theta = \dfrac{x}{1} = x$ radians

or a central angle of x radians on the same circle (if x is positive we go counterclockwise, and if x is negative we go clockwise). Noting that the point on the terminal end of the arc of x units is also on the terminal side of the angle of x radians, we have the following very useful relationships between the previously defined trigonometric functions with angle domains and the circular functions with real number domains.

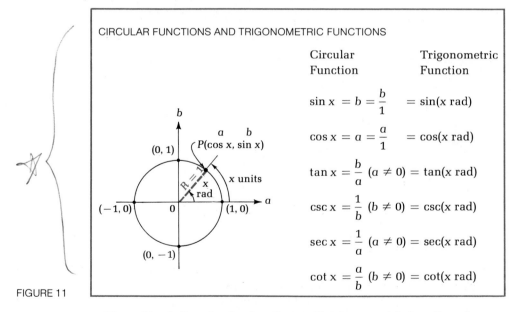

CIRCULAR FUNCTIONS AND TRIGONOMETRIC FUNCTIONS

	Circular Function	Trigonometric Function
	$\sin x = b = \dfrac{b}{1}$	$= \sin(x \text{ rad})$
	$\cos x = a = \dfrac{a}{1}$	$= \cos(x \text{ rad})$
	$\tan x = \dfrac{b}{a} \; (a \neq 0)$	$= \tan(x \text{ rad})$
	$\csc x = \dfrac{1}{b} \; (b \neq 0)$	$= \csc(x \text{ rad})$
	$\sec x = \dfrac{1}{a} \; (a \neq 0)$	$= \sec(x \text{ rad})$
	$\cot x = \dfrac{a}{b} \; (b \neq 0)$	$= \cot(x \text{ rad})$

FIGURE 11

Figure 11, relating circular functions with trigonometric functions, is very useful and should be understood and memorized. We will use it to develop a number of properties that simultaneously hold for the circular and the trigonometric functions.

EVALUATING CIRCULAR FUNCTIONS   Because circular functions are related to the trigonometric functions (as indicated in Figure 11), we can take advantage of the procedures used to evaluate trigonometric functions in the last three sections to evaluate circular functions.

EXAMPLE 12   Find:

(A) $\sin \dfrac{8\pi}{3}$ exactly   (B) $\tan 7\pi$ exactly

(C) $\sec(-1.087)$ to four significant digits using a calculator

Solutions   (A) $\sin \dfrac{8\pi}{3} \boxed{= \sin\left(\dfrac{8\pi}{3} \text{rad}\right)} = \dfrac{\sqrt{3}}{2}$ Evaluate as in Section 2.4

(B) $\tan 7\pi = 0$ Evaluate as in Section 2.4

(C) $\sec(-1.087) \boxed{= \sec(-1.087 \text{ rad})}$

$= 2.150$ $\boxed{1.087}$ $\boxed{+/-}$ $\boxed{\cos}$ $\boxed{1/x}$

PROBLEM 12   Find:

(A) $\cos\left(-\dfrac{7\pi}{6}\right)$ exactly   (B) $\csc \dfrac{3\pi}{2}$ exactly

(C) $\cot 4.793$ to four significant digits using a calculator

PERIODIC PROPERTIES   Since a unit circle has a circumference of 2π, we find for a given value of x (see Figure 11) that we will come back to the terminal point of x if we add any integer multiple of 2π to x.[†] Thus, for x any real number and for k any integer,

$$\sin(x + 2k\pi) = \sin x \qquad \cos(x + 2k\pi) = \cos x$$

Functions with this kind of repetitive behavior are called **periodic functions.** In general:

PERIODIC FUNCTIONS

A function f is **periodic** if there exists a positive real number p such that

$$f(x + p) = f(x)$$

for all x in the domain of f. The smallest such positive p, if it exists, is called the **period of f.**

Both the sine and cosine functions are periodic with period 2π. We will have more to say about the periodic properties of the trigonometric functions in the next chapter when we consider their graphs.

[†]Think of a point P moving around the unit circle in either direction. Then every time P covers a distance of 2π, the circumference of the circle, it will be back at the point where it started.

BASIC IDENTITIES

Returning to the definition of the circular functions and noting that

$$\sin x = b \quad \text{and} \quad \cos x = a$$

we can obtain the following useful relationships among the six functions:

$$\csc x = \frac{1}{b} = \frac{1}{\sin x} \tag{1}$$

$$\sec x = \frac{1}{a} = \frac{1}{\cos x} \tag{2}$$

$$\cot x = \frac{a}{b} = \frac{1}{b/a} = \frac{1}{\tan x} \tag{3}$$

$$\tan x = \frac{b}{a} = \frac{\sin x}{\cos x} \tag{4}$$

$$\cot x = \frac{a}{b} = \frac{\cos x}{\sin x} \tag{5}$$

Because the terminal points of x and $-x$ are symmetric with respect to the horizontal axis (see Figure 12), we have the following sign properties:

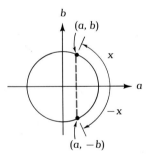

FIGURE 12

$$\sin(-x) = -b = -\sin x \tag{6}$$

$$\cos(-x) = a = \cos x \tag{7}$$

$$\tan(-x) = \frac{-b}{a} = -\frac{b}{a} = -\tan x \tag{8}$$

Finally, because $(a, b) = (\cos x, \sin x)$ is on the unit circle $a^2 + b^2 = 1$, it follows that

$$(\cos x)^2 + (\sin x)^2 = 1$$

which is usually written in the form

$$\sin^2 x + \cos^2 x = 1 \tag{9}$$

where $\sin^2 x$ and $\cos^2 x$ are concise ways of writing $(\sin x)^2$ and $(\cos x)^2$, respectively.

The equations (1) through (9) are called **basic identities.** They hold true for all replacements of x by real numbers, or angles in degree or radian measure, for which both sides of an equation are defined. Some of these identities will be used in the next chapter as aids to graphing some of the trigonometric functions. A detailed discussion of identities is found in Chapter 4.

ANSWERS TO
MATCHED PROBLEMS

12. (A) $-\sqrt{3}/2$ (B) -1 (C) -0.08079

EXERCISE 2.6

A *How does the indicated function vary as x varies over the indicated interval? (Refer to Figure 11.)*

1. $\sin x$, $0 \leq x \leq \pi/2$
2. $\cos x$, $0 \leq x \leq \pi/2$
3. $\sin x$, $\pi/2 \leq x \leq \pi$
4. $\cos x$, $\pi/2 \leq x \leq \pi$
5. $\sin x$, $\pi \leq x \leq 3\pi/2$
6. $\cos x$, $\pi \leq x \leq 3\pi/2$
7. $\sin x$, $3\pi/2 \leq x \leq 2\pi$
8. $\cos x$, $3\pi/2 \leq x \leq 2\pi$

B *Find the values of x from the indicated interval that satisfy the indicated equation or condition. (Refer to Figure 11.)*

9. $\sin x = 1$, $0 \leq x \leq 4\pi$
10. $\cos x = 1$, $0 \leq x \leq 4\pi$
11. $\sin x = 0$, $0 \leq x \leq 4\pi$
12. $\cos x = 0$, $0 \leq x \leq 4\pi$
13. $\tan x = 0$, $0 \leq x \leq 4\pi$
14. $\cot x = 0$, $0 \leq x \leq 4\pi$
15. $\cos x = 1$, $-2\pi \leq x \leq 2\pi$
16. $\sin x = 1$, $-2\pi \leq x \leq 2\pi$
17. $\cos x = 0$, $-2\pi \leq x \leq 2\pi$
18. $\sin x = 0$, $-2\pi \leq x \leq 2\pi$
19. $\tan x$ not defined, $0 \leq x \leq 4\pi$
20. $\cot x$ not defined, $0 \leq x \leq 4\pi$
21. $\csc x$ not defined, $0 \leq x \leq 4\pi$
22. $\sec x$ not defined, $0 \leq x \leq 4\pi$

Evaluate to four significant digits using a calculator.

23. $\sin(-0.2103)$
24. $\cos 14.78$
25. $\sec 1.432$
26. $\cot(-7.809)$
27. $\tan 4.704$
28. $\csc(-3.109)$

Evaluate exactly.

29. $\cos \dfrac{3\pi}{4}$
30. $\sin \dfrac{2\pi}{3}$
31. $\csc \left(-\dfrac{\pi}{4}\right)$
32. $\tan \left(-\dfrac{7\pi}{6}\right)$
33. $\csc \dfrac{7\pi}{2}$
34. $\cos \left(-\dfrac{5\pi}{2}\right)$

Fill in the blanks in the "Reason" column below with the appropriate identity, (1) through (9).

35.   Statement   Reason

$$\tan^2 x + 1 = \left(\frac{\sin x}{\cos x}\right)^2 + 1$$ (A) _____

$$= \frac{\sin^2 x}{\cos^2 x} + 1$$ Algebra

$$= \frac{\sin^2 x + \cos^2 x}{\cos^2 x}$$ Algebra

$$= \frac{1}{\cos^2 x}$$ (B) _____

$$= \left(\frac{1}{\cos x}\right)^2$$ Algebra

$$= \sec^2 x$$ (C) _____

36.   Statement   Reason

$$\cos^2 x + 1 = \left(\frac{\cos x}{\sin x}\right)^2 + 1$$ (A) _____

$cot^2 x$

$$= \frac{\cos^2 x}{\sin^2 x} + 1$$ Algebra

$$= \frac{\cos^2 x + \sin^2 x}{\sin^2 x}$$ Algebra

$$= \frac{1}{\sin^2 x}$$ (B) _____

$$= \left(\frac{1}{\sin x}\right)^2$$ Algebra

$$= \csc^2 x$$ (C) _____

C   **37.** What is the period of the cosecant function?
38. What is the period of the secant function?
39. What is the range of the sine function?
40. What is the range of the cosine function?

_____

2.7   TRIGONOMETRIC APPROXIMATIONS (OPTIONAL)

omit

There are situations in theory and applications where an approximation to a trigonometric value is useful. A well-chosen trigonometric approximation will often simplify an expression immensely. In this section we will approximate the sine and tangent functions with expressions that may be easily evaluated using any four function ($+$, $-$, $\times$, $\div$) calculator, or, in some cases, without a calculator at all.

APPROXIMATION OF
SINE AND TANGENT

A particularly simple approximation for both the sine and tangent functions may be observed by forming the following table using a calculator (set in radian mode) or by using Table III:

TRIGONOMETRIC FUNCTIONS—RADIANS

⤸	sine	tan
.00	.0000	.0000
.10	.0998	.1003
.20	.1987	.2027
.30	.2955	.3093
.40	.3894	.4228
.50	.4794	.5463
.60	.5646	.6841
.70	.6442	.8423

The table suggests that for a small angle θ, the radian measure of θ, $\sin \theta$, and $\tan \theta$ are all approximately equal. Let us look at these relationships through right triangle ratios.

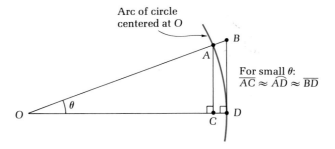

FIGURE 13

In Figure 13 the angles OCA and ODB are right angles, and AD is an arc of a circle centered at O. Let $\overline{OA}$, $\overline{OC}$, $\overline{AC}$, and $\overline{BD}$ be the lengths of the corresponding line segments, and let AD be the length of the arc AD. Then for $0 \leq \theta < \pi/2$,

$$\sin \theta = \frac{\overline{AC}}{\overline{OA}} = \frac{\overline{AC}}{\overline{OD}}$$

OA and OD are radii of the same circle; hence, $\overline{OA} = \overline{OD}$

$$\tan \theta = \frac{\overline{BD}}{\overline{OD}}$$

$$\theta \text{ (in rad)} = \frac{\overset{\frown}{AD}}{\overline{OD}}$$

From definition of radian measure (see Section 2.2)

For small positive θ

$$\overline{AC} \approx \overset{\frown}{AD} \approx \overline{BD}$$

Thus,

$$\sin \theta \approx \theta \approx \tan \theta$$

Following the same line of reasoning as above for negative θ near 0, we arrive at the same result. Thus, we have the following useful trigonometric approximations:

TRIGONOMETRIC APPROXIMATIONS

For θ in radian measure and $|\theta|$ small:
(1) $\sin \theta \approx \theta$
(2) $\tan \theta \approx \theta$
Note: The approximations are accurate to within 1% for angles up to about 0.17 radian (near 10°).

EXAMPLE 13   Radar indicates an approaching plane is 11 km from the airport at an angle of elevation of 0.087 radian.
(A)   Estimate the height of the airplane without using a calculator or table.
(B)   Compute the height of the airplane using a calculator (or table) without the use of an approximation formula.

Solutions   First draw a figure:

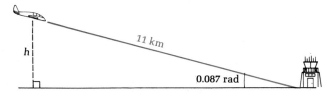

(A) $\sin 0.087 = \dfrac{h}{11}$

$\qquad\qquad h = 11 \sin 0.087 \qquad\quad \sin \theta \approx \theta \quad$ for small θ

$\qquad\qquad\quad \approx 11(0.087)$

$\qquad\qquad\quad \approx 0.96 \text{ km} \qquad\qquad$ To two significant digits

(B) $\sin 0.087 = \dfrac{h}{11}$

$\qquad\qquad h = 11 \sin 0.087 \qquad\quad$ Use a calculator or table

$\qquad\qquad\quad = 0.96 \text{ km}$

PROBLEM 13   At 400 ft from the base of a tree (on level ground) the angle of elevation to the top of the tree is 0.14 radian.

(A)   Estimate the height of the tree without using a calculator or table.

(B)   Compute the height of the tree using a calculator (or table) without the use of an approximation formula.

In the analysis of pendulum motion (considered in most physics courses), it is shown that the restoring force acting on a bob weighing W lb [tending to return it to the equilibrium position (straight down)] is given by (see Figure 14):

$$F = -W \sin \theta \qquad (1)$$

(More will be said about forces in Section 6.3.)

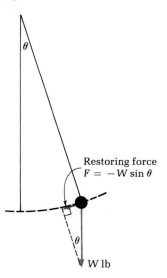

FIGURE 14   Simple pendulum

If the length of the pendulum string is long relative to the arc through which the bob swings, then θ will be small and $\sin \theta$ in equation (1) can be replaced by θ to yield

$$F \approx -W\theta$$

From this simpler equation physicists derive several other useful equations that describe the motion of the pendulum.

ANSWERS TO   **13.** (A)   56 m   (B)   56 m
MATCHED PROBLEMS

EXERCISE 2.7

A   *In Problems 1 and 2 use a calculator set in radian mode (or Table III) to complete each table to two decimal places.*

1.	θ	.01	.05	.10	.15	.20	.25	.30	.35	.40
	$\sin \theta$									

2.	θ	.01	.05	.10	.15	.20	.25	.30	.35	.40
	$\tan \theta$									

B   3.   A ship is due north of a reef that is 2,000 m west of a lighthouse. The angle formed by the ship, lighthouse, and reef is 0.131 radian. (See the figure.)
(A)   Estimate the distance from the ship to the reef without using a calculator or table.
(B)   Compute the distance using a calculator (or table) without an approximation formula.

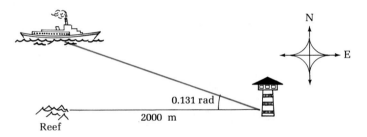

4.   Estimate the height of the Great Pyramid of Egypt if an observer 10,000 cubits from the center of the base views the apex (top) at an angle of elevation of 0.0289 radian. (See the figure.)
(A)   Compute the height with an approximation formula without using a calculator or table.
(B)   Compute the height using a calculator (or table) without an approximation formula.

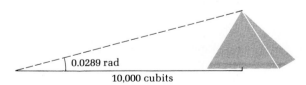

5.   If 110 m of kite string is out at an angle of elevation of 0.138 radian, how high is the kite?
(A)   Compute the height with an approximation formula without using a calculator or table.
(B)   Compute the height using a calculator (or table) without an approximation formula.
6.   Repeat Problem 5 with a kite string of 85 m and an angle of elevation of 0.105 radian.

C   Problems 7–9 refer to the following approximation formulas, which are more accurate than those discussed earlier. These are normally developed in a calculus course. For x in radians (or a real number), and $|x|$ small:

$$\sin x \approx x - \frac{x^3}{6} \qquad \tan x \approx x + \frac{x^3}{3} \qquad \cos x \approx 1 - \frac{x^2}{2}$$

Complete each table to four significant digits. Use a calculator.

7.

x	.05	.15	.25	.35	.45
$x - \dfrac{x^3}{6}$					
$\sin x$					

8.

x	.05	.15	.25	.35	.45
$x + \dfrac{x^3}{3}$					
$\tan x$					

9.

x	.05	.15	.25	.35	.45
$1 - \dfrac{x^2}{2}$					
$\cos x$					

EXERCISE 2.8   CHAPTER REVIEW

A   1. Convert to radian measure in terms of π:
(A) 60°   (B) 45°   (C) 90°
2. Convert to degree measure: (A) $\pi/6$ (B) $\pi/2$ (C) $\pi/4$
3. Find the value of $\sin \theta$ and $\tan \theta$ if the terminal side of θ contains $P(-4, 3)$.

Evaluate Problems 4–6 to four significant digits using a calculator. [Also (optional), evaluate using Tables I, II, or III.]

4. (A)   cot 53°40′   (B)   csc 67°10′
5. (A)   cos 23.5°   (B)   tan 42.3°
6. (A)   cos 0.35   (B)   tan 1.38
7. Evaluate exactly without tables or a calculator:
(A)   sin 60°   (B)   cos($\pi/4$)   (C)   tan 0°

B   **8.** What is the degree measure of a central angle subtended by an arc $\%_{60}$ of the circumference of a circle?

   9. What is the radian measure of a central angle subtended by an arc of length 24 cm if the radius of the circle is 8 cm?

   10. If the radius of a circle is 4 cm, find the length of an arc intercepted by an angle of 1.5 rad.

   11. Convert 212° to radian measure in terms of π.

   12. Convert $\pi/12$ rad to degree measure.

   13. Find the tangent of 0, $\pi/2$, π, and $3\pi/2$.

   14. In which quadrant does the terminal side of each angle lie?
   (A)   732°   (B)   −7 rad

   15. Find reference angles corresponding to:
   (A)   187.4°   (B)   103°20′   (C)   −37°40′

   16. Find reference angles corresponding to:
   (A)   2.39 rad   (B)   5 rad   (C)   −4 rad
   (Use $\pi \approx 3.14$.)

Evaluate Problems 17–28 to four significant digits using a calculator. [Also (optional), evaluate using Tables I, II, or III.]

17. cos 187.4°	**18.** tan 187.4°	**19.** sin 103°20′
20. sec 103°20′	**21.** cot(−37°40′)	**22.** sec(−37°40′)
23. sin 2.39	**24.** cot 2.39	**25.** cos 5
26. tan 5	**27.** sin(−4)	**28.** cot(−4)

   29. Find reference angles in exact form in terms of π for:
   (A) $5\pi/6$ (B) $7\pi/4$ (C) $-4\pi/3$

Find the exact value of each of the following without using tables or a calculator:

30. $\cos \dfrac{5\pi}{6}$	**31.** $\tan \dfrac{5\pi}{6}$	**32.** $\sin \dfrac{7\pi}{4}$
33. $\cot \dfrac{7\pi}{4}$	**34.** $\sin \dfrac{3\pi}{2}$	**35.** $\cos \dfrac{3\pi}{2}$
36. $\sin \dfrac{-4\pi}{3}$	**37.** $\sec \dfrac{-4\pi}{3}$	**38.** $\cos 3\pi$
39. $\cot 3\pi$	**40.** $\cos \dfrac{-11\pi}{6}$	**41.** $\sin \dfrac{-11\pi}{6}$

Use a calculator to evaluate each of the following to five decimal places:

42. sin 384.0314°	**43.** tan(−198°43′6″)
44. cos 26	**45.** cot(−68.005)

46. If $\sin \theta = -\frac{4}{5}$ and the terminal side of θ does not lie in the third quadrant, find the values of $\cos \theta$ and $\tan \theta$ without finding θ.

47. Find the least positive θ in radian measure such that $\sin \theta = -\frac{1}{2}$.

C   **48.** What is the period of the sine function? The cosine function?

49. One of the following is not an identity. Indicate which one.

(A) $\csc x = \dfrac{1}{\sin x}$ (B) $\cot x = \dfrac{1}{\tan x}$

(C) $\tan x = \dfrac{\sin x}{\cos x}$ (D) $\sec x = \dfrac{1}{\sin x}$

(E) $\sin^2 x + \cos^2 x = 1$ (F) $\cot x = \dfrac{\cos x}{\sin x}$

50. Indicate how sin x varies over each interval:
(A) $0 \le x \le \pi/2$ (B) $\pi/2 \le x \le \pi$
(C) $\pi \le x \le 3\pi/2$ (D) $3\pi/2 \le x \le 2\pi$

51. Indicate how cos x varies over each interval:
(A) $0 \le x \le \pi/2$ (B) $\pi/2 \le x \le \pi$
(C) $\pi \le x \le 3\pi/2$ (D) $3\pi/2 \le x \le 2\pi$

52. Through how many radians does a pulley with 10 cm diameter turn when 10 m of rope has been pulled through it without slippage? How many revolutions result?

GRAPHING TRIGONOMETRIC FUNCTIONS

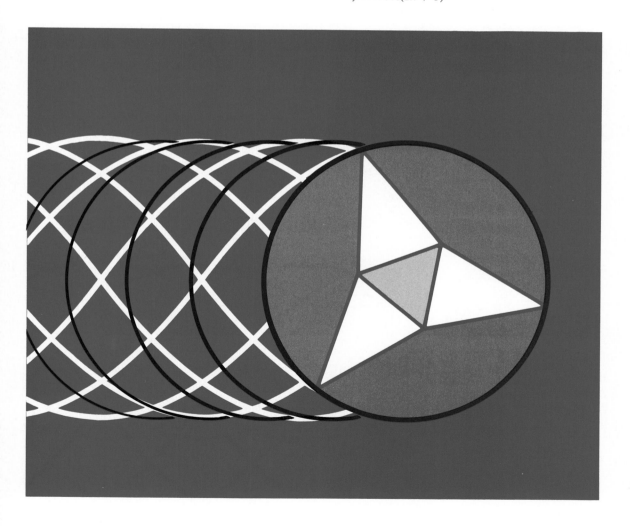

3.1 PERIODIC PHENOMENA

With the trigonometric functions now defined, we are in a position to consider a substantially expanded list of applications and properties. As a brief preview, look at Figure 1. What feature seems to be shared by all the phenomena? All appear to be repetitive—that is, **periodic.** The trigonometric functions, as we will see shortly, can be used to describe such phenomena with remarkable precision.

In this chapter you will learn how to sketch graphs of the trigonometric functions quickly and easily. You will also learn how to recognize certain fundamental and useful properties of these functions.

FIGURE 1

Time of sunrise

Sunrise time (AM)

Ocean waves

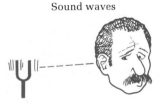

Water level (ft)

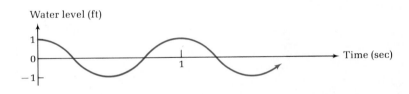

Sound waves

Pressure at ear drum

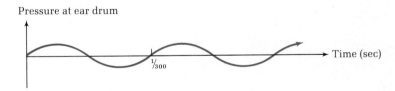

Radio waves

Intensity of magnetic field at antenna

GRAPHS OF
$y = \sin x$
AND $y = \cos x$

Recall that the **graph of an equation in two variables** is the set of all points in a coordinate system with coordinates that satisfy the equation. To graph $y = \sin x$ or $y = \cos x$ for x any real number, we could use a calculator or Table III. However, we can speed up the process by returning to the basic definitions of these functions in terms of a point $P(a, b)$ on the terminal side of the angle with radian measure x. To simplify matters even further, we choose $P(a, b)$ on a unit circle. Then (see Figure 2)

$$\sin x = \frac{b}{R} = \frac{b}{1} = b \qquad \cos x = \frac{a}{R} = \frac{a}{1} = a$$

Let us look at the behavior of sin x and cos x as P moves around the unit circle in Figure 2. Both behave in uniform ways for x in each quadrant, as indicated in Table 1.

Note in Table 1 that after x goes from 0 to 2π, $P(\cos x, \sin x)$ will have completed one revolution. If we let x continue to increase, then Table 1 will repeat every 2π units. In fact, it is a direct consequence of the definition of the sine and cosine functions that for all real x

$$\sin(x + 2\pi) = \sin x \qquad \cos(x + 2\pi) = \cos x$$

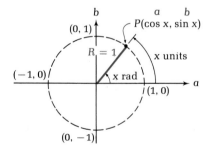

FIGURE 2
Unit circle

TABLE 1

AS x INCREASES FROM	$y = \sin x$	$y = \cos x$
0 to $\pi/2$	Positive Increases from 0 to 1	Positive Decreases from 1 to 0
$\pi/2$ to π	Positive Decreases from 1 to 0	Negative Decreases from 0 to -1
π to $3\pi/2$	Negative Decreases from 0 to -1	Negative Increases from -1 to 0
$3\pi/2$ to 2π	Negative Increases from -1 to 0	Positive Increases from 0 to 1

Recall from Section 2.6 that this important result is called the **periodic property** of the sine and cosine functions. The sine and cosine functions are both periodic with period 2π, which means that their graphs repeat every 2π units.

We are now ready to sketch graphs of

$$y = \sin x \quad \text{and} \quad y = \cos x$$

We determine their graphs for an interval of length 2π, say from 0 to 2π. Then, since both have a period of 2π, we repeat these graphs up and down the x axis as far as we wish. To sketch these graphs, we take advantage of the above information, supplemented with computer, table, or exact values for special angles, to obtain Figures 3 and 4.

FIGURE 3

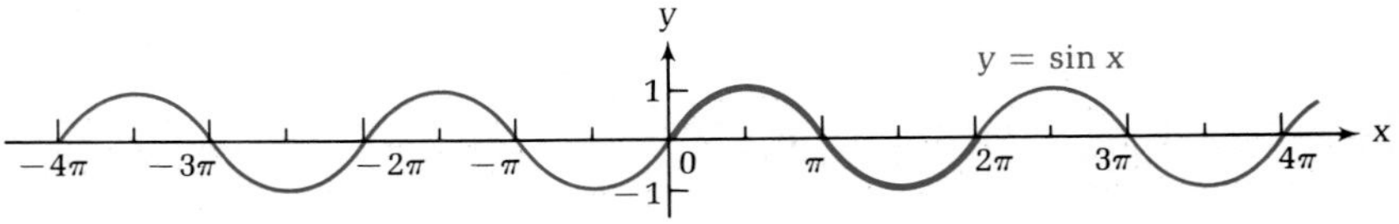

FIGURE 4

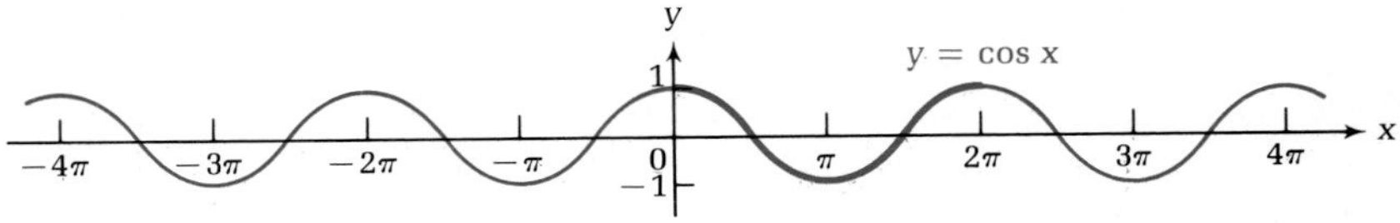

REMARKS

1. In Figures 3 and 4, we indicated multiples of π (special real numbers) on the x axis for convenience and clarity; all other real numbers are also associated with points on this axis.

2. The basic characteristics of the sine and cosine curves should be learned so that the curves can be quickly sketched from memory. In particular, you should be able to answer the following questions:

What is the period of each function (how often does the graph repeat)?

Where does each curve cross the x axis?

Where does each curve cross the y axis?

How far does each curve deviate from the x axis?

Where do the high and low points occur?

GRAPHING
$y = A \sin Bx$
AND $y = A \cos Bx$

Sketching graphs of equations of the form

$$y = A \sin Bx \quad \text{and} \quad y = A \cos Bx \tag{1}$$

is not difficult if we attack the problem step-by-step. Essential to the process, however, is a sound knowledge of the graphs of $y = \sin x$ and $y = \cos x$ in

Figures 3 and 4. We will find that the constants A and B in equation (1) both change the graphs in Figures 3 and 4 in particular ways. Once we know what their effect is, we will be able to modify Figures 3 and 4 to produce graphs of $y = A \sin Bx$ and $y = A \cos Bx$.

Let us first investigate the effect of A by comparing

$$y = \sin x \qquad \text{and} \qquad y = A \sin x$$

Each ordinate (y value) in Figure 3 is multiplied by A. The graph of $y = A \sin x$ will still cross the x axis where $y = \sin x$ crosses the x axis, since A times 0 is 0, but the maximum deviation of $y = \sin x$ from the x axis will change. Since $\sin x$ has period 2π,

$$A \sin(x + 2\pi) = A \sin x$$

Therefore, $A \sin x$ also has period 2π. Compare the graphs of $y = \frac{1}{3} \sin x$ and $y = -3 \sin x$ with the graph of $y = \sin x$ in Figure 5.

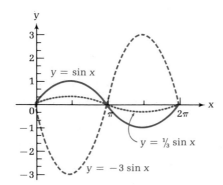

FIGURE 5

The constant $|A|$, the maximum deviation of $A \sin x$ from the x axis, is called the **amplitude** of $y = A \sin x$. Thus, $y = \frac{1}{3} \sin x$ has an amplitude of $|\frac{1}{3}| = \frac{1}{3}$, $y = -3 \sin x$ has an amplitude of $|-3| = 3$, and the amplitude of $y = \sin x$ is 1. The negative sign in $y = -3 \sin x$ turns the graph of $y = 3 \sin x$ upside down. The effect of A is to increase or decrease the ordinates (y values) of $y = \sin x$ without any change in the abscissas (x values).

A similar analysis applies to

$$y = A \cos x$$

and we conclude that this function also has period 2π and an amplitude of $|A|$.

Next, we compare

$$y = \sin x \qquad \text{and} \qquad y = \sin Bx \qquad B > 0$$

Both of these functions have amplitudes of 1, but how do their periods compare? To gain further insight, let us look at three particular cases with $B = 1$, $B = 2$, and $B = \frac{1}{2}$, respectively:

$$y = \sin x \qquad y = \sin 2x \qquad y = \sin \frac{x}{2}$$

Critical points for $y = \sin x$ (see Figure 3)

Maximum of $+1$ occurs at $x = \pi/2$

Minimum of -1 occurs at $x = 3\pi/2$

Curve crosses the x axis at $x = 0, \pi, 2\pi$

The period of $f(x) = \sin x$ is 2π, since

$$f(x + 2\pi) = \sin(x + 2\pi) = \sin x = f(x)$$

We use the results of this first case to give us information about the other two cases.

Critical points for $y = \sin 2x$

Maximum of $+1$ occurs at $2x = \pi/2$, or $x = \pi/4$

Minimum of -1 occurs at $2x = 3\pi/2$, or $x = 3\pi/4$

Curve crosses the x axis at $2x = 0, \pi, 2\pi$, or $x = 0, \pi/2, \pi$

The period of $f(x) = \sin 2x$ is π, since

$$f(x + \pi) = \sin[2(x + \pi)] = \sin(2x + 2\pi) = \sin 2x = f(x)$$

Critical points for $y = \sin(x/2)$

Maximum of $+1$ occurs at $x/2 = \pi/2$, or $x = \pi$

Minimum of -1 occurs at $x/2 = 3\pi/2$, or $x = 3\pi$

Curve crosses the x axis at $x/2 = 0, \pi, 2\pi$, or $x = 0, 2\pi, 4\pi$

The period of $f(x) = \sin(x/2)$ is 4π, since

$$f(x + 4\pi) = \sin\left(\frac{x + 4\pi}{2}\right) = \sin\left(\frac{x}{2} + 2\pi\right) = \sin\frac{x}{2} = f(x)$$

From the analysis above, we conclude that $\sin 2x$ completes one full cycle in an interval of π (half that for $\sin x$), while $\sin(x/2)$ requires an interval of 4π to complete one full cycle (twice that for $\sin x$). Thus, the effect of B is to compress or stretch the basic sine curve; that is, **B changes the period of $\sin x$.** Compare the graphs of $y = \sin 2x$ and $y = \sin(x/2)$ with $y = \sin x$ in Figure 6.

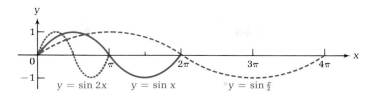

FIGURE 6

$y = \sin 2x \qquad y = \sin x \qquad y = \sin \frac{x}{2}$

A similar analysis applies to $y = \cos Bx$, where $B > 0$. In general,

$$f(x) = \sin Bx \qquad g(x) = \cos Bx$$

each has period $2\pi/B$, since

$$f\left(x + \frac{2\pi}{B}\right) = \sin B\left(x + \frac{2\pi}{B}\right) = \sin(Bx + 2\pi) = \sin Bx = f(x)$$

and

$$g\left(x + \frac{2\pi}{B}\right) = \cos B\left(x + \frac{2\pi}{B}\right) = \cos(Bx + 2\pi) = \cos Bx = g(x)$$

The effect of B is to stretch out the basic sine or cosine curve if $0 < B < 1$, and to compress it if $B > 1$.

Combining the discussions on amplitude and period, we summarize the results as follows:

FOR $y = A \sin Bx$ OR $y = A \cos Bx$, $B > 0$

$$\text{Amplitude} = |A| \qquad \text{Period} = \frac{2\pi}{B}$$

Let us now consider several examples where we will show how graphs of $y = A \sin Bx$ or $y = A \cos Bx$ can be sketched rather quickly.

EXAMPLE 1 State the amplitude and period for $y = -2 \sin x$, and graph the equation for $-2\pi \le x \le 2\pi$.

Solution $$\text{Amplitude} = |A| = |-2| = 2 \qquad \text{Period} = \frac{2\pi}{B} = \frac{2\pi}{1} = 2\pi$$

Since $A = -2$ is negative, the basic curve for $y = \sin x$ is turned upside down. To sketch $y = -2 \sin x$, block out one period from 0 to 2π, divide it into four equal parts, locate high and low points, and locate x intercepts, as shown in Figure 7 (page 78). Then sketch in one period, and extend this period to cover the desired interval, as shown in Figure 8.

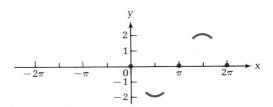

FIGURE 7

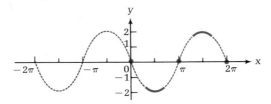

FIGURE 8

PROBLEM 1 State the amplitude and period for $y = \frac{1}{2} \cos x$, and graph the equation for $-2\pi \leq x \leq 2\pi$.

EXAMPLE 2 State the amplitude and period for $y = \cos \pi x$, and graph the equation for $-2 \leq x \leq 2$.

Solution $\quad$ Amplitude $= |A| = |1| = 1 \quad$ Period $= \dfrac{2\pi}{B} = \dfrac{2\pi}{\pi} = 2$

One full cycle of the graph is completed as x goes from 0 to 2. Block out this interval, divide it into four equal parts, locate high and low points, and locate x intercepts, as shown in Figure 9. Then complete the graph as shown in Figure 10. (Be sure to indicate appropriate scales for each axis.)

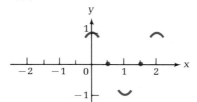

FIGURE 9

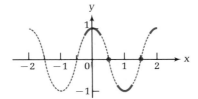

FIGURE 10

PROBLEM 2   State the amplitude and period for $y = \sin(x/4)$, and graph the equation for $-8\pi \le x \le 8\pi$.

EXAMPLE 3   State the period and amplitude for $y = -3 \cos(x/4)$, and graph the equation for $-2\pi \le x \le 12\pi$.

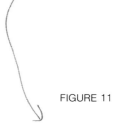

Solution $\text{Amplitude} = |A| = |-3| = 3$ $\text{Period} = \dfrac{2\pi}{B} = \dfrac{2\pi}{\frac{1}{4}} = 8\pi$

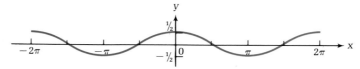

FIGURE 11

PROBLEM 3   State the period and amplitude for $y = 4 \sin \pi x$, and graph the equation for $-1 \le x \le 3$.

ANSWERS TO MATCHED PROBLEMS

1. $\text{Amplitude} = \frac{1}{2}$, $\text{Period} = 2\pi$

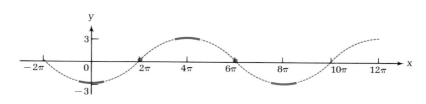

2. $\text{Amplitude} = 1$, $\text{Period} = 8\pi$

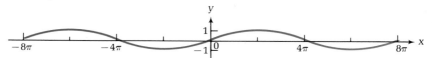

3. $\text{Amplitude} = 4$, $\text{Period} = 2$

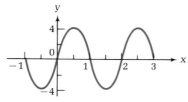

EXERCISE 3.2

A   1. Graph $y = \sin x$, $-2\pi \le x \le 2\pi$, by first plotting multiples of $\pi/6$ from 0 to 2π. Join these points with a smooth curve, and use the periodic property of the sine function to sketch in the rest of the curve.

2. Graph $y = \cos x$, $-2\pi \le x \le 2\pi$, using the procedure described in Problem 1 (above).

3. Use a calculator or Table III to produce an accurate graph of $y = \sin x$, $0 \le x \le 1.6$, using domain values $0, 0.1, 0.2, \ldots, 1.6$.

4. Use a calculator or Table III to produce an accurate graph of $y = \cos x$, $0 \le x \le 1.6$, using domain values $0, 0.1, 0.2, \ldots, 1.6$.

5. Where does $y = \sin x$ cross the x axis for $-2\pi \le x \le 2\pi$?

6. Where does $y = \cos x$ cross the x axis for $-2\pi \le x \le 2\pi$?

7. How far does $y = \sin x$ deviate from the x axis?

8. How far does $y = \cos x$ deviate from the x axis?

Make a rough sketch of each trigonometric function without looking at the text or using a table. Label each point where the graph crosses the x axis.

9. $y = \sin x$, $-\pi \le x \le 3\pi$ **10.** $y = \cos x$, $-\pi \le x \le 3\pi$

B   *State the amplitude and period for each equation, and graph it over the indicated interval.*

11. $y = 3 \cos x$, $-2\pi \le x \le 2\pi$ **12.** $y = 5 \sin x$, $-2\pi \le x \le 2\pi$

13. $y = -2 \sin x$, $0 \le x \le 4\pi$ **14.** $y = -3 \cos x$, $0 \le x \le 4\pi$

15. $y = \dfrac{1}{2} \sin x$, $0 \le x \le 2\pi$ **16.** $y = \dfrac{1}{3} \cos x$, $0 \le x \le 2\pi$

17. $y = \cos 2x$, $-\pi \le x \le \pi$ **18.** $y = \sin 4x$, $-\pi \le x \le \pi$

19. $y = \sin 2\pi x$, $-2 \le x \le 2$ **20.** $y = \cos 4\pi x$, $-1 \le x \le 1$

21. $y = \cos \dfrac{x}{4}$, $0 \le x \le 8\pi$ **22.** $y = \sin \dfrac{x}{2}$, $0 \le x \le 4\pi$

23. $y = 2 \sin 4x$, $-\pi \le x \le \pi$ **24.** $y = 3 \cos 2x$, $-\pi \le x \le \pi$

25. $y = \dfrac{1}{3} \cos 2\pi x$, $-2 \le x \le 2$ **26.** $y = \dfrac{1}{2} \sin 2\pi x$, $-2 \le x \le 2$

27. $y = -\dfrac{1}{4} \sin \dfrac{x}{2}$, $-4\pi \le x \le 4\pi$ **28.** $y = -3 \cos \dfrac{x}{2}$, $-4\pi \le x \le 4\pi$

APPLICATIONS

29. *Electrical circuits*   The voltage E in an electrical circuit is given by

$$E = 110 \sin 120\pi t$$

where t is time in seconds. What are the amplitude and period of the function? What is the **frequency** (cycles completed in 1 sec) of the function? Graph the function for $0 \le t \le \frac{3}{60}$.

30. *Spring–mass system*   The equation

$$y = -4 \cos 8t$$

where t is time in seconds, represents the motion of a weight hanging on a spring after it has been pulled 4 cm below its equilibrium point and released (see the figure). What is the amplitude, period, and frequency (see Problem 29) of the function? (Air resistance and friction are neglected.) Graph the function for $0 \le t \le 3\pi/4$.

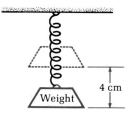

4 cm

Weight

*31. *Rotary and linear motion*   A ferris wheel with a diameter of 40 m rotates counterclockwise at 4 rpm (revolutions per minute) (see the figure). It starts at $\theta = 0$ when time t is 0. At the end of t minutes, $\theta = 8\pi t$. Why? Convince yourself that the position of the person's shadow on the x axis is given by

$$x = 20 \sin 8\pi t$$

Graph this equation for $0 \le t \le 1$.

*32. Repeat Problem 31 for a ferris wheel with a diameter of 30 m rotating at 6 rpm.

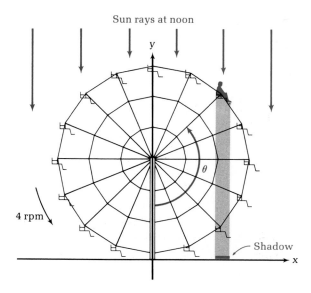

Sun rays at noon

3.3   GRAPHING $y = A \sin(Bx + C)$ AND $y = A \cos(Bx + C)$

We are now ready to consider equations of the more general form

$$y = A \sin(Bx + C) \quad \text{and} \quad y = A \cos(Bx + C) \tag{1}$$

We will find that the graphs of these equations are simply the graphs of

$$y = A \sin Bx \quad \text{or} \quad y = A \cos Bx \tag{2}$$

shifted to the left or right.

Let us start our investigation by comparing the graphs of $y = \sin(x + \pi/2)$ and $y = \sin(x - \pi/2)$ with the graph of $y = \sin x$. Observe the following:

$$f(x) = \sin x \qquad\qquad g(x) = \sin(x + \pi/2)$$
$$f(a) = \sin a \qquad g(a - \pi/2) = \sin(a - \pi/2 + \pi/2)$$
$$= \sin a$$

$$h(x) = \sin(x - \pi/2)$$
$$h(a + \pi/2) = \sin(a + \pi/2 - \pi/2)$$
$$= \sin a$$

Thus, the point with abscissa a on the graph of $y = f(x) = \sin x$ has the same ordinate value, $\sin a$, as the point with abscissa $a - \pi/2$ on the graph of $y = g(x) = \sin(x + \pi/2)$ and the point with abscissa $a + \pi/2$ on the graph of $y = h(x) = \sin(x - \pi/2)$.

Thus also, the graph of $y = \sin(x + \pi/2)$ is simply the graph of $y = \sin x$ shifted $\pi/2$ units to the left, and the graph of $y = \sin(x - \pi/2)$ is the graph of $y = \sin x$ shifted $\pi/2$ units to the right (see Figure 12). This shift of the basic curve is referred to as a **phase shift.** And we say that the graph of $y = \sin(x + \pi/2)$ has a phase shift of $\pi/2$ units to the left and the graph of $y = \sin(x - \pi/2)$ has a phase shift of $\pi/2$ to the right. The shift is just the opposite of what you might expect from the signs of $\pi/2$. The negative sign is associated with a shift to the right and the positive sign with a shift to the left.

Do $y = \sin\left(x - \frac{\pi}{6}\right)$

or HORIZONTAL → SHIFT

ask how to get vert. shift

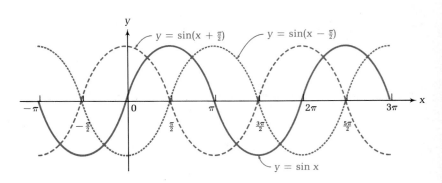

FIGURE 12

In general, for

$$y = A \sin(Bx + C) = A \sin B\left(x + \frac{C}{B}\right)$$

$$y = A \cos(Bx + C) = A \cos B\left(x + \frac{C}{B}\right)$$

the constant C/B is the phase shift (as described above). The graphs of $y = A \sin Bx$ and $y = A \cos Bx$ are shifted to the left C/B units if C/B is positive or to the right $|C/B|$ units if C/B is negative.

Let us summarize these results with those obtained in Section 3.2 in one convenient statement.

FOR $y = A \sin(Bx + C)$ OR $y = A \cos(Bx + C)$, $B > 0$

$$\text{Amplitude} = |A| \qquad \text{Period} = \frac{2\pi}{B}$$

$$\text{Phase shift} = \begin{cases} \left|\dfrac{C}{B}\right| \text{ units to the right} & \text{if } \dfrac{C}{B} < 0 \\[2mm] \dfrac{C}{B} \text{ units to the left} & \text{if } \dfrac{C}{B} > 0 \end{cases}$$

If x is time, then

$$\text{Frequency} = \frac{1}{\text{Period}} = \frac{B}{2\pi}$$

where Frequency = cycles completed in one unit of time.

EXAMPLE 4   State the amplitude, period, and phase shift for $y = 20 \cos(\pi x - \pi/2)$. Graph the equation for $-1 \le x \le 2$.

Solution   Write

$$y = 20 \overset{\overset{A}{\downarrow}}{\cos}\left(\overset{\overset{B}{\downarrow}}{\pi x} - \overset{\overset{C}{\downarrow}}{\frac{\pi}{2}}\right) = 20 \overset{\overset{A}{\downarrow}}{\cos} \overset{\overset{B}{\downarrow}}{\pi}\left(x - \overset{\overset{C/B}{\downarrow}}{\frac{1}{2}}\right)$$

then

$$\text{Amplitude} = |A| = |20| = 20 \qquad \text{Period} = \frac{2\pi}{B} = \frac{2\pi}{\pi} = 2$$

$$\text{Phase shift} = \left|\frac{C}{B}\right| = \frac{1}{2} \qquad \text{Right, since } \frac{C}{B} = -\frac{1}{2} \text{ is negative}$$

To graph the equation, first sketch the graph of $y = 20 \cos \pi x$ over one period from 0 to 2 on scratch paper; then shift the graph $\frac{1}{2}$ unit to the right, then extend and cut off to cover the interval from -1 to 2 (see Figure 13).

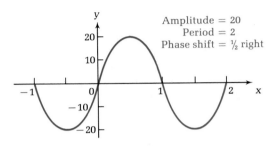

FIGURE 13

PROBLEM 4

Do ⟶

State the amplitude, period, and phase shift for $y = -5 \sin(x/2 + \pi/2)$. Graph the equation for $-3\pi \le x \le 5\pi$.

ANSWERS TO MATCHED PROBLEMS

4. Amplitude = 5,   Period = 4π,   Phase shift = π left;
$y = -5 \sin \frac{1}{2}(x + \pi)$

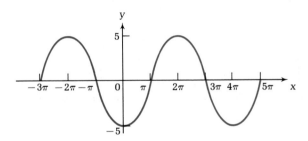

EXERCISE 3.3

A   *Indicate the phase shift for each equation and graph it over the stated interval.*

1. $y = \cos\left(x + \dfrac{\pi}{2}\right)$, $\dfrac{-\pi}{2} \le x \le \dfrac{3\pi}{2}$

2. $y = \cos\left(x - \dfrac{\pi}{2}\right)$, $\dfrac{\pi}{2} \le x \le \dfrac{5\pi}{2}$

3. $y = \sin\left(x - \dfrac{\pi}{4}\right)$, $-\pi \le x \le 2\pi$

4. $y = \cos\left(x + \dfrac{\pi}{4}\right)$, $-\pi \le x \le 2\pi$

5. $y = 3 \sin\left(x - \dfrac{\pi}{2}\right), \quad \dfrac{-\pi}{2} \le x \le \dfrac{5\pi}{2}$

6. $y = \dfrac{1}{4} \cos\left(x + \dfrac{\pi}{2}\right), \quad -\pi \le x \le 2\pi$

B   *State the amplitude, period, and phase shift for each equation and graph it over the indicated interval.*

7. $y = \sin 2\pi\left(x - \dfrac{1}{2}\right), \quad -1 \le x \le 2$

8. $y = \cos \pi(x - 1), \quad -2 \le x \le 3$

9. $y = 4 \cos\left(\pi x + \dfrac{\pi}{4}\right), \quad -1 \le x \le 3$

10. $y = 2 \sin\left(\pi x - \dfrac{\pi}{2}\right), \quad -2 \le x \le 2$

11. $y = -2 \cos(2x + \pi), \quad -\pi \le x \le 3\pi$

12. $y = -3 \sin(4x - \pi), \quad -\pi \le x \le \pi$

C   13. $y = 2 \sin\left(3x - \dfrac{\pi}{2}\right), \quad \dfrac{-2\pi}{3} \le x \le \dfrac{5\pi}{3}$

14. $y = -4 \cos\left(4x + \dfrac{\pi}{2}\right), \quad \dfrac{-\pi}{2} \le x \le 3$

15.   Graph

$$y = \cos\left(x - \dfrac{\pi}{2}\right) \quad \text{and} \quad y = \sin x$$

in the same coordinate system. Conclusion?

16.   Graph

$$y = \sin\left(x + \dfrac{\pi}{2}\right) \quad \text{and} \quad y = \cos x$$

in the same coordinate system. Conclusion?

APPLICATIONS   *Note: Additional applications from several different fields can be found in Chapter 8.*

17.   *Water waves (see Section 8.5)*   At a particular point in the ocean the vertical change in the water due to wave action is given by

$$y = 5 \sin \dfrac{\pi}{6}(t + 3)$$

where y is in meters and t is time in seconds. What is the amplitude, period, and phase shift? Graph the equation for $0 \le t \le 39$.

18. Repeat Problem 17 if the wave equation is

$$y = 8 \cos \frac{\pi}{12}(t - 6) \qquad 0 \leq t \leq 72$$

19. *Electrical circuit*   The current I (in amperes) in an electrical circuit is given by

$$I = 30 \sin(120\pi t - \pi)$$

where t is time in seconds. State the amplitude, period, frequency (cycles per second), and phase shift. Graph the equation for $0 \leq t \leq \frac{3}{60}$.

20. Repeat Problem 19 for

$$I = 110 \cos\left(120\pi t + \frac{\pi}{2}\right), \qquad 0 \leq t \leq \frac{2}{60}$$

*21. *Rotary and linear motion*   If, in Problem 31, Exercise 3.2 (the ferris wheel problem, page 81), we start θ at $\pi/2$ when $t = 0$, find the equation of motion of the shadow of a person. State the amplitude, period, and phase shift. Graph the equation for $0 \leq t \leq 1$.

3.4

TANGENT AND COTANGENT FUNCTIONS

GRAPHS OF OTHER TRIGONOMETRIC FUNCTIONS

We first discuss the graph of $y = \tan x$, then from this graph, because $\cot x = 1/\tan x$, we will be able to get the graph of $y = \cot x$ using reciprocals of ordinates.

To graph $y = \tan x$, we will take advantage of the identity

$$y = \tan x = \frac{\sin x}{\cos x}$$

(see Section 2.6). Let us first graph $y = \sin x$ and $y = \cos x$ on the same coordinate system, then use quotients of ordinates (quotients of y values) from these two graphs to sketch a graph of $y = \tan x$ (see Figure 14).

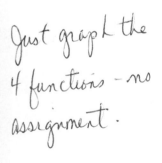

Just graph the 4 functions – no assignment.

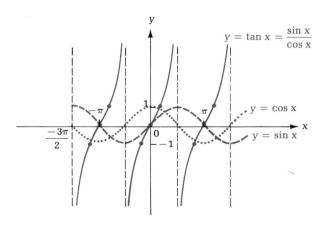

FIGURE 14

In Figure 14 everywhere the graph of $y = \cos x$ crosses the x axis, $\cos x$ is 0 and $\tan x$ is not defined. We draw vertical dashed lines through these points to remind us that the graph of $y = \tan x$ cannot cross these lines. These vertical lines will also be of additional use, as we will see in a moment.

Let us investigate the graph of $y = \tan x$ for the interval from 0 to $\pi/2$. Looking at Figure 14 (and proceeding mentally) we see that when x is 0, $\sin x = 0$ and $\cos x = 1$; thus, $\tan 0 = \sin 0/\cos 0 = 0/1 = 0$. When x is $\pi/4$, $\sin x = \cos x$; thus, $\tan (\pi/4) = \sin x/\cos x = 1$. When $x = \pi/2$, $\sin x = 1$ and $\cos x = 0$; thus, $\tan (\pi/2) = 1/0$ is not defined. What happens to $\tan x$ as x approaches $\pi/2$ from the left? We see that $\sin x$ approaches 1 and $\cos x$ approaches 0, both through positive values. Thus, $\tan x$ increases without bound as x approaches $\pi/2$ from the left.

Continuing in the same way, using quotients of ordinates from the graphs of $y = \sin x$ and $y = \cos x$, we can sketch in the graph of $y = \tan x$ (Figure 14). We see that the graph of $y = \tan x$ crosses the x axis (has x intercepts) everywhere $\sin x$ is 0, and the graph of $y = \tan x$ approaches the vertical lines drawn through the points where $\cos x = 0$. The vertical lines, called **asymptotes,** are useful guide lines for rapid sketching of the graph. We also note that the graph of $y = \tan x$ repeats every π units. Thus, the **tangent function has a period of π;** that is,

$$\tan(x + \pi) = \tan x$$

for all values of x for which both sides are defined.

EXAMPLE 5 *Calculator experiment* Form a table of values of $\tan x$ with x approaching $\pi/2 \approx 1.5707963$ from the left. Any conclusions?

Solution

x	0	0.5	1	1.5	1.57	1.5707	1.5707963
tan x	0	0.5	1.6	14.1	1,256	10,381	37,320,396

Conclusion: As x approaches $\pi/2$ from the left, $\tan x$ appears to increase without bound.

PROBLEM 5 Repeat Example 5, only with x approaching $-\pi/2 \approx -1.5707963$ from the right. Any conclusions?

The graph of $y = \tan x$ is redrawn in Figure 15.

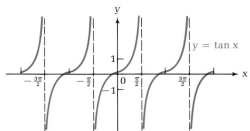

FIGURE 15

Proceeding in a similar manner for the cotangent function, or recalling that

$$\cot x = \frac{1}{\tan x}$$

and taking reciprocals of ordinate values from Figure 15, we obtain the graph of $y = \cot x$ shown in Figure 16. Again, notice the placement of the vertical dashed lines—each place where the graph of $y = \tan x$ crosses the x axis, the graph of $y = \cot x$ has a vertical asymptote.

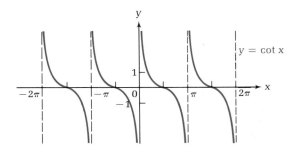

FIGURE 16

Combining these results with the type of analysis we carried out for $y = A \sin(Bx + C)$ and $y = A \cos(Bx + C)$, we arrive at the following general result:

FOR $y = A \tan(Bx + C)$ OR $y = A \cot(Bx + C)$, $B > 0$

$$\text{Period} = \frac{\pi}{B}$$

$$\text{Phase shift} = \begin{cases} \left|\dfrac{C}{B}\right| \text{ to the right} & \text{if } \dfrac{C}{B} < 0 \\[2ex] \dfrac{C}{B} \text{ to the left} & \text{if } \dfrac{C}{B} > 0 \end{cases}$$

Note that the amplitude is not defined for these two functions. Changing A only changes the steepness of the curves, and if A is negative, it reflects the basic curve relative to the x axis.

EXAMPLE 6   Indicate the period of $y = 3 \tan(\pi x/2)$, and sketch its graph for $-3 < x < 3$.

Solution $\text{Period} = \dfrac{\pi}{B} = \dfrac{\pi}{\pi/2} = 2$

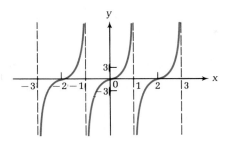

PROBLEM 6   Indicate the period of $y = 2 \cot(x/2)$, and sketch its graph for $-2\pi < x < 2\pi$.

SECANT AND COSECANT FUNCTIONS

We can take advantage of our knowledge of the sine and cosine functions to graph the secant and cosecant functions. Recall from Section 2.2 the reciprocal relationships

$$\sec x = \frac{1}{\cos x} \qquad \csc x = \frac{1}{\sin x}$$

Thus, to graph $y = \sec x$ and $y = \csc x$, we take reciprocals of the ordinate values of the graphs of $y = \cos x$ and $y = \sin x$, respectively. Points on the x axes for which cos x and sin x are zero correspond to vertical asymptotes (vertical guidelines) for the graphs of their respective reciprocal functions. The graphs are shown in Figures 17 and 18 on page 90. Some specific points on the graphs are determined in Table 2.

TABLE 2

x	$\cos x$	$\sec x = \dfrac{1}{\cos x}$	$\sin x$	$\csc x = \dfrac{1}{\sin x}$
0	1	1	0	Not defined
$\pi/6$	$\sqrt{3}/2$	$2/\sqrt{3}$	$1/2$	2
$\pi/3$	$1/2$	2	$\sqrt{3}/2$	$2/\sqrt{3}$
$\pi/2$	0	Not defined	1	1

As with the tangent and cotangent functions, amplitude is not defined for either the secant or the cosecant. The period of each function is the same as that for the sine and cosine, namely, 2π. Since these two functions are not used nearly as often as the other four, we will stop here even though it is a routine matter to consider more general forms.

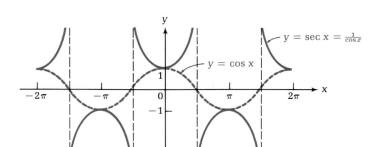

FIGURE 17

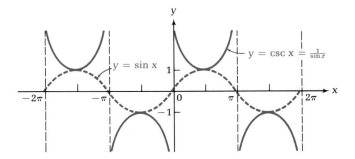

FIGURE 18

ANSWERS TO
MATCHED PROBLEMS

5.

x	0	−0.5	−1	−1.5	−1.57	−1.5707	−1.5707963
tan x	0	−0.5	−1.6	−14.1	−1,256	−10,381	−37,320,396

Conclusion: As x approaches −π/2 from the right, tan x appears to decrease without bound.

6. Period = 2π

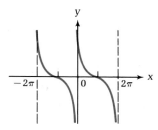

EXERCISE 3.4

A *Make rough sketches of each of the following without looking at the text or using a calculator or table.*

1. y = tan x,   0 ≤ x ≤ 2π
2. y = cot x,   0 < x < 2π
3. y = csc x,   −π < x < π
4. y = sec x,   −π ≤ x ≤ π

B   *Indicate the period of each function, and graph the function over the indicated interval.*

5. $y = 3 \tan 2x$, $-\pi < x < \pi$
6. $y = 2 \cot 4x$, $0 < x < \pi/2$
7. $y = -\frac{1}{2} \cot 2\pi x$, $0 < x < 1$
8. $y = -\frac{1}{4} \tan 8\pi x$, $0 < x < \frac{1}{2}$
9. $y = \sec \pi x$, $-1.5 \le x \le 3.5$
10. $y = \csc (x/2)$, $-3\pi \le x \le 3\pi$
11. $y = \frac{1}{2} \tan (x/2)$, $-\pi < x < 3\pi$
12. $y = \frac{1}{2} \cot (x/2)$, $0 < x < 4\pi$
13. $y = 2 \sec (x/2)$, $0 < x < 8\pi$
14. $y = 2 \sec \pi x$, $-1 < x < 3$

C   *Indicate the period and phase shift, and graph each function.*

15. $y = \tan(x - \pi/2)$, $-\pi < x < \pi$
16. $y = \cot(x + \pi/2)$, $-\pi/2 < x < 3\pi/2$
17. $y = 4 \tan(2x + \pi)$, $-\pi \le x \le \pi$
18. $y = -3 \cot(\pi x - \pi)$, $-2 \le x \le 2$

3.5   ADDITION OF ORDINATES

After having considered the trigonometric functions individually, we now consider them in combination with each other and with other functions. Graphing these combinations is best illustrated through examples.

EXAMPLE 7   Graph $y = (x/2) + \sin x$, $0 \le x \le 2\pi$.

Solution   A simple and fast method of graphing equations involving two or more terms is to graph each term separately on the same coordinate system and then add ordinates. In this case, we form

$$y_1 = \frac{x}{2} \quad \text{and} \quad y_2 = \sin x$$

to

# of squares on graph paper

We sketch the graph of each equation in the same coordinate system, then use a compass, dividers, ruler, or eye to add the ordinates $y_1 + y_2$ (see Figure 19, page 92). The final graph of $y = (x/2) + \sin x$ is shown in Figure 20. Specific points on the graph can also be determined as indicated in Table 3 if greater accuracy is desired.

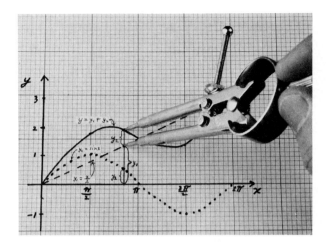

FIGURE 19
Addition of ordinates

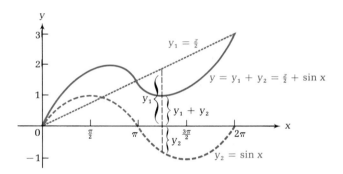

FIGURE 20

TABLE 3

x	$y_1 = \dfrac{x}{2}$	$y_2 = \sin x$	$y = y_1 + y_2 = \dfrac{x}{2} + \sin x$
0	0	0	0
$\dfrac{\pi}{2}$	$\dfrac{\pi}{4}$	1	$\dfrac{\pi}{4} + 1 \approx 1.79$
π	$\dfrac{\pi}{2}$	0	$\dfrac{\pi}{2} + 0 \approx 1.57$
$\dfrac{3\pi}{2}$	$\dfrac{3\pi}{4}$	-1	$\dfrac{3\pi}{4} - 1 \approx 1.36$
2π	π	0	$\pi + 0 \approx 3.14$

PROBLEM 7   Graph $y = (x/2) + \cos x$, $0 \le x \le 2\pi$, using addition of ordinates.

EXAMPLE 8   Graph $y = 3 \sin x + \cos 2x$, $0 \le x \le 3\pi$.

Solution   We use the method of addition of ordinates as in Example 7, letting $y_1 = 3 \sin x$ and $y_2 = \cos 2x$.

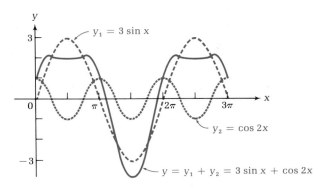

FIGURE 21

PROBLEM 8   Graph $y = 3 \sin x + \sin 3x$, $0 \le x \le 2\pi$, using addition of ordinates.

FOURIER SERIES
(OPTIONAL)

We now point out a significant use of combinations of trigonometric functions, called **Fourier series** (named after the French mathematician Joseph Fourier, 1768–1830). These combinations may be encountered in advanced applied mathematics in the study of heat flow, electrical fields and circuits, spring–mass systems, and sound. The discussion below is included only for illustrative purposes, and the reader is not expected to become proficient in this area at this time.

The following are examples of Fourier series:

$$y = \sin x + \frac{\sin 3x}{3} + \frac{\sin 5x}{5} + \cdots \tag{1}$$

$$y = \sin \pi x + \frac{\sin 2\pi x}{2} + \frac{\sin 3\pi x}{3} + \cdots \tag{2}$$

The three dots at the end of each series indicate that the pattern established in the first three terms continues indefinitely.

If we graph the first term of each series, then the sum of the first two terms, then the sum of the first three terms, and so on, we will obtain a sequence of graphs that will get closer and closer to a **square wave** for (1) and a **sawtooth wave** for (2). The greater the number of terms we take in the

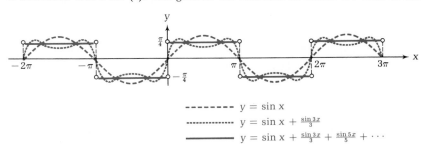

FIGURE 22
Square wave

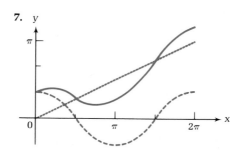

FIGURE 23
Sawtooth wave

- - - - - - - - $y = \sin \pi x$

- - - - - - - - $y = \sin \pi x + \frac{\sin 2\pi x}{2}$

———————— $y = \sin \pi x + \frac{\sin 2\pi x}{2} + \frac{\sin 3\pi x}{3} + \cdots$

series, the more the graph will look like the indicated wave form. Figures 22 and 23 illustrate these phenomena.

ANSWERS TO
MATCHED PROBLEMS

7.

8.

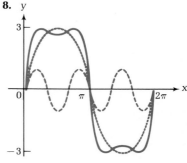

EXERCISE 3.5

A   *Graph each function over the indicated interval.*

1. $y = 2 + \sin x, \quad 0 \le x \le 2\pi$
2. $y = -1 + \cos x, \quad -\pi/2 \le x \le 3\pi/2$
3. $y = x + \cos x, \quad 0 \le x \le 5\pi/2$
4. $y = x + \sin x, \quad 0 \le x \le 2\pi$

B   5. $y = \dfrac{x}{2} + \cos \pi x, \quad 0 \le x \le 3$

6. $y = \dfrac{x}{2} - \sin 2\pi x, \quad 0 \le x \le 2$

7. $y = 3 \cos x + \sin 2x, \quad 0 \le x \le 3\pi$
8. $y = 3 \cos x + \cos 3x, \quad 0 \le x \le 2\pi$

⑨ $y = \sin x + 2 \cos 2x, \quad 0 \le x \le 3\pi$

10. $y = \cos x + 3 \sin 2x, \quad 0 \le x \le 3\pi$

C **11.** $y = \sin x + \dfrac{\sin 3x}{3}, \quad 0 \le x \le 2\pi$

(Compare to the square wave in Figure 22.)

12. $y = \sin \pi x + \dfrac{\sin 2\pi x}{2}, \quad -2 \le x \le 2$

(Compare to the sawtooth wave in Figure 23.)

*__13.__ $y = \dfrac{x}{2} \cos x, \quad 0 \le x \le 3\pi$

*__14.__ $y = \dfrac{x}{4} \sin \pi x, \quad 0 \le x \le 4$

APPLICATIONS **15.** *Pollution* In a large city the amount of sulfur dioxide released into the atmosphere due to the burning of coal and oil for heating purposes varies seasonally. Suppose the number of tons of pollutant released into the atmosphere during the xth week after January 1 is given approximately by

$$P(x) = 1.1 + \cos \frac{\pi x}{26}$$

Graph this function for a 2 year period starting January 1.

16. *Seasonal business cycle* Suppose profit in the sale of swimming suits in a chain department store is given approximately by

$$P(x) = 5 - 5 \cos \frac{\pi x}{26}$$

where *P* is the profit in hundreds of dollars for a week of sales x weeks after January 1. Graph this function for a 2 year period starting January 1.

EXERCISE 3.6 CHAPTER REVIEW

A *Sketch a graph of each function for* $-2\pi \le x \le 2\pi$.

1. $y = \sin x$ **2.** $y = \cos x$ **3.** $y = \tan x$

4. $y = \cot x$ **5.** $y = \sec x$ **6.** $y = \csc x$

Sketch a graph of each function for the indicated interval.

7. $y = 3 \cos \dfrac{x}{2}, \quad -4\pi \le x \le 4\pi$ **8.** $y = \dfrac{1}{2} \sin 2x, \quad -\pi \le x \le \pi$

9. $y = 4 + \cos x, \quad 0 \le x \le 2\pi$

B **10.** $y = -2 \sin \pi x, \quad -2 \le x \le 2$

11. $y = -\dfrac{1}{3} \cos 2\pi x, \quad -2 \le x \le 2$

12. $y = \sin\left(x - \dfrac{\pi}{2}\right), \quad 0 \le x \le 2\pi$

13. $y = \cos(x + \pi), \quad 0 \le x \le 2\pi$

14. $y = \tan 2x, \quad -\pi \le x \le \pi$

15. $y = \cot \pi x, \quad -2 < x < 2$

16. $y = x + \sin \pi x, \quad 0 \le x \le 2$

17. $y = 2 \sin x + \cos 2x, \quad 0 \le x \le 4\pi$

18. What are the period and amplitude of the function in Problem 11?

19. What is the period of the function in Problem 15?

20. What is the phase shift for the function in Problem 12?

21. What are the amplitude, period, and phase shift for $y = -3\cos(\pi x + \pi)$?

C   *Sketch a graph of each function for the indicated interval.*

22. $y = -2 \sin(\pi x - \pi), \quad 0 \le x \le 2$

23. $y = -\dfrac{1}{4}\cos(2x + \pi), \quad 0 \le x \le 2\pi$

*24. $y = 2^x \sin 2\pi x, \quad 0 \le x \le 4$

IDENTITIES 4

4.1 BASIC IDENTITIES AND THEIR USE

Trigonometric functions have many uses. In addition to solving real-world problems, they are used in the development of mathematics—analytic geometry, calculus, and so on. Whatever their use, it is often of value to be able to change a trigonometric expression from one form to an equivalent form. This involves the use of **identities.** An equation in one or more variables is said to be an **identity** if the left side is equal to the right side for all replacements of the variables for which both sides are defined. The equation

$$x^2 - x - 6 = (x - 3)(x + 2)$$

is an identity, while

$$x^2 - x - 6 = 2x$$

is not. The latter is called a **conditional equation,** since it only holds for certain values of x and not for all values for which both sides are defined.

BASIC IDENTITIES

Our first encounter with trigonometric identities was in Section 2.6, where we established 11 basic forms. We restate and name these here for convenient reference. These 11 basic identities will be used very frequently in the work that follows and should be memorized. The second and third

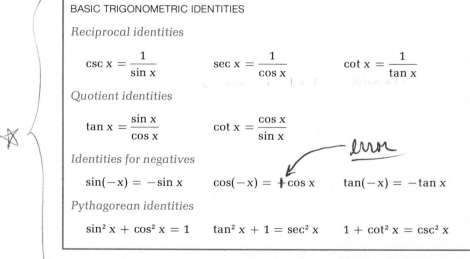

BASIC TRIGONOMETRIC IDENTITIES

Reciprocal identities

$$\csc x = \frac{1}{\sin x} \qquad \sec x = \frac{1}{\cos x} \qquad \cot x = \frac{1}{\tan x}$$

Quotient identities

$$\tan x = \frac{\sin x}{\cos x} \qquad \cot x = \frac{\cos x}{\sin x}$$

Identities for negatives

$$\sin(-x) = -\sin x \qquad \cos(-x) = +\cos x \qquad \tan(-x) = -\tan x$$

Pythagorean identities

$$\sin^2 x + \cos^2 x = 1 \qquad \tan^2 x + 1 = \sec^2 x \qquad 1 + \cot^2 x = \csc^2 x$$

Pythagorean identities were established in Problems 35 and 36 in Exercise 2.6. An easy way to remember them is to note that the second can be obtained from the first by dividing both sides of the first by $\cos^2 x$, and the third can be obtained from the first by dividing both sides of the first by $\sin^2 x$, as shown at the top of the next page.

$$\sin^2 x + \cos^2 x = 1 \qquad \sin^2 x + \cos^2 x = 1$$

$$\frac{\sin^2 x}{\cos^2 x} + \frac{\cos^2 x}{\cos^2 x} = \frac{1}{\cos^2 x} \qquad \frac{\sin^2 x}{\sin^2 x} + \frac{\cos^2 x}{\sin^2 x} = \frac{1}{\sin^2 x}$$

$$\tan^2 x + 1 = \sec^2 x \qquad 1 + \cot^2 x = \csc^2 x$$

ESTABLISHING OTHER IDENTITIES

As indicated earlier, when working with trigonometric expressions, it is often desirable to convert one form to an equivalent form that may be more useful. This section is designed to give you experience in this process. The 11 basic identities listed in the box will be used frequently, so they should be learned before proceeding further. The following examples illustrate some of the techniques used to establish certain identities. To become proficient in the use of identities, it is important that you work out many problems on your own.

EXAMPLE 1   Establish the identity

$$\sin x \cot x = \cos x$$

Do

Proof   Generally, we start with the more complicated side and transform it into the other side using basic identities, algebra, or other established identities. Thus,

$$\sin x \cot x = \sin x \frac{\cos x}{\sin x} \qquad \text{Quotient identity}$$

$$= \cos x \qquad \text{Algebra}$$

PROBLEM 1   Establish the identity

$$\cos x \tan x = \sin x$$

EXAMPLE 2   Establish the identity

$$\csc(-x) = -\csc x$$

Proof $$\csc(-x) = \frac{1}{\sin(-x)} \qquad \text{Reciprocal identity}$$

$$= \frac{1}{-\sin x} \qquad \text{Identity for negatives}$$

$$= -\frac{1}{\sin x} \qquad \text{Algebra}$$

$$= -\csc x \qquad \text{Reciprocal identity}$$

PROBLEM 2   Establish the identity

$$\sec(-x) = \sec x$$

EXAMPLE 3   Establish the identity

Do

$$\tan x \sin x + \cos x = \sec x$$

Proof $\qquad \tan x \sin x + \cos x = \dfrac{\sin x}{\cos x} \sin x + \cos x \qquad$ Quotient identity

$$= \dfrac{\sin^2 x + \cos^2 x}{\cos x} \qquad\qquad \text{Algebra}$$

$$= \dfrac{1}{\cos x} \qquad\qquad \text{Pythagorean identity}$$

$$= \sec x \qquad\qquad \text{Reciprocal identity}$$

PROBLEM 3   Establish the identity

$$\cot x \cos x + \sin x = \csc x$$

There is no fixed procedure that works in the proofs for all identities. Nevertheless, certain steps can be taken that will help in many cases:

SUGGESTED STEPS FOR PROVING IDENTITIES

1. Start with the more complicated side and transform it into the simpler side.

2. Try algebraic operations such as multiplying, factoring, combining fractions, splitting single fractions, and so on.

3. If other steps fail, express each function in terms of sine and cosine functions and then perform appropriate algebraic operations.

4. At each step keep the other side of the identity in mind. This often reveals what one should do in order to get there.

---

5. Use formulas when you see 1's and squares.

EXAMPLE 4   Establish the identity

$$\dfrac{\tan x - \cot x}{\tan x + \cot x} = 1 - 2 \cos^2 x$$

DO

Proof $\qquad \dfrac{\tan x - \cot x}{\tan x + \cot x} = \dfrac{\dfrac{\sin x}{\cos x} - \dfrac{\cos x}{\sin x}}{\dfrac{\sin x}{\cos x} + \dfrac{\cos x}{\sin x}} \qquad$ Change to sines and cosines

$$= \dfrac{(\sin x)(\cos x)\left(\dfrac{\sin x}{\cos x} - \dfrac{\cos x}{\sin x}\right)}{(\sin x)(\cos x)\left(\dfrac{\sin x}{\cos x} + \dfrac{\cos x}{\sin x}\right)}$$

Multiply numerator and denominator by $(\sin x)(\cos x)$ and use algebra to transform complex fraction into simple fraction

$$= \frac{\sin^2 x - \cos^2 x}{\sin^2 x + \cos^2 x} \qquad \text{A simple fraction}$$

$$= \frac{1 - \cos^2 x - \cos^2 x}{1} \qquad \begin{array}{l}\text{Pythagorean identity,}\\ \text{twice}\end{array}$$

$$= 1 - 2\cos^2 x \qquad \text{Algebra}$$

PROBLEM 4   Establish the identity

$\mathcal{D}o$

$$\cot x - \tan x = \frac{2\cos^2 x - 1}{\sin x \cos x}$$

EXAMPLE 5   Establish the identity

$$\frac{1 + \cos x}{\sin x} + \frac{\sin x}{1 + \cos x} = 2 \csc x$$

Proof

$$\frac{1 + \cos x}{\sin x} + \frac{\sin x}{1 + \cos x} = \frac{(1 + \cos x)^2 + \sin^2 x}{(\sin x)(1 + \cos x)} \qquad \text{Algebra}$$

$$= \frac{1 + 2\cos x + \cos^2 x + \sin^2 x}{(\sin x)(1 + \cos x)} \qquad \text{Algebra}$$

$$= \frac{1 + 2\cos x + 1}{(\sin x)(1 + \cos x)} \qquad \begin{array}{l}\text{Pythagorean}\\ \text{identity}\end{array}$$

$$= \frac{2 + 2\cos x}{(\sin x)(1 + \cos x)} \qquad \text{Algebra}$$

$$= \frac{2(1 + \cos x)}{(\sin x)(1 + \cos x)} \qquad \text{Algebra}$$

$$= \frac{2}{\sin x} \qquad \text{Algebra}$$

$$= 2 \csc x \qquad \begin{array}{l}\text{Reciprocal}\\ \text{identity}\end{array}$$

PROBLEM 5   Establish the identity

$\mathcal{D}o$

$$\frac{1 + \sin x}{\cos x} + \frac{\cos x}{1 + \sin x} = 2 \sec x$$

PROOFS FOR MATCHED PROBLEMS

1. $\cos x \tan x = \cos x \dfrac{\sin x}{\cos x} = \sin x$

2. $\sec(-x) = \dfrac{1}{\cos(-x)} = \dfrac{1}{\cos x} = \sec x$

3. $\cot x \cos x + \sin x = \dfrac{\cos^2 x}{\sin x} + \sin x = \dfrac{\cos^2 x + \sin^2 x}{\sin x} = \dfrac{1}{\sin x} = \csc x$

4. $\cot x - \tan x = \dfrac{\cos x}{\sin x} - \dfrac{\sin x}{\cos x} = \dfrac{\cos^2 x - \sin^2 x}{\sin x \cos x}$

$$= \dfrac{\cos^2 x - (1 - \cos^2 x)}{\sin x \cos x} = \dfrac{2 \cos^2 x - 1}{\sin x \cos x}$$

5. $\dfrac{1 + \sin x}{\cos x} + \dfrac{\cos x}{1 + \sin x} = \dfrac{(1 + \sin x)^2 + \cos^2 x}{(\cos x)(1 + \sin x)}$

$$= \dfrac{1 + 2 \sin x + \sin^2 x + \cos^2 x}{(\cos x)(1 + \sin x)}$$

$$= \dfrac{2 + 2 \sin x}{(\cos x)(1 + \sin x)} = \dfrac{2}{\cos x} = 2 \sec x$$

EXERCISE 4.1

A *Match each function in Problems 1–12 with one of the following to form a fundamental identity or a minor variation of one. Do not look at the list of 11 identities that was given in this section.*

(A) $\dfrac{1}{\csc x}$ (B) $\cos x$ (C) 1 (D) $\tan x$

(E) $\dfrac{1}{\sec x}$ (F) $\sin^2 x$ (G) $\tan^2 x$ (H) $\cot x$

(I) $-\sin x$ (J) $1 - \sin^2 x$ (K) $1 - \cos^2 x$ (L) $\sec^2 x$

1. $\sin x$ **2.** $\cos(-x)$ **3.** $\cos x$

4. $\dfrac{\sin x}{\cos x}$ **5.** $\dfrac{\cos x}{\sin x}$ **6.** $\sin^2 x + \cos^2 x$

7. $\cos^2 x$ **8.** $\dfrac{1}{\cot x}$ **9.** $\tan^2 x + 1$

10. $\sin(-x)$ **11.** $\sin^2 x$ **12.** $\sec^2 x - 1$

Establish each identity.

13. $\cos x \sec x = 1$ **14.** $\sin x \csc x = 1$

15. $\tan x \cos x = \sin x$ **16.** $\cot x \sin x = \cos x$

17. $\tan x = \sin x \sec x$ **18.** $\cot x = \cos x \csc x$

19. $\tan(-x) = -\tan x$ **20.** $\cot(-x) = -\cot x$

21. $\dfrac{\sin x}{\csc x} + \dfrac{\cos x}{\sec x} = 1$ **22.** $\dfrac{1}{\sec^2 x} + \dfrac{1}{\csc^2 x} = 1$

B **23.** $\dfrac{\sin^2 x}{\cos x} + \cos x = \sec x$ **24.** $\dfrac{\cos^2 x}{\sin x} + \sin x = \csc x$

25. $\dfrac{1 - \cos^2 y}{\cos^2 y} = \tan^2 y$ **26.** $\dfrac{1 - (\sin x - \cos x)^2}{\sin x} = 2 \cos x$

27. $\tan x + \cot x = \sec x \csc x$ **28.** $\dfrac{\csc x}{\cot x + \tan x} = \cos x$

Discuss
cross-mult.

29. $\dfrac{1 - \csc x}{1 + \csc x} = \dfrac{\sin x - 1}{\sin x + 1}$

30. $\dfrac{1 - \cos x}{1 + \cos x} = \dfrac{\sec x - 1}{\sec x + 1}$

31. $\sin^2 x - \cos^2 x = 1 - 2 \cos^2 x$

32. $\cos^2 x - \sin^2 x = 1 - 2 \sin^2 x$

33. $(\sin x + \cos x)^2 - 1 = 2 \sin x \cos x$

34. $\sec x - 2 \sin x = \dfrac{(\sin x - \cos x)^2}{\cos x}$

35. $\sin^4 x - \cos^4 x = 1 - 2 \cos^2 x$

36. $\sin^4 x + 2 \sin^2 x \cos^2 x + \cos^4 x = 1$

37. $\csc u - \dfrac{\sin u}{1 + \cos u} = \cot u$

38. $\sec u - \dfrac{\cos u}{1 + \sin u} = \tan u$

39. $\dfrac{\cos x}{1 - \sin x} + \dfrac{\cos x}{1 + \sin x} = 2 \sec x$

40. $\dfrac{\cos x}{\csc x + 1} + \dfrac{\cos x}{\csc x - 1} = 2 \tan x$

41. $\dfrac{\cos^2 n - 3 \cos n + 2}{\sin^2 n} = \dfrac{2 - \cos n}{1 + \cos n}$

42. $\dfrac{\sin^2 n + 4 \sin n + 3}{\cos^2 n} = \dfrac{3 + \sin n}{1 - \sin n}$

43. $\dfrac{1 - \tan^2 y}{\cot^2 y - 1} = \tan^2 y$

44. $\dfrac{\tan y}{\sin y - 2 \tan y} = \dfrac{1}{\cos y - 2}$

45. $\dfrac{1 - \cot^2 x}{\tan^2 x - 1} = \cot^2 x$

46. $\dfrac{\tan^2 x - 1}{1 - \cot^2 x} = \tan^2 x$

47. $\sec^2 x \cdot + \csc^2 x = \sec^2 x \csc^2 x$

48. $\tan^2 x - \sin^2 x = \tan^2 x \sin^2 x$

49. $(\sec x - \tan x)^2 = \dfrac{1 - \sin x}{1 + \sin x}$

50. $\dfrac{1 - \cos x}{1 + \cos x} = (\cot x - \csc x)^2$

51. $\dfrac{1 + \sin t}{\cos t} = \dfrac{\cos t}{1 - \sin t}$

52. $\dfrac{\sin t}{1 - \cos t} = \dfrac{1 + \cos t}{\sin t}$

Show that each of the following equations is not an identity by finding a value for which both sides are defined but are not equal to each other. For example, sin x = cos x is not an identity, since the two sides are not equal for x = 0.

53. $\tan x = \cot x$ **54.** $\sin^2 x + \sin x = 1$

55. $\sin^2 x - \cos^2 x = 1$ **56.** $\cot x = 2 \sin x \cos x$

57. $\cot^2 x + \cos x = \sin^2 x$ **58.** $\cos^2 x + 2 \cos x - 8 = 0$

C   *Prove that the statements in Problems 59–62 are identities.*

59. $\dfrac{3 \cos^2 m + 5 \sin m - 5}{\cos^2 m} = \dfrac{3 \sin m - 2}{1 + \sin m}$

60. $\dfrac{2 \sin^2 z + 3 \cos z - 3}{\sin^2 z} = \dfrac{2 \cos z - 1}{1 + \cos z}$

61. $\dfrac{\tan y + \sin y}{\tan y - \sin y} = \dfrac{\sec y + 1}{\sec y - 1}$

62. $\dfrac{\sin x \cos y + \cos x \sin y}{\cos x \cos y - \sin x \sin y} = \dfrac{\tan x + \tan y}{1 - \tan x \tan y}$

Each of the following is an identity in certain quadrants. Indicate which quadrants.

63. $\sqrt{1 - \cos^2 x} = \sin x$ **64.** $\sqrt{1 - \sin^2 x} = \cos x$

65. $\sqrt{1 - \sin^2 x} = -\cos x$ **66.** $\sqrt{1 - \cos^2 x} = -\sin x$

67. $\sqrt{1 - \sin^2 x} = |\cos x|$ **68.** $\sqrt{1 - \cos^2 x} = |\sin x|$

69. $\dfrac{\sin x}{\sqrt{1 - \sin^2 x}} = \tan x$ **70.** $\dfrac{\sin x}{\sqrt{1 - \sin^2 x}} = -\tan x$

4.2

SUM AND DIFFERENCE
IDENTITIES FOR COSINE

[handwritten: Show $\cos(x-y) \neq \cos x - \cos y$ ✱ w/ $x = 60°$, $y = 30°$]

SUM, DIFFERENCE, AND COFUNCTION IDENTITIES

The basic identities discussed in Section 4.1 involve only one variable. We will now consider an important identity, called a difference identity for cosine, which involves two variables:

$$\cos(x - y) = \cos x \cos y + \sin x \sin y \tag{1}$$

Many other useful identities can be readily established from this particular one.

    We will sketch a proof of (1) in which we assume that x and y are restricted as follows: $0 < y < x < 2\pi$. Identity (1) holds, however, for all real numbers and angles in radian or degree measure.

    We associate x and y with arcs and angles on a unit circle as indicated in Figure 1a. Using the definitions of the circular functions in Section 2.6, the terminal points of x and y are labeled as indicated in Figure 1a.

    Now if we rotate the triangle *AOB* clockwise about the origin until the

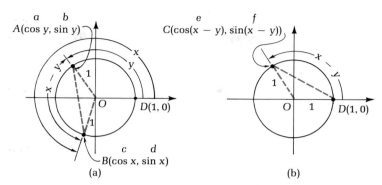

FIGURE 1   (a)   (b)

terminal point A coincides with $D(1, 0)$, then terminal point B will be at C (see Figure 1b). Thus, since rotation preserves lengths,

$$d(A, B) = d(C, D)$$
$$\sqrt{(c - a)^2 + (d - b)^2} = \sqrt{(1 - e)^2 + (0 - f)^2}$$
$$(c - a)^2 + (d - b)^2 = (1 - e)^2 + f^2$$
$$c^2 - 2ac + a^2 + d^2 - 2db + b^2 = 1 - 2e + e^2 + f^2$$
$$(c^2 + d^2) + (a^2 + b^2) - 2ac - 2db = 1 - 2e + (e^2 + f^2) \qquad (2)$$

Since $c^2 + d^2 = 1$, $a^2 + b^2 = 1$, and $e^2 + f^2 = 1$ (Why?), equation (2) becomes

$$e = ac + bd \qquad (3)$$

Replacing e, a, c, b, and d with $\cos(x - y)$, $\cos y$, $\cos x$, $\sin y$, and $\sin x$, respectively (see Figure 1), we obtain:

$$\cos(x - y) = \cos y \cos x + \sin y \sin x$$
$$= \cos x \cos y + \sin x \sin y \qquad (4)$$

If we replace y with $-y$ in (4) and use the identities for negatives, we obtain the sum identity for cosine:

$$\cos(x + y) = \cos x \cos y - \sin x \sin y \qquad (5)$$

COFUNCTION
IDENTITIES

To obtain sum and difference identities for the sine and tangent functions, we first derive **cofunction** identities directly out of (1), the difference identity for cosine.

$$\cos(x - y) = \cos x \cos y + \sin x \sin y \qquad \text{Let } x = \pi/2$$
$$\cos\left(\frac{\pi}{2} - y\right) = \cos\frac{\pi}{2}\cos y + \sin\frac{\pi}{2}\sin y$$
$$= (0)\cos y + (1)\sin y$$
$$= \sin y$$

Thus,

$$\cos\left(\frac{\pi}{2} - y\right) = \sin y \qquad (6)$$

Handwritten annotations: DO · this is where we get cosine · osine : the complement of sine for y any real number or angle in radian measure. If y is in degree measure, replace $\pi/2$ with 90°.

Now, if in (6) we let $y = \pi/2 - x$, then we have

$$\cos\left[\frac{\pi}{2} - \left(\frac{\pi}{2} - x\right)\right] = \sin\left(\frac{\pi}{2} - x\right)$$

$$\cos x = \sin\left(\frac{\pi}{2} - x\right)$$

or

$$\sin\left(\frac{\pi}{2} - x\right) = \cos x \qquad (7)$$

where x is any real number or angle in radian measure. If x is in degree measure, replace $\pi/2$ with 90°.

Finally, we state the cofunction identity for tangent and leave its derivation to Problem 9 in Exercise 4.2:

$$\tan\left(\frac{\pi}{2} - x\right) = \cot x \qquad (8)$$

for x any real number or angle in radian measure. If x is in degree measure, replace $\pi/2$ with 90°.

REMARK   If $0 < x < 90°$, then x and $90° - x$ are complementary angles. Originally, *cosine, cotangent,* and *cosecant* meant, respectively, "complements sine," "complements tangent," and "complements secant." Now we simply refer to cosine, cotangent, and cosecant as **cofunctions** of sine, tangent, and secant, respectively.

SUM AND DIFFERENCE   To derive a difference identity for sine, we use (7), (1), and (6) as follows:
FORMULAS FOR SINE
AND TANGENT

$$\sin(x - y) = \cos\left[\frac{\pi}{2} - (x - y)\right] \qquad \text{Use (6)}$$

$$= \cos\left[\left(\frac{\pi}{2} - x\right) - (-y)\right] \qquad \text{Algebra}$$

$$= \cos\left(\frac{\pi}{2} - x\right)\cos(-y) + \sin\left(\frac{\pi}{2} - x\right)\sin(-y) \quad \text{Use (1)}$$

$$= \sin x \cos y - \cos x \sin y \qquad \text{Use (6), (7), and identities for negatives}$$

The same result is obtained by replacing $\pi/2$ with 90°. Thus,

$$\sin(x - y) = \sin x \cos y - \cos x \sin y \qquad (9)$$

Now, if we replace y with $-y$ (a good exercise to do), we obtain

$$\sin(x + y) = \sin x \cos y + \cos x \sin y \qquad (10)$$

It is not difficult to derive sum and difference identities for the tangent function. See if you can supply the reason for each step.

$$\tan(x - y) = \frac{\sin(x - y)}{\cos(x - y)}$$

$$= \frac{\sin x \cos y - \cos x \sin y}{\cos x \cos y + \sin x \sin y}$$

$$= \frac{\dfrac{\sin x \cos y}{\cos x \cos y} - \dfrac{\cos x \sin y}{\cos x \cos y}}{\dfrac{\cos x \cos y}{\cos x \cos y} + \dfrac{\sin x \sin y}{\cos x \cos y}}$$

$$= \frac{\tan x - \tan y}{1 + \tan x \tan y}$$

Thus, for all angles or real numbers x and y,

$$\tan(x - y) = \frac{\tan x - \tan y}{1 + \tan x \tan y} \tag{11}$$

And if we replace y in (9) with −y (another good exercise to do), we obtain

$$\tan(x + y) = \frac{\tan x + \tan y}{1 - \tan x \tan y} \tag{12}$$

← Write this ; ask st. to derive

SUMMARY AND USE

Before proceeding with examples illustrating the use of these new identities, let us list them for convenient reference.

SUMMARY OF IDENTITIES

Sum identities

$$\sin(x + y) = \sin x \cos y + \cos x \sin y$$
$$\cos(x + y) = \cos x \cos y - \sin x \sin y$$

$$\tan(x + y) = \frac{\tan x + \tan y}{1 - \tan x \tan y}$$

Difference identities

$$\sin(x - y) = \sin x \cos y - \cos x \sin y$$
$$\cos(x - y) = \cos x \cos y + \sin x \sin y$$

$$\tan(x - y) = \frac{\tan x - \tan y}{1 + \tan x \tan y}$$

Cofunction identities   (Replace $\pi/2$ with 90° if x is in degree measure.)

$$\sin\left(\frac{\pi}{2} - x\right) = \cos x \qquad \tan\left(\frac{\pi}{2} - x\right) = \cot x \qquad \sec\left(\frac{\pi}{2} - x\right) = \csc x$$

EXAMPLE 6   Simplify $\sin(x - \pi)$ using a difference identity.

Solution   Use the difference identity for sine, replacing y with π.

$$\sin(x - y) = \sin x \cos y - \cos x \sin y$$
$$\sin(x - \pi) = \sin x \cos \pi - \cos x \sin \pi$$
$$= (\sin x)(-1) - (\cos x)(0)$$
$$= -\sin x$$

DO

DISCUSS   Q III - sin is neg.

PROBLEM 6   Simplify $\cos(x + 3\pi/2)$ using a sum identity.

EXAMPLE 7   Write $\sin 75°$ in the form $\cos \theta$, $0 \le \theta \le 90°$.

Solution   Use $\sin x = \cos(90° - x)$. Thus,

$$\sin 75° = \cos(90° - 75°) = \cos 15°$$

DO

PROBLEM 7   Write $\cos 37°$ in the form $\sin \theta$, $0 \le \theta \le 90°$.

EXAMPLE 8   Find the value of $\tan 75°$ in exact radical form.

Solution   Since we can write $75° = 45° + 30°$, the sum of two special angles, we can use the sum identity for tangents with $x = 45°$ and $y = 30°$.

DO

$$\tan(x + y) = \frac{\tan x + \tan y}{1 - \tan x \tan y}$$

$$\tan(45° + 30°) = \frac{\tan 45° + \tan 30°}{1 - \tan 45° \tan 30°}$$

$$= \frac{1 + \dfrac{1}{\sqrt{3}}}{1 - 1 \cdot \dfrac{1}{\sqrt{3}}}$$ Multiply numerator and denominator by $\sqrt{3}$

$$= \frac{\sqrt{3} + 1}{\sqrt{3} - 1}$$ Rationalize denominator

$$= 2 + \sqrt{3}$$

PROBLEM 8   Find the value of $\cos 15°$ in exact radical form.

EXAMPLE 9   Find the exact value of $\cos(x + y)$, given $\sin x = \frac{3}{5}$, $\cos y = \frac{4}{5}$, x in quadrant II, and y in quadrant I. Do not use a calculator or table.

Solution   We start with the sum identity for cosine:

$$\cos(x + y) = \cos x \cos y - \sin x \sin y$$

We know $\sin x$ and $\cos y$, but not $\sin y$ and $\cos x$. We find the latter two values by using reference triangles and the Pythagorean theorem:

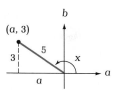
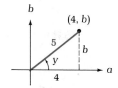

$$a = -\sqrt{5^2 - 3^2} = -4 \qquad\qquad b = \sqrt{5^2 - 4^2} = 3$$

$$\cos x = -\frac{4}{5} \qquad\qquad\qquad \sin y = \frac{3}{5}$$

Thus,

$$\cos(x + y) = \cos x \cos y - \sin x \sin y$$

$$= \left(-\frac{4}{5}\right)\left(\frac{4}{5}\right) - \left(\frac{3}{5}\right)\left(\frac{3}{5}\right) = \frac{-25}{25} = -1$$

Now ask what (x+y) = ?

PROBLEM 9   Find the exact value of $\sin(x - y)$, given $\sin x = -\frac{2}{3}$, $\cos y = \sqrt{5}/3$, x in quadrant III, and y in quadrant IV.

EXAMPLE 10   Establish the identity

$$\tan x + \cot y = \frac{\cos(x - y)}{\cos x \sin y}$$

Solution
$$\frac{\cos(x - y)}{\cos x \cos y} = \frac{\cos x \cos y + \sin x \sin y}{\cos x \sin y} \qquad \text{Difference identity for cosine}$$

$$= \frac{\cos x \cos y}{\cos x \sin y} + \frac{\sin x \sin y}{\cos x \sin y} \qquad \text{Algebra}$$

$$= \cot y + \tan x \qquad\qquad \text{Quotient identities}$$

$$= \tan x + \cot y$$

PROBLEM 10   Establish the identity

Do →

$$\cot y - \cot x = \frac{\sin(x - y)}{\sin x \sin y}$$

ANSWERS TO MATCHED PROBLEMS
6. $\sin x$ **7.** $\sin 53°$ **8.** $\dfrac{1 + \sqrt{3}}{2\sqrt{2}}$ **9.** $\dfrac{-4\sqrt{5}}{9}$

10. $\dfrac{\sin(x - y)}{\sin x \sin y} = \dfrac{\sin x \cos y - \cos x \sin y}{\sin x \sin y}$

$$= \frac{\sin x \cos y}{\sin x \sin y} - \frac{\cos x \sin y}{\sin x \sin y}$$

$$= \cot y - \cot x$$

EXERCISE 4.2

A   We can use sum identities to establish periodic properties for the trigonometric functions. Establish the following identities using sum identities.

1. $\cos(x + 2\pi) = \cos x$ 2. $\sin(x + 2\pi) = \sin x$
3. $\cot(x + \pi) = \cot x$ 4. $\tan(x + \pi) = \tan x$
5. $\sin(x + 2k\pi) = \sin x$,   6. $\cos(x + 2k\pi) = \cos x$,
    k an integer k an integer
7. $\tan(x + k\pi) = \tan x$,   8. $\cot(x + k\pi) = \cot x$,
    k an integer k an integer

Establish each identity using cofunction identities for sine and cosine and
the basic identities discussed in the last section.

9. $\tan\left(\dfrac{\pi}{2} - x\right) = \cot x$ 10. $\cot\left(\dfrac{\pi}{2} - x\right) = \tan x$

11. $\sec\left(\dfrac{\pi}{2} - x\right) = \csc x$ 12. $\csc\left(\dfrac{\pi}{2} - x\right) = \sec x$

Convert to forms involving sin x, cos x, and/or tan x using sum or difference
identities.

13. $\sin(x - 45°)$ 14. $\sin(30° - x)$
15. $\cos(x + 180°)$ 16. $\sin(180° - x) = \sin x$

17. $\tan\left(\dfrac{\pi}{4} - x\right)$ 18. $\tan\left(x + \dfrac{\pi}{3}\right)$

B   Find, using appropriate identities, exact values for each of the following.
Do not use a calculator or table.

19. $\sin 75°$ 20. $\sec 75°$

21. $\cos \dfrac{\pi}{12}$ 22. $\sin \dfrac{7\pi}{12}$

$\left[Hint: \ \dfrac{\pi}{12} = \dfrac{\pi}{4} - \dfrac{\pi}{6}\right]$ $\left[Hint: \ \dfrac{7\pi}{12} = \dfrac{\pi}{3} + \dfrac{\pi}{4}\right]$

23. $\sin 22° \cos 38° + \cos 22° \sin 38°$
24. $\cos 74° \cos 44° + \sin 74° \sin 44°$

25. $\dfrac{\tan 110° - \tan 50°}{1 + \tan 110° \tan 50°}$ 26. $\dfrac{\tan 27° + \tan 18°}{1 - \tan 27° \tan 18°}$

Find sin(x − y) and tan(x + y) exactly without a calculator or table using
the information given and appropriate identities.

27. $\sin x = \frac{2}{3}$, $\cos y = -\frac{1}{4}$, x in quadrant II, and y in quadrant III

28. $\sin x = -\frac{3}{5}$, $\sin y = \sqrt{8}/3$, x in quadrant IV, and y in quadrant I
29. $\cos x = -\frac{1}{3}$, $\tan y = \frac{1}{2}$, x in quadrant II, and y in quadrant III
30. $\tan x = \frac{3}{4}$, $\tan y = -\frac{1}{2}$, x in quadrant III, and y in quadrant IV

Establish each identity.

Give hint ⟶
$2x = x + x$

31. $\sin 2x = 2 \sin x \cos x$

32. $\cos 2x = \cos^2 x - \sin^2 x$

33. $\cot(x - y) = \dfrac{\cot x \cot y + 1}{\cot y - \cot x}$

34. $\cot(x + y) = \dfrac{\cot x \cot y - 1}{\cot x + \cot y}$

35. $\cot 2x = \dfrac{\cot^2 x - 1}{2 \cot x}$

36. $\tan 2x = \dfrac{2 \tan x}{1 - \tan^2 x}$

37. $\tan x - \tan y = \dfrac{\sin(x - y)}{\cos x \cos y}$

38. $\cot x - \tan y = \dfrac{\cos(x + y)}{\sin x \cos y}$

39. $\tan(x + y) = \dfrac{\cot x + \cot y}{\cot x \cot y - 1}$

40. $\tan(x - y) = \dfrac{\cot y - \cot x}{\cot x \cot y + 1}$

41. $\dfrac{\sin(x + h) - \sin x}{h} = \sin x \left(\dfrac{\cos h - 1}{h}\right) + \cos x \left(\dfrac{\sin h}{h}\right)$

42. $\dfrac{\cos(x + h) - \cos x}{h} = \cos x \left(\dfrac{\cos h - 1}{h}\right) - \sin x \left(\dfrac{\sin h}{h}\right)$

C   *Establish the identities in Problems 43 and 44. Hint:* $\sin(x + y + z) = \sin[(x + y) + z]$.

43. $\sin(x + y + z) = \sin x \cos y \cos z + \cos x \sin y \cos z$
$$+ \cos x \cos y \sin z - \sin x \sin y \sin z$$

44. $\cos(x + y + z) = \cos x \cos y \cos z - \sin x \sin y \cos z$
$$- \sin x \cos y \sin z - \cos x \sin y \sin z$$

APPLICATIONS   45.   Use the information in the figure to show that

$$\tan(\theta_2 - \theta_1) = \dfrac{m_2 - m_1}{1 + m_1 m_2}$$

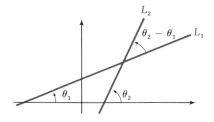

$\tan \theta_1 = $ Slope of $L_1 = m_1$
$\tan \theta_2 = $ Slope of $L_2 = m_2$

46.   Find the acute angle of intersection between the two lines $y = 3x + 1$ and $y = \frac{1}{2}x - 1$. (Use the results of Problem 45.)

Evaluate both sides of the difference identity for sine and the sum identity for tangent for the indicated values of x and y. Evaluate to four significant digits using a calculator.

EXAMPLE   Use a calculator to evaluate, to four significant digits, the identity

$$\cos(x - y) = \cos x \cos y + \sin x \sin y$$

for $x = 2.317$ and $y = 1.583$.

Solution   Set calculator in radian mode.

$$\cos(2.317 - 1.583) = 0.7425$$

A: [2.317] [−] [1.583] [=] [cos]

P: [2.317] [ENTER] [1.583] [−] [cos]

$$\cos 2.317 \cos 1.583 + \sin 2.317 \sin 1.583 = 0.7425$$

A: [2.317] [cos] [×] [1.583] [cos] [=] [+] [(] [2.317] [sin] [×] [1.583] [sin] [)] [=]

P: [2.317] [cos] [ENTER] [1.583] [cos] [×] [2.317] [sin] [ENTER] [1.583] [sin] [×] [+]

47. $x = 3.042$, $y = 2.384$ **48.** $x = 5.288$, $y = 1.769$

49. $x = 128.3°$, $y = 25.62°$ **50.** $x = 42.08°$, $y = 68.37°$

4.3   DOUBLE-ANGLE AND HALF-ANGLE IDENTITIES

We now develop another important set of identities called **double-angle** and **half-angle** identities. We can obtain these identities directly from the sum and difference identities that were found in Section 4.2. In spite of names involving the word "angle," the new identities hold for real numbers as well.

DOUBLE-ANGLE
IDENTITIES

If we start with the sum identity for sine,

$$\sin(x + y) = \sin x \cos y + \cos x \sin y$$

and let $y = x$, we obtain

$$\sin(x + x) = \sin x \cos x + \cos x \sin x$$

or

$$\sin 2x = 2 \sin x \cos x \tag{1}$$

Similarly, if we start with the sum identity for cosine,

$$\cos(x + y) = \cos x \cos y - \sin x \sin y$$

and let $y = x$, we obtain

$$\cos(x + x) = \cos x \cos x - \sin x \sin x$$

or

$$\cos 2x = \cos^2 x - \sin^2 x \tag{2}$$

Now, using the Pythagorean identities in the two forms

$$\cos^2 x = 1 - \sin^2 x \tag{3}$$
$$\sin^2 x = 1 - \cos^2 x \tag{4}$$

and substituting (3) into (2), we obtain

$$\cos 2x = 1 - \sin^2 x - \sin^2 x$$
$$\cos 2x = 1 - 2 \sin^2 x \tag{5}$$

Substituting (4) into (2), we obtain

$$\cos 2x = \cos^2 x - (1 - \cos^2 x)$$
$$\cos 2x = 2 \cos^2 x - 1 \tag{6}$$

A double-angle identity can be developed for the tangent function in the same way by starting with the sum identity for tangent. This is left as an exercise for you to do. We list these double-angle identities for convenient reference.

DOUBLE-ANGLE IDENTITIES

For x any real number or angle for which both sides are defined:

$$\sin 2x = 2 \sin x \cos x$$
$$\cos 2x = \cos^2 x - \sin^2 x$$
$$= 1 - 2 \sin^2 x$$
$$= 2 \cos^2 x - 1$$
$$\tan 2x = \frac{2 \tan x}{1 - \tan^2 x}$$

EXAMPLE 11   Establish the identity

$$\sin 2x = \frac{2 \tan x}{1 + \tan^2 x}$$

Proof   We start with the right side:

Do this way?

$$\frac{2\tan x}{1+\tan^2 x} = \frac{2\left(\dfrac{\sin x}{\cos x}\right)}{1+\dfrac{\sin^2 x}{\cos^2 x}}$$ Quotient identity

$$\frac{2\tan x}{1+\tan^2 x} = \frac{2\tan x}{\sec^2 x}$$

$$= \frac{2\left(\dfrac{\sin x}{\cos x}\right)}{\dfrac{1}{\cos^2 x}}\cdot\cos^2 x$$

$$= \frac{2\sin x\cos x}{\cos^2 x + \sin^2 x}$$ Multiply numerator and denominator by $\cos^2 x$

$$= \frac{\sin 2x}{1}$$ Double-angle and Pythagorean identities

$$= 2\sin x\cos x$$

$$= \sin 2x$$

$$= \sin 2x$$

PROBLEM 11   Establish the identity

Do

$$\cos 2x = \frac{1-\tan^2 x}{1+\tan^2 x}$$

EXAMPLE 12   Find the exact value of cos 2x and tan 2x if sin x = ⁴⁄₅, $\pi/2 < x < \pi$.

Solution   First draw a reference triangle in the second quadrant and find cos x and
tan x:

Do

$$a = -\sqrt{5^2 - 4^2} = -3$$

$$\sin x = \frac{4}{5}$$

$$\cos x = -\frac{3}{5}$$

$$\tan x = -\frac{4}{3}$$

$$\cos 2x = 1 - 2\sin^2 x$$ Use double-angle identity and the results above

$$= 1 - 2\left(\frac{4}{5}\right)^2$$

$$= -\frac{7}{25}$$

$$\tan 2x = \frac{2\tan x}{1-\tan^2 x}$$ Use double-angle identity and the results above

$$= \frac{2(-⁴⁄₃)}{1 - (-⁴⁄₃)^2}$$

$$= \frac{24}{7}$$

PROBLEM 12   Find the exact value of sin 2x and cos 2x if tan x = −³⁄₄, $-\pi/2 < x < 0$.

HALF-ANGLE IDENTITIES

Half-angle identities are simply double-angle identities stated in an alternate form. Let us start with the double-angle identity for cosine in the form

$$\cos 2u = 1 - 2\sin^2 u$$ ← *Just leave as u and solve for sin u*

and let $u = x/2$. Then

$$\cos x = 1 - 2\sin^2 \frac{x}{2}$$

Now solve for $\sin x/2$ to obtain a half-angle formula for the sine function:

$$2\sin^2 \frac{x}{2} = 1 - \cos x$$

$$\sin^2 \frac{x}{2} = \frac{1 - \cos x}{2}$$

$$\sin \frac{x}{2} = \pm\sqrt{\frac{1 - \cos x}{2}} \qquad (7)$$

In equation (7) the choice of the sign is determined by the quadrant in which $x/2$ lies.

Now let us start with the double-angle identity for cosine in the form

$$\cos 2u = 2\cos^2 u - 1$$

and let $u = x/2$. We then obtain a half-angle formula for the cosine function:

$$\cos x = 2\cos^2 \frac{x}{2} - 1$$

$$2\cos^2 \frac{x}{2} = 1 + \cos x$$

$$\cos^2 \frac{x}{2} = \frac{1 + \cos x}{2}$$

$$\cos \frac{x}{2} = \pm\sqrt{\frac{1 + \cos x}{2}} \qquad (8)$$

In equation (8) the choice of the sign is again determined by the quadrant in which $x/2$ lies.

To obtain a half-angle identity for the tangent function, we can use the quotient identity and the half-angle formulas for sine and cosine:

$$\tan \frac{x}{2} = \frac{\sin\left(\frac{x}{2}\right)}{\cos\left(\frac{x}{2}\right)} = \frac{\pm\sqrt{\dfrac{1 - \cos x}{2}}}{\pm\sqrt{\dfrac{1 + \cos x}{2}}} = \pm\sqrt{\frac{1 - \cos x}{1 + \cos x}} \qquad (9)$$

where the sign is determined by the quadrant in which $x/2$ lies.

We now list all of the half-angle identities for convenient reference. Two of the half-angle identities for tangent are left to Problems 17 and 18 in Exercise 4.3.

HALF-ANGLE IDENTITIES

$$\sin \frac{x}{2} = \pm \sqrt{\frac{1 - \cos x}{2}}$$

$$\cos \frac{x}{2} = \pm \sqrt{\frac{1 + \cos x}{2}}$$

$$\tan \frac{x}{2} = \pm \sqrt{\frac{1 - \cos x}{1 + \cos x}} = \frac{\sin x}{1 + \cos x} = \frac{1 - \cos x}{\sin x}$$

where the sign is determined by the quadrant in which $x/2$ lies.

EXAMPLE 13   Find $\cos 165°$ exactly by means of a half-angle identity.

Solution
$$\cos 165° = \cos \frac{330°}{2} = -\sqrt{\frac{1 + \cos 330°}{2}}$$

The negative square root is used since $165°$ is in the second quadrant and cosine is negative there. We complete the evaluation by noting that the reference triangle for $330°$ is a $30°$–$60°$ triangle in the fourth quadrant.

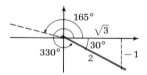

$$\cos 330° = \cos 30°$$
$$= \frac{\sqrt{3}}{2}$$

Thus,

$$\cos 165° = -\sqrt{\frac{1 + \sqrt{3}/2}{2}} = -\frac{\sqrt{2 + \sqrt{3}}}{2}$$

Find sin 30°

PROBLEM 13   Find the exact value of $\sin 165°$ using a half-angle identity.

EXAMPLE 14   Find the exact value of $\sin(x/2)$, $\cos(x/2)$, and $\tan(x/2)$ if $\sin x = -\frac{3}{5}$, $\pi < x < 3\pi/2$.

Solution   Draw a reference triangle in the third quadrant and find $\cos x$:

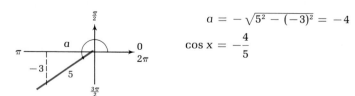

$$a = -\sqrt{5^2 - (-3)^2} = -4$$

$$\cos x = -\frac{4}{5}$$

If $\pi < x < 3\pi/2$, then

$$\frac{\pi}{2} < \frac{x}{2} < \frac{3\pi}{4} \qquad \text{Divide each member of } \pi < x < 3\pi/2 \text{ by 2}$$

Thus, $x/2$ is in the second quadrant, where sine is positive and cosine and tangent are negative. Using half-angle identities, we obtain

$$\sin\frac{x}{2} = \sqrt{\frac{1 - \cos x}{2}}$$

$$= \sqrt{\frac{1 - (-\tfrac{4}{5})}{2}}$$

$$= \sqrt{\frac{9}{10}} \quad \text{or} \quad \frac{3\sqrt{10}}{10}$$

$$\cos\frac{x}{2} = -\sqrt{\frac{1 + \cos x}{2}}$$

$$= -\sqrt{\frac{1 + (-\tfrac{4}{5})}{2}}$$

$$= -\sqrt{\frac{1}{10}} \quad \text{or} \quad \frac{-\sqrt{10}}{10}$$

$$\tan\frac{x}{2} = \frac{\sin(x/2)}{\cos(x/2)}$$

$$= \frac{3\sqrt{10}/10}{-\sqrt{10}/10} = -3$$

PROBLEM 14   Find the exact value for $\sin(x/2)$, $\cos(x/2)$, and $\tan(x/2)$ if $\cot x = -\tfrac{4}{3}$, $\pi/2 < x < \pi$.

EXAMPLE 15   Establish the identity

$$\cos^2\frac{x}{2} = \frac{\tan x + \sin x}{2 \tan x}$$

Proof $\cos^2\dfrac{x}{2} = \dfrac{1 + \cos x}{2}$ Square both sides of the half-angle identity for cosine

$$= \frac{\tan x}{\tan x} \cdot \frac{1 + \cos x}{2} \qquad \text{Algebra}$$

$$= \frac{\tan x + \tan x \cos x}{2 \tan x} \qquad \text{Algebra}$$

$$= \frac{\tan x + \sin x}{2 \tan x} \qquad \text{Quotient identity and algebra}$$

PROBLEM 15   Establish the identity

$$\sin^2 \frac{x}{2} = \frac{\tan x - \sin x}{2 \tan x}$$

ANSWERS TO   **11.** $\dfrac{1 - \tan^2 x}{1 + \tan^2 x} = \dfrac{1 - \dfrac{\sin^2 x}{\cos^2 x}}{1 + \dfrac{\sin^2 x}{\cos^2 x}} = \cos^2 x - \sin^2 x = \cos 2x$
MATCHED PROBLEMS

12. $\sin 2x = -\dfrac{24}{25}, \quad \cos 2x = \dfrac{7}{25}$

13. $\dfrac{\sqrt{2 - \sqrt{3}}}{2}$

14. $\sin(x/2) = 3\sqrt{10}/10, \quad \cos(x/2) = \sqrt{10}/10, \quad \tan(x/2) = 3$

15. $\sin^2 \dfrac{x}{2} = \dfrac{1 - \cos x}{2} = \dfrac{\tan x}{\tan x} \cdot \dfrac{1 - \cos x}{2} = \dfrac{\tan x - \sin x}{2 \tan x}$

EXERCISE 4.3

A   *Evaluate each side of the indicated identity for x = 60° (thus verifying it for one particular case).*

1. $\sin 2x = 2 \sin x \cos x$ **2.** $\cos 2x = \cos^2 x - \sin^2 x$

3. $\tan 2x = \dfrac{2 \tan x}{1 - \tan^2 x}$ **4.** $\sin \dfrac{x}{2} = \pm \sqrt{\dfrac{1 - \cos x}{2}}$

Use half-angle identities to find the exact value of Problems 5–8. Do not use a calculator or table.

5. $\sin 105°$ **6.** $\cos 105°$ **7.** $\tan 15°$ **8.** $\tan 75°$

B   *Establish each of the following identities.*

9. $\dfrac{2 \tan x}{\sin 2x} = \sec^2 x$ **10.** $(\sin x + \cos x)^2 = 1 + \sin 2x$

11. $\sin 2x = (\tan x)(1 + \cos 2x)$ **12.** $1 - \cos 2x = \tan x \sin 2x$

13. $2 \sin^2 \dfrac{x}{2} = \dfrac{\sin^2 x}{1 + \cos x}$ **14.** $2 \cos^2 \dfrac{x}{2} = \dfrac{\sin^2 x}{1 - \cos x}$

15. $\tan 2x = \dfrac{2 \tan x}{1 - \tan^2 x}$ **16.** $\sin 2x = \dfrac{2 \tan x}{1 + \tan^2 x}$

17. $\tan \dfrac{x}{2} = \dfrac{\sin x}{1 + \cos x}$ **18.** $\tan \dfrac{x}{2} = \dfrac{1 - \cos x}{\sin x}$

19. $\sec^2 x = (\sec 2x)(2 - \sec^2 x)$ **20.** $2 \csc 2x = \dfrac{1 + \tan^2 x}{\tan x}$

21. $\cos 2x = \dfrac{\cot x - \tan x}{\cot x + \tan x}$ **22.** $\cos x = \dfrac{1 - \tan^2(x/2)}{1 + \tan^2(x/2)}$

Find the exact value of sin 2x, cos 2x, and tan 2x for the information given in Problems 23–28. Do not use a calculator or a table.

23. $\sin x = \frac{3}{5}$, $0° < x < 90°$ 24. $\cos x = \frac{4}{5}$, $0° < x < 90°$

25. $\cos x = -\frac{4}{5}$, $\pi/2 < x < \pi$ 26. $\sin x = \frac{3}{5}$, $\pi/2 < x < \pi$

27. $\cot x = -\frac{5}{12}$, $-\pi/2 < x < 0$ 28. $\tan x = -\frac{5}{12}$, $-\pi/2 < x < 0$

Find the exact value of sin(x/2) and cos(x/2) for the information given in Problem 29–34. Do not use a calculator or a table.

29. $\cos x = \frac{1}{3}$, $0° < x < 90°$ 30. $\sin x = \frac{4}{5}$, $0° < x < 90°$

31. $\sin x = -\frac{1}{3}$, $\pi < x < 3\pi/2$ 32. $\cos x = -\frac{1}{4}$, $\pi < x < 3\pi/2$

33. $\cot x = \frac{3}{4}$, $-\pi < x < -\pi/2$ 34. $\tan x = \frac{3}{4}$, $-\pi < x < -\pi/2$

C   Establish each identity.

    $\Rightarrow \sin(x + 2x)$

35. $\sin 3x = 3 \sin x - 4 \sin^3 x$ 36. $\cos 3x = 4 \cos^3 x - 3 \cos x$

37. $\sin 4x = (\cos x)(4 \sin x - 8 \sin^3 x)$

38. $\cos 4x = 8 \cos^4 x - 8 \cos^2 x + 1$

39. $\tan 3x = \dfrac{3 \tan x - \tan^3 x}{1 - 3 \tan^2 x}$

40. $4 \sin^4 x = 1 - 2 \cos 2x + \cos^2 2x$

CALCULATOR PROBLEMS   Verify each of the following identities for the indicated value of x. Compute values to five significant digits using a calculator.

(A) $\tan 2x = \dfrac{2 \tan x}{1 - \tan^2 x}$ (B) $\cos \dfrac{x}{2} = \pm \sqrt{\dfrac{1 + \cos x}{2}}$ (choose correct sign)

41. $x = 252.06°$ 42. $x = 72.358°$

43. $x = 0.93457$ 44. $x = 4$

APPLICATIONS   45. *Projectile distance*   In physics it can be shown that the distance d a javelin will travel (see figure below) is given approximately by

$$d = \frac{2v_0^2 \sin \theta \cos \theta}{32 \text{ ft/sec}^2} \;\dot=\; \frac{v_0^2 \sin 2\theta}{g}$$

where v_0 is the initial velocity of the javelin in feet per second. Write the formula in terms of sine only by using a suitable identity.

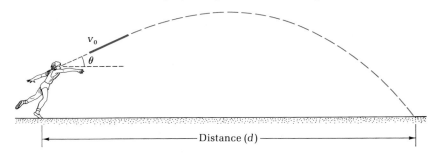

Distance (d)

(46.) *Projectile distance*   Using the resulting equation in Problem 45, determine the angle θ that will produce the maximum distance d for a given initial velocity v_0. [*Hint:* For what value θ, $0 < \theta < 90°$, will sin 2θ be maximum?] This result is an important consideration for javelin, shot-put, and discus throwers.

4.4   PRODUCT AND FACTOR IDENTITIES

PRODUCT IDENTITIES

The product and factor identities are easily derived from the sum and difference identities developed in Section 4.2. To obtain a product identity, let us add, left side to left side and right side to right side, the sum and difference identities for sine:

$$\sin(x + y) = \sin x \cos y + \cos x \sin y$$
$$\underline{\sin(x - y) = \sin x \cos y - \cos x \sin y}$$
$$\sin(x + y) + \sin(x - y) = 2 \sin x \cos y$$

or

$$\sin x \cos y = \frac{1}{2}[\sin(x + y) + \sin(x - y)]$$

Similarly, by adding or subtracting appropriate sum and difference identities, we can obtain three other product identities. These identities are listed below for convenient reference.

Mention –
then omit

PRODUCT IDENTITIES

$$\sin x \cos y = \frac{1}{2}[\sin(x + y) + \sin(x - y)]$$

$$\cos x \sin y = \frac{1}{2}[\sin(x + y) - \sin(x - y)]$$

$$\sin x \sin y = \frac{1}{2}[\cos(x - y) - \cos(x + y)]$$

$$\cos x \cos y = \frac{1}{2}[\cos(x + y) + \cos(x - y)]$$

EXAMPLE 16   Write the product cos $3t$ sin t as a sum or difference.

Solution

$$\cos x \sin y = \frac{1}{2}[\sin(x + y) - \sin(x - y)] \qquad \text{Let } x = 3t \text{ and } y = t$$

$$\cos 3t \sin t = \frac{1}{2}[\sin(3t + t) - \sin(3t - t)]$$

$$= \frac{1}{2} \sin 4t - \frac{1}{2} \sin 2t$$

PROBLEM 16   Write the product $\cos 5\theta \cos 2\theta$ as a sum or difference.

EXAMPLE 17   Evaluate $\sin 105° \sin 15°$ exactly using an appropriate product identity.

Solution
$$\sin x \sin y = \frac{1}{2}[\cos(x - y) - \cos(x + y)]$$

$$\sin 105° \sin 15° = \frac{1}{2}[\cos(105° - 15°) - \cos(105° + 15°)]$$

$$= \frac{1}{2}[\cos 90° - \cos 120°]$$

$$= \frac{1}{2}\left[0 - \left(-\frac{1}{2}\right)\right] = \frac{1}{4} \text{ or } 0.25$$

PROBLEM 17   Evaluate $\cos 165° \sin 75°$ exactly using an appropriate product identity.

FACTOR IDENTITIES

The product identities above can be transformed into equivalent forms called **factor identities.** These identities are used to express sums and differences involving sines and cosines as products involving sines and cosines. We illustrate the transformation for one identity. The other three identities can be obtained by following the same procedure.

Let us start with the product identity

$$\sin \alpha \cos \beta = \frac{1}{2}[\sin(\alpha + \beta) + \sin(\alpha - \beta)] \tag{1}$$

We would like

$$\alpha + \beta = x \qquad \alpha - \beta = y$$

Solving this system, we have

$$\alpha = \frac{x + y}{2} \qquad \beta = \frac{x - y}{2} \tag{2}$$

By substituting (2) into (1) and simplifying, we obtain

$$\sin x + \sin y = 2 \sin \frac{x + y}{2} \cos \frac{x - y}{2}$$

All four factor identities are listed below for convenient reference.

FACTOR IDENTITIES

$$\sin x + \sin y = 2 \sin \frac{x + y}{2} \cos \frac{x - y}{2} \qquad\qquad \cos x + \cos y = 2 \cos \frac{x + y}{2} \cos \frac{x - y}{2}$$

$$\sin x - \sin y = 2 \cos \frac{x + y}{2} \sin \frac{x - y}{2} \qquad\qquad \cos x - \cos y = -2 \sin \frac{x + y}{2} \sin \frac{x - y}{2}$$

EXAMPLE 18 Write the difference $\sin 7\theta - \sin 3\theta$ as a product.

Solution

$$\sin x - \sin y = 2 \cos \frac{x + y}{2} \sin \frac{x - y}{2}$$

$$\sin 7\theta - \sin 3\theta = 2 \cos \frac{7\theta + 3\theta}{2} \sin \frac{7\theta - 3\theta}{2}$$

$$= 2 \cos 5\theta \sin 2\theta$$

PROBLEM 18 Write the sum $\cos 3t + \cos t$ as a product.

EXAMPLE 19 Find the exact value of $\sin 105° - \sin 15°$ using an appropriate factor identity.

Solution

$$\sin x - \sin y = 2 \cos \frac{x + y}{2} \sin \frac{x - y}{2}$$

$$\sin 105° - \sin 15° = 2 \cos \frac{105° + 15°}{2} \sin \frac{105° - 15°}{2}$$

$$= 2 \cos 60° \sin 45°$$

$$= 2 \left(\frac{1}{2}\right)\left(\frac{\sqrt{2}}{2}\right) = \frac{\sqrt{2}}{2}$$

PROBLEM 19 Find the exact value of $\cos 165° - \cos 75°$ by using an appropriate factor identity.

ANSWERS TO MATCHED PROBLEMS

16. $\cos 5\theta \cos 2\theta = \dfrac{1}{2} \cos 7\theta + \dfrac{1}{2} \cos 3\theta$

17. $\dfrac{-\sqrt{3} - 2}{4}$

18. $\cos 3t + \cos t = 2 \cos 2t \cos t$

19. $-\sqrt{6}/2$

EXERCISE 4.4

A *Write each product as a sum or difference involving sines and cosines.*

1. $\cos 7A \cos 5A$
2. $\sin 3m \cos m$
3. $\cos 2\theta \sin 3\theta$
4. $\sin u \sin 3u$

Write each difference or sum as a product involving sines and cosines.

5. $\cos 7\theta + \cos 5\theta$
6. $\sin 3t + \sin t$
7. $\sin u - \sin 5u$
8. $\cos 5w - \cos 9w$

B *Evaluate each of the following exactly using an appropriate identity.*

9. cos 75° sin 15° 10. sin 195° cos 75°

11. sin 105° sin 165° 12. cos 15° cos 75°

Evaluate each of the following exactly using an appropriate identity.

13. sin 195° + sin 105° 14. cos 285° + cos 195°

15. sin 75° − sin 165° 16. cos 15° − cos 105°

Use sum and difference identities to establish:

17. $\sin x \sin y = \dfrac{1}{2}[\cos(x - y) - \cos(x + y)]$

18. $\cos x \cos y = \dfrac{1}{2}[\cos(x + y) + \cos(x - y)]$

Use appropriate substitutions in the product identities to obtain:

19. $\sin x - \sin y = 2 \cos \dfrac{x + y}{2} \sin \dfrac{x - y}{2}$

20. $\cos x - \cos y = -2 \sin \dfrac{x + y}{2} \sin \dfrac{x - y}{2}$

Establish each identity below.

21. $\dfrac{\cos t - \cos 3t}{\sin t + \sin 3t} = \tan t$

22. $\dfrac{\sin 2t + \sin 4t}{\cos 2t - \cos 4t} = \cot t$

23. $\dfrac{\sin x + \sin y}{\cos x + \cos y} = \tan \dfrac{x + y}{2}$

24. $\dfrac{\sin x - \sin y}{\cos x - \cos y} = -\cot \dfrac{x + y}{2}$

25. $\dfrac{\cos x - \cos y}{\sin x + \sin y} = -\tan \dfrac{x - y}{2}$

26. $\dfrac{\cos x + \cos y}{\sin x - \sin y} = \cot \dfrac{x - y}{2}$

27. $\dfrac{\sin x + \sin y}{\sin x - \sin y} = \dfrac{\tan \dfrac{1}{2}(x + y)}{\tan \dfrac{1}{2}(x - y)}$

28. $\dfrac{\cos x + \cos y}{\cos x - \cos y} = \cot \dfrac{x + y}{2} \cot \dfrac{x - y}{2}$

C 29. $\sin x \sin y \sin z = \dfrac{1}{4}[\sin(x + y - z) + \sin(y + z - x)$
$+ \sin(z + x - y) - \sin(x + y + z)]$

30. $\cos x \cos y \cos z = \dfrac{1}{4}[\cos(x + y - z) + \cos(y + z - x)$
$+ \cos(z + x - y) + \cos(x + y + z)]$

CALCULATOR PROBLEMS

Verify each of the following identities for the indicated values of x and y. Evaluate each side to five significant digits.

(A) $\cos x \sin y = \dfrac{1}{2}[\sin(x + y) - \sin(x - y)]$

(B) $\cos x + \cos y = 2 \cos \dfrac{x + y}{2} \cos \dfrac{x - y}{2}$

31. $x = 50.137°, \quad y = 18.044°$ **32.** $x = 172.63°, \quad y = 20.177°$

33. $x = 0.03917, \quad y = 0.61052$ **34.** $x = 1.1255, \quad y = 3.6014$

4.5   FROM $M \sin Bt + N \cos Bt$ TO $A \sin(Bt + C)$

omit

In the process of solving certain kinds of problems that require the use of more advanced mathematics—problems dealing with heat flow, electrical circuits, spring–mass systems, and so on—we are led naturally to the form

$$y = M \sin Bt + N \cos Bt \tag{1}$$

With a little ingenuity and the use of the sum identity for sine,

$$\sin(x + y) = \sin x \cos y + \cos x \sin y \tag{2}$$

we can convert (1) into the form

$$y = A \sin(Bt + C) \tag{3}$$

This form of a solution is often more convenient than (1), since from it we can easily determine amplitude, period, frequency, and phase shift (and a graph if necessary).

  How do we proceed? We start by trying to get the right side of (1) to look like the right side of (2). Then we use (2), from right to left, to obtain (3) Getting (1) to look like (2) requires constants M and N to be replaced by $\cos C$ and $\sin C$, respectively, where C is an appropriate number. It turns out that C can be any angle (in radians if t is real or in radians) having $P(M, N)$ on its terminal side (see Figure 2). In (1), to be able to replace M with $\cos C$ and N with $\sin C$, M and N must each be divided by $\sqrt{M^2 + N^2}$. We accomplish this as follows:

$$\sin C = \frac{N}{\sqrt{M^2 + N^2}}$$

$$\cos C = \frac{M}{\sqrt{M^2 + N^2}}$$

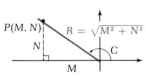

FIGURE 2

$$
\begin{aligned}
y &= M \sin Bt + N \cos Bt \\
&= \frac{\sqrt{M^2 + N^2}}{\sqrt{M^2 + N^2}} (M \sin Bt + N \cos Bt) \\
&= \sqrt{M^2 + N^2} \left(\frac{M}{\sqrt{M^2 + N^2}} \sin Bt + \frac{N}{\sqrt{M^2 + N^2}} \cos Bt \right) \\
&= \sqrt{M^2 + N^2} (\cos C \sin Bt + \sin C \cos Bt) \\
&= \sqrt{M^2 + N^2} (\sin Bt \cos C + \cos Bt \sin C) \\
&= \sqrt{M^2 + N^2} \sin(Bt + C) \qquad\qquad \text{Using (2)}
\end{aligned}
$$

We thus have form (3) with $A = \sqrt{M^2 + N^2}$. We summarize the results in the following box for convenient reference.

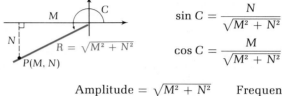

FROM $M \sin Bt + N \cos Bt$ TO $A \sin(Bt + C)$

$$y = M \sin Bt + N \cos Bt = \sqrt{M^2 + N^2} \sin(Bt + C)$$

where C is any angle (in radians if t is real) having $P(M, N)$ on its terminal side.

$$\sin C = \frac{N}{\sqrt{M^2 + N^2}}$$

$$R = \sqrt{M^2 + N^2} \qquad \cos C = \frac{M}{\sqrt{M^2 + N^2}}$$

$$\text{Amplitude} = \sqrt{M^2 + N^2} \qquad \text{Frequency} = \frac{B}{2\pi}$$

$$\text{Period} = \frac{2\pi}{B} \qquad \text{Phase shift} = \left|\frac{C}{B}\right|$$

In converting (1) to (3) you can either go through the steps that led up to the boxed material or use the boxed material directly, whichever is easier. Let us consider several examples.

EXAMPLE 20 Write

$$y = -\sqrt{3} \sin 2t + \cos 2t$$

in the form $y = A \sin(Bt + C)$ and indicate its amplitude, period, frequency, and phase shift. (Choose C as small as possible, but positive.)

Solution $\qquad M = -\sqrt{3} \qquad \text{and} \qquad N = 1$

Locate $P(M, N) = P(-\sqrt{3}, 1)$ to determine C:

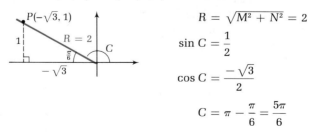

$$R = \sqrt{M^2 + N^2} = 2$$

$$\sin C = \frac{1}{2}$$

$$\cos C = \frac{-\sqrt{3}}{2}$$

$$C = \pi - \frac{\pi}{6} = \frac{5\pi}{6}$$

(Reference triangle is a special 30°–60° triangle.)
Thus,

$$y = -\sqrt{3} \sin 2t + \cos 2t = 2 \sin\left(2t + \frac{5\pi}{6}\right)$$

$$\text{Amplitude} = 2 \qquad \text{Frequency} = \frac{2}{2\pi} = \frac{1}{\pi}$$

$$\text{Period} = \frac{2\pi}{2} = \pi \qquad \text{Phase shift} = \frac{5\pi/6}{2} = \frac{5\pi}{12} \, (\text{left})$$

PROBLEM 20   Repeat Example 20 for $y = \sin \pi t - \sqrt{3} \cos \pi t$.

EXAMPLE 21   Write

$$y = \sin \frac{t}{2} - \sqrt{3} \cos \frac{t}{2}$$

in the form $y = A \sin(Bt + C)$. Indicate amplitude, period, frequency, and phase shift. Choose C (positive or negative) so that $|C|$ is minimum.

Solution $M = 1 \quad$ and $\quad N = -\sqrt{3}$

Locate $P(M, N) = P(1, -\sqrt{3})$ to determine C.

$$R = \sqrt{M^2 + N^2} = 2$$

$$\sin C = \frac{-\sqrt{3}}{2}$$

$$\cos C = \frac{1}{2}$$

$$C = -\frac{\pi}{3} \qquad |C| \text{ is minimum for this choice}$$

Thus,

$$y = 2 \sin\left(\frac{t}{2} - \frac{\pi}{3}\right)$$

$$\text{Amplitude} = 2 \qquad \text{Frequency} = \frac{1}{4\pi}$$

$$\text{Period} = \frac{2\pi}{1/2} = 4\pi \qquad \text{Phase shift} = \left|\frac{\pi/3}{1/2}\right| = \frac{2\pi}{3} \, (\text{right})$$

PROBLEM 21   Repeat Example 21 for $y = -\sqrt{3} \sin 4t - \cos 4t$.

EXAMPLE 22   Write

$$y = -3 \sin 2\pi t - 4 \cos 2\pi t$$

in the form $y = A \sin(Bt + C)$, where C is chosen so that $|C|$ is minimum. Compute C to two decimal places.

Solution   Let $M = -3$ and $N = -4$; then locate $P(M, N) = P(-3, -4)$ to determine C.

$$R = \sqrt{(-3)^2 + (-4)^2} = 5$$

$$\sin C = \frac{-4}{5} = -0.8 \qquad \cos C = \frac{-3}{5} = -0.6$$

Find the reference angle α. Then

$$C = \alpha - \pi$$

$$\sin \alpha = \frac{4}{5} = 0.8$$

$$\alpha \approx 0.93 \qquad \text{From Table III or calculator}$$

Thus,

$$C \approx 0.93 - \pi \approx -2.21 \qquad \text{Use } \pi \approx 3.14$$

and $|C|$ is minimum for this choice of C.
    We can now write

$$y = 5 \sin(2\pi t - 2.21)$$

Amplitude $= 5$ \qquad\qquad Frequency $= 1$

$$\text{Period} = \frac{2\pi}{2\pi} = 1 \qquad \text{Phase shift} = \left|\frac{2.21}{2\pi}\right| \approx 0.35 \text{ (right)}$$

PROBLEM 22   Repeat Example 22 for $y = -4 \sin(t/2) + 3 \cos(t/2)$.

ANSWERS TO **20.** $y = 2 \sin(\pi t + 5\pi/3)$; Amplitude $= 2$, Period $= 2$, Frequency $= \frac{1}{2}$,
MATCHED PROBLEMS Phase shift $= \frac{5}{3}$ (left)
21. $y = 2 \sin(4t - 5\pi/6)$; Amplitude $= 2$, Period $= \pi/2$,
Frequency $= 2/\pi$, Phase shift $= 5\pi/24$ (right)
22. $y = 5 \sin\left(\frac{t}{2} + 2.50\right)$; Amplitude $= 5$, Period $= 4\pi$,

Frequency $= \dfrac{1}{4\pi}$, Phase shift $= 5$ (left)

EXERCISE 4.5

A   Write each equation in the form $y = A \sin(Bt + C)$. Indicate amplitude, period, frequency, and phase shift. (Choose the least positive C.)

1. $y = \sin t + \cos t$ \qquad\qquad **2.** $y = -\sin t + \cos t$
3. $y = -\sqrt{3} \sin t + \cos t$ \qquad **4.** $y = \sin t - \sqrt{3} \cos t$

B   Continue as above, but choose C, either positive or negative, so that $|C|$ is minimum.

5. $y = \sqrt{3} \sin 2t - \cos 2t$ \qquad **6.** $y = \sin \pi t - \cos \pi t$

7. $y = -\sin \dfrac{t}{2} - \cos \dfrac{t}{2}$ \qquad **8.** $y = -\sin 4t - \sqrt{3} \cos 4t$

Continue as above, but compute C to two decimal places using Table III or a calculator. A calculator will be especially helpful for Problems 11 and 12.

C **9.** $y = 4 \sin \pi t - 3 \cos \pi t$ **10.** $y = -3 \sin \dfrac{t}{4} + 4 \cos \dfrac{t}{4}$

11. $y = -5 \sin 3t + 3 \cos 3t$ **12.** $y = 2 \sin 8t - 5 \cos 8t$

APPLICATIONS

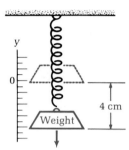

Downward initial
velocity = 24 cm/sec

*13.   A weight suspended from a spring, with stiffness constant 12, is pulled 4 cm below its equilibrium position and is then given a downward thrust to produce an initial downward velocity of 24 cm per second. In more advanced mathematics (differential equations) the equation of motion (neglecting friction and air resistance) is found to be given approximately by

$$y = -3 \sin 8t - 4 \cos 8t$$

where y is the position on the scale in the figure at time t (y is in centimeters and t is in seconds). Write this equation in the form

$$y = A \sin(Bt + C)$$

and indicate the amplitude, period, frequency, and phase shift of the motion. (Choose the least positive C and keep A positive.) [*Note:* The unit that is now generally used for frequency is the hertz (Hz); 1 Hz is 1 cycle per second; 50 Hz is 50 cycles per second, and so on.]

EXERCISE 4.6   CHAPTER REVIEW

A   *Choose the function on the right that will make the equation an identity.*

1. $\sin x = \begin{cases} \dfrac{1}{\sec x} \\ \dfrac{1}{\csc x} \end{cases}$ **2.** $\cot x = \begin{cases} \dfrac{\sin x}{\cos x} \\ \dfrac{\cos x}{\sin x} \end{cases}$

3. $\sin^2 x = \begin{cases} 1 - \cos^2 x \\ 1 + \cos^2 x \end{cases}$ **4.** $\cos(-x) = \begin{cases} -\cos x \\ \cos x \end{cases}$

Establish each identity without looking at a table of identities.

5. $\csc x \sin x = \sec x \cos x$ **6.** $\cot x \sin x = \cos x$

7. $\tan x = -\tan(-x)$ **8.** $\dfrac{\sin^2 x}{\cos x} + \cos x = \sec x$

9. $\dfrac{\sin^2 x}{\cos x} = \sec x - \cos x$ **10.** $\dfrac{\csc x}{\cos x} = \tan x + \cot x$

11. $\cot^2 x \cos^2 x = \cot^2 x - \cos^2 x$

12. Using $\cos(x + y) = \cos x \cos y - \sin x \sin y$, show that

$$\cos(x + 2\pi) = \cos x$$

13. Using $\sin(x + y) = \sin x \cos y + \cos x \sin y$, show that

$$\sin(x + \pi) = -\sin x$$

Verify each identity for the indicated value(s).

14. $\cos 2x = 1 - 2 \sin^2 x$, $x = 30°$ 15. $\sin \dfrac{x}{2} = \pm \sqrt{\dfrac{1 - \cos x}{2}}$, $x = \dfrac{\pi}{2}$

16. $2 \sin x \cos y = \sin(x + y) + \sin(x - y)$, $x = 60°$, $y = 30°$

B   *Establish the identities in Problems 17–23. Use the list of identities inside the front cover if necessary.*

17. $\dfrac{\sin x}{1 - \cos x} = (\csc x)(1 + \cos x)$ 18. $\dfrac{1 - \tan^2 x}{1 - \cot^2 x} = 1 - \sec^2 x$

19. $\tan(x + \pi) = \tan x$ 20. $\dfrac{\sin 2x}{\cot x} = 1 - \cos 2x$

21. $\dfrac{2 \tan x}{1 + \tan^2 x} = \sin 2x$ 22. $2 \csc 2x = \tan x + \cot x$

23. $\csc x = \dfrac{\cot(x/2)}{1 + \cos x}$

24. Write in factored form: $\sin 3x - \sin x$
25. Write as a sum: $2 \cos 5x \cos 2x$
26. Write $y = -\sqrt{3} \sin 6t - \cos 6t$ in the form of $y = A \sin(Bt + C)$. Indicate amplitude, period, frequency, and phase shift. Choose C so that $|C|$ is minimum.

C   27. Use the definition of sine, cosine, and tangent on a unit circle to prove that

$$\tan x = \frac{\sin x}{\cos x}$$

28. Prove that the cosine function has a period of 2π.
29. Prove that the cotangent function has a period of π.
30. By letting

$$x + y = u \qquad x - y = v$$

in $\sin x \sin y = \frac{1}{2}[\cos(x - y) - \cos(x + y)]$, show that

$$\cos v - \cos u = 2 \sin \frac{u + v}{2} \sin \frac{u - v}{2}$$

31. Show that $\sin 3x = 3 \sin x - 4 \sin^3 x$ is an identity.

Use ə instead ŋ

INVERSE RELATIONS AND FUNCTIONS; TRIGONOMETRIC EQUATIONS

5.1

INVERSE SINE
AND COSINE RELATIONS

*first draw line
and its inverse*

INVERSE SINES AND COSINES

If we start with the sine and cosine functions

$$y = \sin x \quad \text{and} \quad y = \cos x \tag{1}$$

and interchange the role of the variables, we obtain the **inverse sine and cosine relations**

$$x = \sin y \quad \text{and} \quad x = \cos y \tag{2}$$

These are also called, respectively, the **arcsine and arccosine relations,** and are denoted by

$$y = \arcsin x \qquad\qquad y = \arccos x$$
$$\text{or} \tag{3}$$
$$y = \sin^{-1} x \qquad\qquad y = \cos^{-1} x$$

Both arcsin x and sin⁻¹ x are used to represent the arcsine relation and arccos x and cos⁻¹ x are used to represent the arccosine relation. It is important to note that $\sin^{-1} x$ and $\cos^{-1} x$ do not mean $1/\sin x$ and $1/\cos x$, respectively. The -1 is part of the function symbol.

Comparing (2) and (3), we see that arcsin x and sin⁻¹ x both represent a number or angle with sine equal to x. Similarly, arccos x and cos⁻¹ x both represent a number or angle with cosine equal to x. We summarize the above discussion as follows:

INVERSE SINE AND COSINE RELATIONS   *Use θ instead of y*

$$y = \arcsin x$$
are equivalent to $\sin y = x$
$$y = \sin^{-1} x$$

(arcsin x is any angle or number with sine equal to x)

$$y = \arccos x$$
are equivalent to $\cos y = x$
$$y = \cos^{-1} x$$

(arccos x is any angle or number with cosine equal to x)

If we graph the sine function and its inverse relation in the same coordinate system and the cosine function and its inverse relation in another coordinate system, we obtain the graphs shown in Figure 1.

Recall that domains and ranges are interchanged when going from a given function to its inverse. Thus, we have the domain and range relationships listed in Table 1.

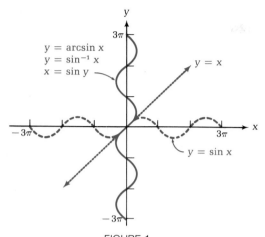

FIGURE 1
Inverse sine and cosine
relations

TABLE 1

RELATION	DOMAIN	RANGE
$y = \sin x$	All real numbers	$-1 \le y \le 1$
$y = \sin^{-1} x$	$-1 \le x \le 1$	All real numbers
$y = \cos x$	All real numbers	$-1 \le y \le 1$
$y = \cos^{-1} x$	$-1 \le x \le 1$	All real numbers

Neither the sine nor the cosine functions are one-to-one; hence, the inverse relations are not functions. The fact that neither inverse relation is a function is also obvious from Figure 1. Later, we will restrict the domains of the sine and cosine functions to obtain one-to-one functions so that their inverses will be functions. First, however, let us evaluate the inverse relations for some values of x, since this process will be useful to us when we solve trigonometric equations in Section 5.3.

EXAMPLE 1   Evaluate as real numbers: (A)   arcsin 1   (B) $\cos^{-1} \frac{1}{2}$

Solutions   (A) $y = \arcsin 1$ is equivalent to $\sin y = 1$ (y is any number with sine equal to 1). There are infinitely many y values that have this property. Choose all solutions for one period, say from 0 to 2π, and then add multiples of 2π to these. Thus, since $\pi/2$ is the only solution from 0 to 2π, all solutions are given by

$$y = \frac{\pi}{2} + 2k\pi \qquad k \text{ any integer}$$

That is,

$$\arcsin 1 = \frac{\pi}{2} + 2k\pi \qquad k \text{ any integer}$$

(Interpret these results relative to Figure 1.)

(B) $y = \cos^{-1} \frac{1}{2}$ is equivalent to $\cos y = \frac{1}{2}$ (y is any number with cosine equal to $\frac{1}{2}$). Since $\frac{1}{2}$ is positive, the reference triangle associated with y is either in the first or fourth quadrants. The only two solutions from the period $-\pi$ to π are $-\pi/3$ and $\pi/3$ (if we had chosen the period from 0 to 2π, then the solutions would be $\pi/3$ and $5\pi/3$). Thus, all solutions are given by

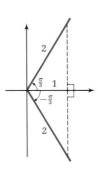

$$y = \pm\frac{\pi}{3} + 2k\pi \qquad k \text{ any integer}$$

That is,

$$\cos^{-1}\frac{1}{2} = \pm\frac{\pi}{3} + 2k\pi \qquad k \text{ any integer}$$

and we have all values with cosine equal to $\frac{1}{2}$. If we had chosen the solutions for the period from 0 to 2π, then we would write

$$\cos^{-1}\frac{1}{2} = \begin{cases} \dfrac{\pi}{3} + 2k\pi \\[2mm] \dfrac{5\pi}{3} + 2k\pi \end{cases} \qquad k \text{ any integer}$$

which would also represent all values with cosine equal to $\frac{1}{2}$.

PROBLEM 1 Evaluate as real numbers starting with the period from 0 to 2π:

(A) $\sin^{-1}\frac{1}{2}$ (B) $\arccos(-1)$

INVERSE SINE AND COSINE FUNCTIONS

Can we form functions out of the inverse sine and cosine relations? Yes, if we restrict the domains of sine and cosine so that the functions become one-to-one. This can be done in infinitely many ways. The generally accepted restrictions are indicated in Figure 2. Using these restrictions, we obtain the important **inverse sine and cosine functions** with the graphs given in Figure 3.

We use capital letters to represent inverse functions and lowercase letters to represent inverse relations.

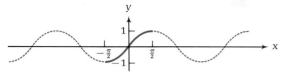

(a) $y = \sin x$ is one-to-one for $-\frac{\pi}{2} \leq x \leq \frac{\pi}{2}$

FIGURE 2

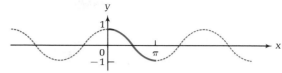

(b) $y = \cos x$ is one-to-one for $0 \leq x \leq \pi$

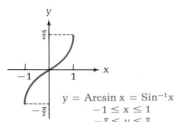

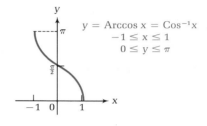

$y = \text{Arccos } x = \text{Cos}^{-1}x$
$-1 \leq x \leq 1$
$0 \leq y \leq \pi$

FIGURE 3
Inverse sine and cosine
functions

$y = \text{Arcsin } x = \text{Sin}^{-1}x$
$-1 \leq x \leq 1$
$-\frac{\pi}{2} \leq y \leq \frac{\pi}{2}$

Thus, Arcsin x and Sin^{-1} x are single-valued functions, whereas arcsin x and sin^{-1} x are multivalued relations. We summarize the above discussion in the box.

INVERSE SINE AND COSINE FUNCTIONS

$y = \text{Arcsin } x$

$y = \text{Sin}^{-1} x$

are equivalent to $\sin y = x$ where $\begin{cases} -\dfrac{\pi}{2} \leq y \leq \dfrac{\pi}{2} \\ -1 \leq x \leq 1 \end{cases}$

$y = \text{Arccos } x$

$y = \text{Cos}^{-1} x$

are equivalent to $\cos y = x$ where $\begin{cases} 0 \leq y \leq \pi \\ -1 \leq x \leq 1 \end{cases}$

Arcsin x and Arccos x are often referred to as **principal values** of the corresponding inverse relations.

Again, we emphasize that domains and ranges are interchanged when going from a given function to its inverse (see Table 2).

TABLE 2

FUNCTION	DOMAIN	RANGE
$y = \sin x$	$-\dfrac{\pi}{2} \leq x \leq \dfrac{\pi}{2}$†	$-1 \leq y \leq 1$
$y = \text{Sin}^{-1} x$	$-1 \leq x \leq 1$	$-\dfrac{\pi}{2} \leq y \leq \dfrac{\pi}{2}$
$y = \cos x$	$0 \leq x \leq \pi$†	$-1 \leq y \leq 1$
$y = \text{Cos}^{-1} x$	$-1 \leq x \leq 1$	$0 \leq y \leq \pi$

†Restricted to obtain one-to-one functions.

It follows from the general properties of functions and their inverses that:

$$\sin(\mathrm{Sin}^{-1}\, x) = x,\ -1 \le x \le 1 \qquad \mathrm{Sin}^{-1}(\sin x) = x,\ -\frac{\pi}{2} \le x \le \frac{\pi}{2}$$

$$\cos(\mathrm{Cos}^{-1}\, x) = x,\ -1 \le x \le 1 \qquad \mathrm{Cos}^{-1}(\cos x) = x,\ 0 \le x \le \pi$$

Thus, for example,

$$\sin(\mathrm{Sin}^{-1}\, 0.8) = 0.8 \qquad \text{but} \qquad \mathrm{Cos}^{-1}(\cos 35) \ne 35 \ (\text{why?})$$

EXAMPLE 2 Evaluate as real numbers: (A) $\mathrm{Cos}^{-1}\, \tfrac{1}{2}$ (B) Arcsin 1

Solutions (A) $y = \mathrm{Cos}^{-1}\, \tfrac{1}{2}$ is equivalent to $\cos y = \tfrac{1}{2}$ and $0 \le y \le \pi$. What y between 0 and π has cosine $\tfrac{1}{2}$? We use two methods to answer the question:

Method 1   Use reference triangles.

$\cos y = \tfrac{1}{2}$ $y = \tfrac{\pi}{3}$

Reference triangle is
a special 30°–60° triangle

Thus,

$$\mathrm{Cos}^{-1}\frac{1}{2} = \frac{\pi}{3}$$

($\pi/3$ is the only number between 0 and π with cosine equal to $\tfrac{1}{2}$).

Method 2   Use a calculator.

Calculators that have the capability of evaluating inverse sines and cosines do so according to the restrictions indicated in Table 2. To evaluate directly as a real number, put the calculator in radian mode, enter 0.5, and push the button(s) that yields $\mathrm{Cos}^{-1}\, 0.5$ to obtain

$$\mathrm{Cos}^{-1}\, 0.5 = 1.0472 \qquad \boxed{0.5}\ \boxed{\cos^{-1}} \quad \text{or} \quad \boxed{0.5}\ \boxed{\text{inv}}\ \boxed{\cos}$$

(The actual button-pushing process varies among the various types of calculators. Read your instruction booklet for your particular calculator.)

(B) $y = $ Arcsin 1 is equivalent to $\sin y = 1$ and $-\pi/2 \le y \le \pi/2$. What y between $-\pi/2$ and $\pi/2$ has sine 1? $\pi/2$ is the only value; hence,

$$\text{Arcsin } 1 = \frac{\pi}{2}$$

If we use a calculator, we obtain

$$\text{Arcsin } 1 = 1.5708 \qquad \boxed{1}\ \boxed{\sin^{-1}} \quad \text{or} \quad \boxed{1}\ \boxed{\text{inv}}\ \boxed{\sin}$$

to four decimal places.

PROBLEM 2 Evaluate as real numbers: (A) Arccos 1 (B) $\text{Sin}^{-1}\ \tfrac{1}{2}$

EXAMPLE 3 Evaluate as real numbers: (A) $\text{Sin}^{-1}(-\sqrt{3}/2)$ (B) Arccos 3

Solutions (A) $y = \text{Sin}^{-1}(-\sqrt{3}/2)$ is equivalent to $\sin y = -\sqrt{3}/2$ and $-\pi/2 \le y \le \pi/2$. What y between $-\pi/2$ and $\pi/2$ has sine $-\sqrt{3}/2$? y must be negative and in the fourth quadrant:

Do

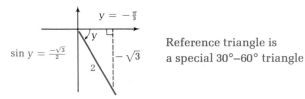

Reference triangle is a special 30°–60° triangle

Thus,

$$\text{Sin}^{-1}\!\left(-\frac{\sqrt{3}}{2}\right) = -\frac{\pi}{3}$$

[*Note:* y cannot be $5\pi/3$. Why?]

(B) Arccos 3 is not defined, since 3 is not in the domain of this function.

PROBLEM 3 Evaluate as real numbers: (A) $\text{Arcsin}(-2)$ (B) $\text{Cos}^{-1}(-\sqrt{3}/2)$

EXAMPLE 4 Find $\cos[\text{Sin}^{-1}(-0.2)]$:
(A) Without a table or calculator (B) With a calculator

Solutions (A) Let $y = \text{Sin}^{-1}(-0.2)$; then

Do

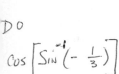

$$\sin y = -0.2 \qquad \text{and} \qquad -\frac{\pi}{2} \le y \le \frac{\pi}{2}$$

y must be negative and in the fourth quadrant. Draw the reference triangle, find the third side, and then determine $\cos y$ from the triangle:

$$\sin y = -0.2 = \frac{-0.2}{1} \qquad\qquad a = \sqrt{1^2 - (-0.2)^2}$$
$$= \sqrt{0.96}$$

Thus,

$$\cos[\text{Sin}^{-1}(-0.2)] = \cos y = \sqrt{0.96}$$

(B)   To find $\cos[\text{Sin}^{-1}(-0.2)]$ using a calculator we can use either degree or radian mode (think about this!). Enter -0.2, push the Sin^{-1} button (or its equivalent), and then push the cos button to obtain

$$\cos[\text{Sin}^{-1}(-0.2)] = 0.98 \qquad \boxed{-0.2}\ \boxed{\sin^{-1}}\ \boxed{\cos}$$

PROBLEM 4   Find $\sin[\text{Cos}^{-1}(-0.3)]$:
(A)   Without a table or calculator   (B)   With a calculator

EXAMPLE 5   Evaluate $\cos(\text{Sin}^{-1}\tfrac{3}{5} - \text{Cos}^{-1}\tfrac{4}{5})$ without using tables or a calculator.

Solution   We use the difference identity for cosine and the procedure outlined in Example 4(A) to obtain

$$\cos\left(\text{Sin}^{-1}\frac{3}{5} - \text{Cos}^{-1}\frac{4}{5}\right)$$

$$= \cos\left(\text{Sin}^{-1}\frac{3}{5}\right)\cos\left(\text{Cos}^{-1}\frac{4}{5}\right) + \sin\left(\text{Sin}^{-1}\frac{3}{5}\right)\sin\left(\text{Cos}^{-1}\frac{4}{5}\right)$$

$$= \quad\left(\frac{4}{5}\right)\quad\cdot\quad\left(\frac{4}{5}\right)\quad + \quad\left(\frac{3}{5}\right)\quad\cdot\quad\left(\frac{3}{5}\right)$$

$$= 1$$

PROBLEM 5   Evaluate $\sin(2\ \text{Cos}^{-1}\tfrac{3}{5})$ without using tables or a calculator.

ANSWERS TO MATCHED PROBLEMS

1.   (A) $\sin^{-1}\dfrac{1}{2} = \begin{cases} \pi/6 + 2k\pi \\ 5\pi/6 + 2k\pi \end{cases}$ k any integer

     (B) $\pi + 2k\pi$, k any integer
2.   (A)   0   (B) $\pi/6$ 3.   (A)   Not defined   (B) $5\pi/6$
4.   (A) $\sqrt{0.91}$ (B)   0.954   5. $\tfrac{24}{25}$

EXERCISE 5.1

A   *Find each of the following exactly as real numbers without a table or calculator:*

1. $\text{Cos}^{-1} 0$ 2. $\text{Sin}^{-1} 0$ 3.   Arccos 1

4.   Arcsin 1   5. $\text{Sin}^{-1}\dfrac{1}{2}$ 6. $\text{Cos}^{-1}\dfrac{1}{2}$

7. $\text{Arcsin}\dfrac{1}{\sqrt{2}}$ 8. $\text{Arccos}\dfrac{\sqrt{3}}{2}$ 9. $\text{Cos}^{-1}\dfrac{1}{\sqrt{2}}$

10. $\text{Sin}^{-1}\dfrac{\sqrt{3}}{2}$ 11. $\text{Cos}^{-1}(-1)$ 12. $\text{Sin}^{-1}(-1)$

13. $\text{Arcsin}\left(\dfrac{-1}{2}\right)$ 14. $\text{Arccos}\left(\dfrac{-1}{2}\right)$ 15. $\text{Cos}^{-1}\left(\dfrac{-1}{\sqrt{2}}\right)$

16. $\text{Sin}^{-1}\left(-\dfrac{\sqrt{3}}{2}\right)$

B   17. $\sin[\text{Sin}^{-1}(-0.7)]$ 18. $\text{Cos}^{-1}(\cos 0.43)$ 19. $\cos(\text{Cos}^{-1}\, 0.37)$

     20. $\text{Sin}^{-1}[\sin(-1.4)]$ 21. $\sin(\text{Cos}^{-1}\, 1)$ 22. $\cos(\text{Sin}^{-1}\, 0)$

     23. $\cos\left(\text{Sin}^{-1}\dfrac{1}{2}\right)$ 24. $\sin\left(\text{Cos}^{-1}\dfrac{\sqrt{3}}{2}\right)$ 25. $\sin\left[\text{Cos}^{-1}\left(-\dfrac{3}{5}\right)\right]$

     26. $\cos\left[\text{Sin}^{-1}\left(-\dfrac{4}{5}\right)\right]$ 27. $\sin(\text{Cos}^{-1}\, 0.3)$ 28. $\cos(\text{Sin}^{-1}\, 0.6)$

     29. $\cos[\text{Sin}^{-1}(-0.8)]$ 30. $\sin[\text{Cos}^{-1}(-0.9)]$

Find all real values exactly, starting with values from 0 to 2π.

     31. $\arcsin\dfrac{1}{2}$ 32. $\arccos\dfrac{1}{2}$ 33. $\sin^{-1}\dfrac{1}{\sqrt{2}}$

     34. $\cos^{-1}\dfrac{1}{\sqrt{2}}$ 35. $\cos^{-1}\left(-\dfrac{1}{2}\right)$ 36. $\sin^{-1}\left(-\dfrac{1}{2}\right)$

     37. $\arcsin\left(-\dfrac{1}{\sqrt{2}}\right)$ 38. $\arccos\left(-\dfrac{\sqrt{3}}{2}\right)$

Find each of the following as real numbers using a calculator (compute answers to three decimal places):

     39. $\text{Cos}^{-1}\, 0.613$ 40. $\text{Sin}^{-1}\, 0.839$

     41. $\text{Arcsin}(-0.103)$ 42. $\text{Arccos}(-0.417)$

     43. $\sin[\text{Cos}^{-1}(-0.237)]$ 44. $\cos[\text{Sin}^{-1}(-0.305)]$

     45. $\text{Sin}^{-1}(\cos 23.5)$ 46. $\text{Cos}^{-1}[\sin(-5.013)]$

C   *Evaluate exactly as real numbers without the use of tables or a calculator.*

     47. $\sin\left[\text{Cos}^{-1}\left(-\dfrac{4}{5}\right) + \text{Sin}^{-1}\left(-\dfrac{3}{5}\right)\right]$ 48. $\cos\left[\text{Sin}^{-1}\left(-\dfrac{3}{5}\right) + \text{Cos}^{-1}\dfrac{4}{5}\right]$

     49. $\sin\left[\text{Arccos}\dfrac{1}{2} + \text{Arcsin}(-1)\right]$

     50. $\cos\left[\text{Cos}^{-1}\left(-\dfrac{\sqrt{3}}{2}\right) - \text{Sin}^{-1}\left(-\dfrac{1}{2}\right)\right]$

     51. $\sin\left[2\,\text{Sin}^{-1}\left(-\dfrac{4}{5}\right)\right]$ 52. $\sin\left[2\,\text{Cos}^{-1}\left(-\dfrac{3}{5}\right)\right]$

     53. Show that $\sin(\text{Cos}^{-1} x) = \sqrt{1 - x^2}$ for $-1 \leq x \leq 1$.

     54. Show that $\cos(\text{Sin}^{-1} x) = \sqrt{1 - x^2}$ for $-1 \leq x \leq 1$.

     55. Show that $\tan(\text{Sin}^{-1} x) = x/\sqrt{1 - x^2}$ for $-1 < x < 1$.

     56. Show that $\cot(\text{Cos}^{-1} x) = x/\sqrt{1 - x^2}$ for $-1 < x < 1$.

5.2

INVERSE RELATIONS

INVERSES OF OTHER TRIGONOMETRIC FUNCTIONS

Paralleling the development for the sine and cosine functions, we define the inverse relations for the other four trigonometric functions as follows:

INVERSE TANGENT, COTANGENT, SECANT, AND COSECANT RELATIONS

$y = \arctan x = \tan^{-1} x$ are equivalent to $\tan y = x$

$y = \text{arccot } x = \cot^{-1} x$ are equivalent to $\cot y = x$

$y = \text{arcsec } x = \sec^{-1} x$ are equivalent to $\sec y = x$

$y = \text{arccsc } x = \csc^{-1} x$ are equivalent to $\csc y = x$

EXAMPLE 6

Evaluate as real numbers:

(A) $\arctan 1$ (B) $\cot^{-1}(1/\sqrt{3})$ (C) $\text{arcsec } 2$

Solutions

(A) $y = \arctan 1$ is equivalent to $\tan y = 1$ (y is any number with tangent equal to 1). Choose all solutions over one period, say from $-\pi/2$ to $\pi/2$, and then add multiples of π (the period of the tangent function). Drawing a reference triangle sometimes helps:

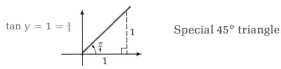

$\tan y = 1 = \frac{1}{1}$ Special 45° triangle

$$y = \frac{\pi}{4} + k\pi \qquad k \text{ any integer}$$

That is,

$$\arctan 1 = \frac{\pi}{4} + k\pi \qquad k \text{ any integer}$$

(B) $y = \cot^{-1}(1/\sqrt{3})$ is equivalent to $\cot y = 1/\sqrt{3}$ (y is any number with cotangent equal to $1/\sqrt{3}$). Choose all solutions over one period, say from 0 to π, and then add multiples of π (the period of the cotangent function). Drawing a reference triangle helps:

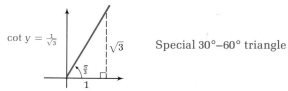

$\cot y = \frac{1}{\sqrt{3}}$ Special 30°–60° triangle

$$y = \frac{\pi}{3} + k\pi \qquad k \text{ any integer}$$

That is,

$$\cot^{-1}\frac{1}{\sqrt{3}} = \frac{\pi}{3} + k\pi \qquad k \text{ any integer}$$

(C) $y = \text{arcsec } 2$ is equivalent to sec $y = 2$ (y is any number with secant equal to 2). Choose all solutions over one period, say from 0 to 2π, and then add multiples of 2π (the period of the secant function). We start by drawing reference triangles:

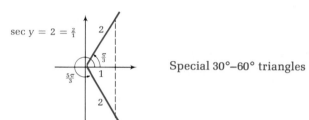

Special 30°–60° triangles

Thus,

$$y = \begin{cases} \dfrac{\pi}{3} + 2k\pi \\[2mm] \dfrac{5\pi}{3} + 2k\pi \end{cases} \qquad k \text{ any integer}$$

That is,

$$\text{arcsec } 2 = \begin{cases} \dfrac{\pi}{3} + 2k\pi \\[2mm] \dfrac{5\pi}{3} + 2k\pi \end{cases} \qquad k \text{ any integer}$$

PROBLEM 6   Evaluate as real numbers:

(A) $\tan^{-1}\sqrt{3}$ (B) $\text{arccot } 1$ (C) $\csc^{-1}\sqrt{2}$

INVERSE FUNCTIONS   Of course, the inverse relations just considered are not functions. However, if we restrict the domains of the four functions so that they become one-to-one, then the inverses will be functions. This is done as shown in the box on page 142 (again, we use capital letters to represent the inverse functions and lowercase letters to represent the inverse relations).

INVERSE TANGENT, COTANGENT, SECANT, AND COSECANT FUNCTIONS

$y = \text{Tan}^{-1}\, x$ is equivalent to $\tan y = x$ where $-\dfrac{\pi}{2} < y < \dfrac{\pi}{2}$ and x is any real number

$y = \text{Cot}^{-1}\, x$ is equivalent to $\cot y = x$ where $0 < y < \pi$ and x is any real number

$y = \text{Sec}^{-1}\, x$ is equivalent to $\sec y = x$ where $0 \le y \le \pi$, $y \ne \dfrac{\pi}{2}$,   and $x \le -1$ or $x \ge 1$

$y = \text{Csc}^{-1}\, x$ is equivalent to $\csc y = x$ where $-\dfrac{\pi}{2} \le y \le \dfrac{\pi}{2}$, $y \ne 0$,   and $x \le -1$ or $x \ge 1$

The graphs of these four functions are given in Figure 4.

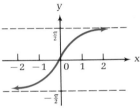

Domain: All real numbers
Range: $-\frac{\pi}{2} < y < \frac{\pi}{2}$

(a) $y = \text{Tan}^{-1}\, x$

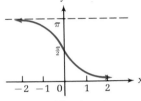

Domain: All real numbers
Range: $0 < y < \pi$

(b) $y = \text{Cot}^{-1}\, x$

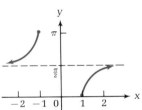

Domain: $x \le -1$ or $x \ge 1$
Range: $0 \le y \le \pi$, $y \ne \frac{\pi}{2}$

(c) $y = \text{Sec}^{-1}\, x$

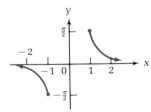

Domain: $x \le -1$ or $x \ge 1$
Range: $-\frac{\pi}{2} \le y \le \frac{\pi}{2}$, $y \ne 0$

(d) $y = \text{Csc}^{-1}\, x$

FIGURE 4
Inverse tangent, cotangent,
secant, and cosecant
functions

It follows from the general properties of functions and their inverses that

for T a trigonometric function and T^{-1} its inverse, and if the domain and range of T and T^{-1} are suitably restricted (see above), then

$$T[T^{-1}(x)] = x \qquad \text{and} \qquad T^{-1}[T(x)] = x$$

Thus, $\sec[\text{Sec}^{-1}(-3.05)] = -3.05$ but $\text{Cot}^{-1}(\cot 403.5) \ne 403.5$. (Why?)

EXAMPLE 7   Evaluate as real numbers:

(A) $\text{Tan}^{-1}(-1/\sqrt{3})$ (B) $\text{Arccot}(-1)$ (C) $\text{Sec}^{-1}(2/\sqrt{3})$

Solutions   (A) $y = \text{Tan}^{-1}(-1/\sqrt{3})$ is equivalent to $\tan y = -1/\sqrt{3}$, $-\pi/2 < y < \pi/2$.
What number between $-\pi/2$ and $\pi/2$ has tangent $-1/\sqrt{3}$? y must be
negative and in the fourth quadrant. A reference triangle is helpful:

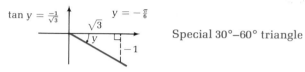

Special 30°–60° triangle

Thus,

$$\text{Tan}^{-1}\left(-\frac{1}{\sqrt{3}}\right) = -\frac{\pi}{6}$$

[*Note:* y cannot be $11\pi/6$. Why?]

(B) $y = \text{Arccot}(-1)$ is equivalent to $\cot y = -1$ and $0 < y < \pi$. What number between 0 and π has cotangent -1? y must be positive and in the second quadrant:

cot $y = -1 = \frac{-1}{1}$ $\alpha = \frac{\pi}{4}$ $y = \frac{3\pi}{4}$

Thus,

$$\text{Arccot}(-1) = \frac{3\pi}{4}$$

(C) $y = \text{Sec}^{-1}(2/\sqrt{3})$ is equivalent to $\sec y = 2/\sqrt{3}$ and $0 \le y \le \pi$, $y \ne \pi/2$.
What number between 0 and π has secant $2/\sqrt{3}$? y is positive and in the first quadrant. We draw a reference triangle:

sec $y = \frac{2}{\sqrt{3}}$ $y = \frac{\pi}{6}$

Thus,

$$\text{Sec}^{-1}\frac{2}{\sqrt{3}} = \frac{\pi}{6}$$

PROBLEM 7   Evaluate as real numbers:

(A) $\text{Tan}^{-1}\sqrt{3}$ (B) $\text{Arccot}(-\sqrt{3})$ (C) $\text{Csc}^{-1}(-2)$

EXAMPLE 8   Evaluate $\tan[\text{Sec}^{-1}(-3)]$ exactly without a calculator or table.

Solution   Let $y = \text{Sec}^{-1}(-3)$; then

$$\sec y = -3 \qquad 0 \le y \le \pi, \quad y \ne \frac{\pi}{2}$$

y is positive and in the second quadrant. Draw a reference triangle, find the third side, and then determine $\tan y$ from the triangle:

Same as before

$$\sec y = -3 = \tfrac{3}{-1}$$

$$b = \sqrt{3^2 - (-1)^2}$$
$$= \sqrt{8} = 2\sqrt{2}$$

Thus,

$$\tan[\text{Sec}^{-1}(-3)] = \tan y = \frac{b}{a} = \frac{2\sqrt{2}}{-1} = -2\sqrt{2}$$

PROBLEM 8   Evaluate $\cot[\text{Csc}^{-1}(-\tfrac{5}{3})]$ exactly without a calculator or table.

CALCULATOR
EVALUATION

Many hand calculators have sin, cos, tan, Sin^{-1}, Cos^{-1}, Tan^{-1}, or their equivalents. To find sec x, csc x, and cot x using a calculator, we use the reciprocal identities

$$\sec x = \frac{1}{\cos x} \qquad \csc x = \frac{1}{\sin x} \qquad \cot x = \frac{1}{\tan x}$$

How can we evaluate $\text{Sec}^{-1} x$, $\text{Csc}^{-1} x$, and $\text{Cot}^{-1} x$ using a calculator with only Sin^{-1}, Cos^{-1}, and Tan^{-1}? The following inverse identities are made to order for this purpose:

INVERSE TRIGONOMETRIC IDENTITIES

$$\text{Cot}^{-1} x = \begin{cases} \text{Tan}^{-1}\dfrac{1}{x}, & x > 0 \\[3mm] \pi + \text{Tan}^{-1}\dfrac{1}{x}, & x < 0 \end{cases}$$

$$\text{Sec}^{-1} x = \text{Cos}^{-1}\frac{1}{x}, \qquad x \ge 1 \quad \text{or} \quad x \le -1$$

$$\text{Csc}^{-1} x = \text{Sin}^{-1}\frac{1}{x}, \qquad x \ge 1 \quad \text{or} \quad x \le -1$$

We will establish the first part of the inverse cotangent identity. The inverse secant and cosecant identities are left to you to do (see Problems 53 and 54, Exercise 5.2). Let

$$y = \text{Cot}^{-1} x \qquad x > 0$$

Then

$$\cot y = x \qquad 0 < y < \frac{\pi}{2} \qquad \text{Definition of Cot}^{-1}$$

$$\frac{1}{\tan y} = x \qquad 0 < y < \frac{\pi}{2} \qquad \text{Reciprocal identity}$$

$$\tan y = \frac{1}{x} \qquad 0 < y < \frac{\pi}{2} \qquad \text{Algebra}$$

$$y = \text{Tan}^{-1} \frac{1}{x} \qquad 0 < y < \frac{\pi}{2} \qquad \text{Definition of Tan}^{-1}$$

Thus,

$$\text{Cot}^{-1} x = \text{Tan}^{-1} \frac{1}{x} \qquad \text{for} \quad x > 0$$

EXAMPLE 9   Use a calculator to evaluate as real numbers to three decimal places:

(A) $\text{Tan}^{-1}(-35.2)$ (B) $\text{Cot}^{-1} 4.05$ (C) $\text{Csc}^{-1}(-12)$

Solutions   (A)   Put the calculator in radian mode, enter -35.2, and push the Tan^{-1} button (or its equivalent) to obtain

$$\text{Tan}^{-1}(-35.2) = -1.542 \qquad \boxed{35.2} \; \boxed{+/-} \; \boxed{\tan^{-1}}$$

(B)   Put the calculator in radian mode, enter 4.05, take its reciprocal, and then push the Tan^{-1} button (or its equivalent) to obtain

$$\text{Cot}^{-1} 4.05 = \text{Tan}^{-1} \frac{1}{4.05} \qquad \boxed{4.05} \; \boxed{1/x} \; \boxed{\tan^{-1}}$$

$$= 0.242$$

(C)   Put the calculator in radian mode, enter -12, take its reciprocal, and then push the Sin^{-1} button (or its equivalent) to obtain

$$\text{Csc}^{-1}(-12) = \text{Sin}^{-1} \left(-\frac{1}{12} \right) \qquad \boxed{12} \; \boxed{+/-} \; \boxed{1/x} \; \boxed{\sin^{-1}}$$

$$= -0.083$$

PROBLEM 9   Use a calculator to evaluate as real numbers to three decimal places:

(A) $\text{Tan}^{-1} 5.13$ (B) $\text{Cot}^{-1} 2.314$ (C) $\text{Sec}^{-1}(-1.549)$

6. (A) $\pi/3 + k\pi$, k any integer   (B) $\pi/4 + k\pi$, k any integer
   (C) $\pi/4 + 2k\pi$, $3\pi/4 + 2k\pi$, k any integer
7. (A) $\pi/3$ (B) $5\pi/6$ (C) $-\pi/6$ **8.** $-\frac{4}{5}$
9. (A) 1.378   (B) 0.408   (C) 2.273

EXERCISE 5.2

A   *Find each of the following exactly as real numbers without a table or calculator:*

1. $\text{Tan}^{-1} 0$ **2.** $\text{Cot}^{-1} 0$ **3.** $\text{Arccot } \sqrt{3}$
4. $\text{Arctan } \sqrt{3}$ **5.** $\text{Sec}^{-1} \sqrt{2}$ **6.** $\text{Csc}^{-1} 2$

7. $\sin(\text{Tan}^{-1} 1)$ **8.** $\cos(\text{Cot}^{-1} 1)$ **9.** $\tan\left(\text{Csc}^{-1} \frac{5}{4}\right)$

10. $\cot\left(\text{Sec}^{-1} \frac{5}{3}\right)$

B   **11.** $\text{Cot}^{-1}(-1)$ **12.** $\text{Tan}^{-1}(-1)$
   13. $\text{Arcsec}(-2)$ **14.** $\text{Arccsc}(-\sqrt{2})$
   15. $\text{Arctan}(-\sqrt{3})$ **16.** $\text{Arccot}(-\sqrt{3})$

17. $\text{Csc}^{-1} \frac{1}{2}$ **18.** $\text{Sec}^{-1}\left(-\frac{1}{2}\right)$

19. $\cot\left[\text{Tan}^{-1}\left(-\frac{1}{\sqrt{3}}\right)\right]$ **20.** $\tan\left[\text{Cot}^{-1}\left(-\frac{1}{\sqrt{3}}\right)\right]$

21. $\cos\left[\text{Tan}^{-1}\left(-\frac{3}{4}\right)\right]$ **22.** $\sin\left[\text{Cot}^{-1}\left(-\frac{3}{4}\right)\right]$

23. $\csc\left[\text{Tan}^{-1}\left(-\frac{2}{3}\right)\right]$ **24.** $\sec\left[\text{Cot}^{-1}\left(-\frac{1}{2}\right)\right]$

25. $\cos[\text{Sec}^{-1}(-2)]$ **26.** $\sin[\text{Csc}^{-1}(-2)]$
27. $\tan[\text{Tan}^{-1} 33.4]$ **28.** $\sec[\text{Sec}^{-1}(-44)]$
29. $\csc[\text{Csc}^{-1}(-4)]$ **30.** $\cot[\text{Cot}^{-1}(-7.3)]$

Find all real values exactly.

31. $\arctan \frac{1}{\sqrt{3}}$ **32.** $\text{arccot} \frac{1}{\sqrt{3}}$ **33.** $\sec^{-1} \frac{2}{\sqrt{3}}$

34. $\csc^{-1} 2$ **35.** $\cot^{-1}\left(-\frac{1}{\sqrt{3}}\right)$ **36.** $\tan^{-1}\left(-\frac{1}{\sqrt{3}}\right)$

37. $\text{arccsc}(-2)$ **38.** $\text{arcsec}(-2)$ **39.** $\sin(\csc^{-1} 2)$

40. $\cos\left(\sec^{-1} \frac{2}{\sqrt{3}}\right)$

Use a calculator to evaluate the following as real numbers to three decimal places:

41. $\text{Tan}^{-1} 2.035$ **42.** $\text{Tan}^{-1}(-2.087)$ **43.** $\text{Cot}^{-1} 3.065$

44. $\mathrm{Cot}^{-1}\, 7.306$ **45.** $\mathrm{Sec}^{-1}(-1.963)$ **46.** $\mathrm{Sec}^{-1}\, 2.041$

47. $\mathrm{Csc}^{-1}\, 1.172$ **48.** $\mathrm{Csc}^{-1}(-1.938)$

C   *Evaluate exactly without using tables or a calculator.*

49. $\tan\left[\mathrm{Csc}^{-1}\left(-\dfrac{5}{3}\right) + \mathrm{Tan}^{-1}\dfrac{1}{4}\right]$ **50.** $\tan[\mathrm{Tan}^{-1}\, 4 - \mathrm{Sec}^{-1}(-\sqrt{5})]$

51. $\tan\left[2\,\mathrm{Cot}^{-1}\left(-\dfrac{3}{4}\right)\right]$ **52.** $\tan[2\,\mathrm{Sec}^{-1}(-\sqrt{5})]$

53. Show that $\mathrm{Sec}^{-1} x = \mathrm{Cos}^{-1}(1/x)$ for $x \geq 1$ and $x \leq -1$.

54. Show that $\mathrm{Csc}^{-1} x = \mathrm{Sin}^{-1}(1/x)$ for $x \geq 1$ and $x \leq -1$.

Establish the following identities for all real x:

55. $\sin(\mathrm{Tan}^{-1} x) = \dfrac{x}{\sqrt{1 + x^2}}$ **56.** $\cos(\mathrm{Cot}^{-1} x) = \dfrac{x}{\sqrt{1 + x^2}}$

57. $\sin(2\,\mathrm{Tan}^{-1} x) = \dfrac{2x}{1 + x^2}$ **58.** $\cos(2\,\mathrm{Cot}^{-1} x) = \dfrac{x^2 - 1}{x^2 + 1}$

59. $\mathrm{Tan}^{-1} x + \mathrm{Tan}^{-1}(-x) = 0$ **60.** $\mathrm{Cot}^{-1} x + \mathrm{Cot}^{-1}(-x) = \pi$

61. Evaluate $\mathrm{Cot}^{-1} x$ and $\mathrm{Tan}^{-1}(1/x)$ without a calculator or table for $x = -1$.

APPLICATIONS   ***62.** A painting 4 ft high and 8 ft from the floor is being observed in an art gallery (see the figure). If the observer is standing x ft from the wall and his eye is 5 ft from the floor, show that

$$\theta = \mathrm{Arctan}\,\frac{4x}{x^2 + 21}$$

[*Hint:*   Use the sum identity, $\tan(\alpha + \theta) = (\tan \alpha + \tan \theta)/(1 - \tan \alpha \tan \theta)$.]

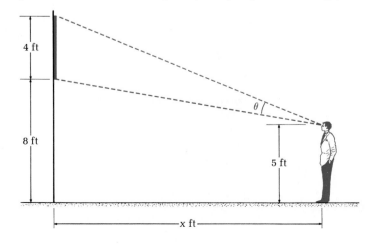

63. The figure below represents a circular courtyard surrounded by a high stone wall. A flood light, located at E, shines into the courtyard. If a person walks x ft away from the center along DC, show that his or her shadow will move a distance given by

$$d = 2R\theta = 2R \, \text{Tan}^{-1} \frac{x}{R}$$

where θ is in radians. [*Hint:* Draw a line from A to C.]

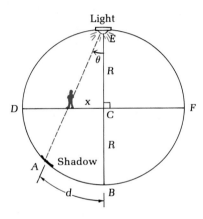

5.3   TRIGONOMETRIC EQUATIONS

In Chapter 4 we considered trigonometric equations called identities. These equations were true for all replacements of the variable(s) for which both sides were defined. Now we will consider another class of trigonometric equations, called **conditional equations,** which are true for some replacement of the variable but false for other replacements. For example,

$$\sin x = \cos x$$

is a conditional equation, since it is true for $x = \pi/4$ and false for $x = 0$.

In this section we will consider some standard techniques for solving conditional trigonometric equations. Even though there is no step-by-step process that will work for all equations, some guidelines that will be helpful in many cases are listed in the box at the top of the next page.

Solve by :
① graph
② ÷ by cos x
③ sq. both sides
(CK. answers!)

EXAMPLE 10   Solve $2 \sin x - \sqrt{3} = 0, 0 \leq x \leq \pi/2$.

GUIDELINES FOR SOLVING TRIGONOMETRIC EQUATIONS

1.   If only one of the six trigonometric functions is present in an equation, try to solve for that function; then solve for the variable.

2.   In the process of solving for a particular trigonometric function, try rewriting the equation so that only zero remains to the right of the equals sign; then try solving by factoring or use of the quadratic formula.

3.   If two or more different trigonometric functions are present in an equation, try using identities to transform the equation into one with only one trigonometric function, or rewrite the equation so that only zero remains on the right and try factoring the left side.

4.   If multiple-angle forms are present, try using identities to reduce to single-angle forms.

Solution

$$2 \sin x - \sqrt{3} = 0 \qquad \text{Solve for } \sin x$$

$$\sin x = \frac{\sqrt{3}}{2} \qquad \text{Solve for } x$$

$$x = \text{Sin}^{-1} \frac{\sqrt{3}}{2}$$

$$x = \frac{\pi}{3}$$

PROBLEM 10   Solve $\sqrt{3} \tan x - 1 = 0, 0° \le x \le 90°$.

EXAMPLE 11   Solve $\sqrt{2} \cos x - 1 = 0$ for all real numbers and angles in degrees.

Solution   *Step 1*   *Solve for cos x.*

$$\sqrt{2} \cos x - 1 = 0$$

$$\cos x = \frac{1}{\sqrt{2}}$$

Step 2   *Draw reference triangles where cosine is positive and find reference angle α (recall, α is never negative).*

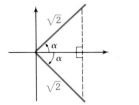

$$\alpha = \text{Cos}^{-1} \frac{1}{\sqrt{2}}$$

$$= \frac{\pi}{4} \quad \text{or} \quad 45°$$

Step 3   Find all solutions for one period, say from 0 to 2π.

$$\frac{\pi}{4} \; (45°) \qquad 2\pi - \frac{\pi}{4} = \frac{7\pi}{4} \; (315°)$$

Step 4   Add multiples of 2π ($360°$) to solutions from one period to obtain all solutions.

$$\text{Real solutions:} \quad x = \begin{cases} \dfrac{\pi}{4} + 2k\pi \\[2mm] \dfrac{7\pi}{4} + 2k\pi \end{cases} \quad k \text{ any integer}$$

$$\text{Degree solutions:} \quad x = \begin{cases} 45° + k360° \\ 315° + k360° \end{cases} \quad k \text{ any integer}$$

PROBLEM 11   Repeat Example 11 for $\cot x - \sqrt{3} = 0$.

EXAMPLE 12   Solve $8 \sin^2 x = \sin x$, $0° \le x \le 90°$. Use decimal degrees to two decimal places.

Solution

$$8 \sin^2 x = \sin x$$
$$8 \sin^2 x - \sin x = 0 \qquad \text{Factor left side}$$
$$(\sin x)(8 \sin x - 1) = 0 \qquad \text{Solve for } \sin x$$

Therefore,

$$\sin x = 0 \qquad \text{or} \qquad 8 \sin x - 1 = 0$$
$$x = \text{Sin}^{-1} 0 \qquad\qquad \sin x = \frac{1}{8}$$

Degree mode

$$x = 0° \qquad\qquad\qquad x = \text{Sin}^{-1} \frac{1}{8} \qquad \boxed{8} \; \boxed{1/x} \; \boxed{\sin^{-1}}$$
$$= 7.18°$$

Thus, $x = 0°$ or $x = 7.18°$.

PROBLEM 12   Repeat Example 12 for $4 \cos^2 x = \cos x$, $0° \le x \le 90°$.

EXAMPLE 13   Solve $\cos^2 x = \cos x + \sin^2 x$, $0 \le x \le 2\pi$.

Solution

$$\cos^2 x = \cos x + \sin^2 x \qquad \text{Use Pythagorean identity}$$
$$\cos^2 x = \cos x + 1 - \cos^2 x \qquad \text{Solve for } \cos x$$
$$2 \cos^2 x - \cos x - 1 = 0 \qquad \text{Quadratic in } \cos x$$
$$(\cos x - 1)(2 \cos x + 1) = 0$$

Therefore,

$$\cos x - 1 = 0$$

$$\cos x = 1$$

$$x = 0, \quad 2\pi$$

or

$$2 \cos x + 1 = 0$$

$$\cos x = -\frac{1}{2}$$

$$x = \begin{cases} \pi - \dfrac{\pi}{3} = \dfrac{2\pi}{3} \\[2mm] \pi + \dfrac{\pi}{3} = \dfrac{4\pi}{3} \end{cases}$$

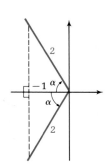

Reference angle:

$$\alpha = \frac{\pi}{3}$$

Thus, $x = 0, 2\pi/3, 4\pi/3,$ or 2π.

PROBLEM 13   Solve $\sin^2 x = \cos^2 x - \sin x, 0 \le x \le 2\pi$.

$\mathcal{D}o$

EXAMPLE 14   Solve $\cos 2x = 2(\sin x - 1), 0 \le x \le 360°$. Use decimal degrees to two decimal places.

Solution

$$\cos 2x = 2(\sin x - 1) \qquad \text{Use double-angle identity}$$

$$1 - 2 \sin^2 x = 2 \sin x - 2 \qquad \text{Solve for } \sin x$$

$$2 \sin^2 x + 2 \sin x - 3 = 0 \qquad \text{Use quadratic formula}$$

$$\sin x = \frac{-2 \pm \sqrt{4 - 4(2)(-3)}}{2(2)}$$

$$\sin x = -1.823 \quad \text{or} \quad 0.823$$

Thus,

$$\sin x = -1.823 \qquad \text{No solution. Why?}$$

Or

$$\sin x = 0.823$$

$$x = \sin^{-1} 0.823 \qquad 0° \le x \le 360°$$

Find reference angle:

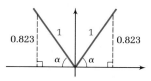

$$\sin \alpha = 0.823$$
$$\alpha = \text{Sin}^{-1}\, 0.823$$
$$\alpha = 55.39°$$

Degree mode
$\boxed{0.823}$ $\boxed{\sin^{-1}}$

Thus,

$$x = \begin{cases} 55.39° \\ 180° - 55.39° = 124.61° \end{cases}$$

PROBLEM 14   Solve $\cos 2x = 4 \cos x - 2$, $0 \le x \le 2\pi$, to two decimal places.

Do

EXAMPLE 15   Solve $\sin x + \cos x = 0$ exactly for all real x.

Solution

$$\sin x + \cos x = 0$$

$$\frac{\sin x}{\cos x} + \frac{\cos x}{\cos x} = \frac{0}{\cos x} \qquad \cos x \ne 0$$

$$\tan x + 1 = 0$$

$$\tan x = -1$$

$$x = \tan^{-1}(-1)$$

$$x = \frac{3\pi}{4} + k\pi \qquad k \text{ any integer}$$

PROBLEM 15   Repeat Example 15 for $\sin x - \sqrt{3} \cos x = 0$.

EXAMPLE 16   Solve $\csc x + \cot x = 1$, $0 \le x \le 2\pi$, exactly without a calculator or table.

Solution   To solve this equation, we square both sides after subtracting $\cot x$ from each side. We can then use Pythagorean identities to transform the equation into one involving only one type of trigonometric function. The squaring process can create extraneous solutions, so all solutions must be checked in the original equation to see if any must be discarded.

$$\csc x = 1 - \cot x$$
$$\csc^2 x = 1 - 2 \cot x + \cot^2 x$$
$$1 + \cot^2 x = 1 - 2 \cot x + \cot^2 x$$
$$0 = -2 \cot x$$
$$\cot x = 0$$
$$x = \pi/2,\, 3\pi/2$$

Checking these in the original equation, we find that $3\pi/2$ must be discarded. Hence, $x = \pi/2$ is the only solution.

Do   **PROBLEM 16**   Solve $1 + \cos x = \sin x$, $0 \le x \le 2\pi$, exactly without using a calculator or table.

ANSWERS TO
MATCHED PROBLEMS

10. $30°$ **11.** $\pi/6 + k\pi$, $30° + k180°$, k any integer
12. $75.52°$, $90°$ **13.** $\pi/6$, $5\pi/6$, $3\pi/2$ **14.** 1.27, 5.01
15. $\pi/3 + k\pi$, k any integer   **16.** $\pi/2$, π

EXERCISE 5.3

A   *Solve the equations exactly for x, $0 \le x \le 90°$.*

1. $1 - \cos x = 0$ **2.** $1 - \sin x = 0$
3. $2 \sin x = \sqrt{3}$ **4.** $2 \cos x = 1$
5. $\sqrt{3} - \tan x = 0$ **6.** $1 - \cot x = 0$

Solve the equations exactly for x, $0 \le x \le \pi/2$.

7. $1 - 2 \sin x = 0$ **8.** $1 - \sqrt{2} \cos x = 0$
9. $1 - \tan x = 0$ **10.** $1 - \sqrt{3} \cot x = 0$
11. $2 - \csc x = 0$ **12.** $2 - \sqrt{3} \sec x = 0$

Solve the equations exactly for x, $0° \le x \le 360°$.

13. $1 - 2 \sin x = 0$ **14.** $1 - 2 \cos x = 0$
15. $(\cos x)(2 \sin x - 1) = 0$ **16.** $(\sin x)(\sqrt{2} \sin x + 1) = 0$
17. $\sin^2 x = \sin x$ **18.** $\cos^2 x = \cos x$
19. $4 \sin^2 x - 3 = 0$ **20.** $2 \cos^2 x = 1$

Solve the equations exactly for x, $0 \le x \le 2\pi$.

21. $2 \sin x \cos x = \sin x$ **22.** $\cos x = 2 \sin x \cos x$
23. $\cot^2 x = 3$ **24.** $4 \cos^2 x - 3 = 0$

B   **25.** $2 \sin 2x = \sqrt{3}$ **26.** $2 \cos 2x = 1$

27. $\tan \dfrac{x}{2} - 1 = 0$ **28.** $\sec \dfrac{x}{2} + 2 = 0$

29. $(\sin x - 3)(\sin x + 1) = 0$ **30.** $(2 \cos x + 1)(\cos x - 4) = 0$
31. $2 \sin^2 x = 3 \sin x - 1$ **32.** $2 \cos^2 x + \cos x = 1$
33. $\sin^2 x + 2 \cos x = -2$ **34.** $2 \cos^2 x + 3 \sin x = 0$
35. $\cos 2x + \sin^2 x = 0$ **36.** $\sin 2x = \sin x$
37. $\csc^2 x = \csc x + 2$ **38.** $\tan^2 x + 1 = 2 \tan x$
39. $\sin x = \cos x$ **40.** $\sqrt{3} \sin x - \cos x = 0$
41. $2 \sin^2(x/2) - 3 \sin(x/2) + 1 = 0$
42. $4 \cos^2 2x - 4 \cos 2x + 1 = 0$

Find all solutions for:

43. Problem 13	**44.** Problem 14	**45.** Problem 9
46. Problem 10	**47.** Problem 21	**48.** Problem 22
49. Problem 29	**50.** Problem 30	

C   *Find all real solutions and decimal degree solutions to two decimal places; x is restricted to first quadrant values only. Use a calculator or tables.*

51. $3 - 4\cos x = 0$ **52.** $4 - 5\tan x = 0$

53. $\sin x = \dfrac{3 - \sqrt{2}}{2}$ **54.** $\cos x = \dfrac{-2 + \sqrt{5}}{2}$

Find all real x, $0 \le x \le 2\pi$, to two decimal places. Use a calculator or tables.

55. $2\sin^2 x = 1 - 2\sin x$ **56.** $\cos^2 x = 3 - 5\cos x$
57. $2\sin x = \cos 2x$ **58.** $\sec^2 x = 2(1 - \tan x)$

Find all real solutions exactly for $0 \le x \le 2\pi$. Be careful of extraneous roots.

59. $\sin x = 1 - \cos x$ **60.** $\cos x - \sin x = 1$
  [*Hint:*   Square both sides.]
61. $\sec x + \tan x = 1$ **62.** $\tan x - \sec x = 1$

Find simultaneous solutions for each system of equations for $0 \le \theta \le 360°$ (these are polar equations, which will be discussed in Chapter 7).

63. $r = 2\sin\theta$ **64.** $r = 2\sin\theta$
  $r = 2(1 - \sin\theta)$ $r = \sin 2\theta$

APPLICATIONS   **65.** *Electrical current*   An alternating current generator produces a current given by

  $I = 30\sin 120\pi t$

where t is time in seconds and I is in amperes. Find the least positive t to four significant digits such that $I = 25$ amperes.
66. *Electrical current*   Find the least positive t in Problem 65, to four significant digits, such that $I = -10$ amperes.
67. *Astronomy*   The planet Mercury travels around the sun in an elliptical orbit given approximately by

  $r = \dfrac{3.44 \times 10^7}{1 - 0.206\cos\theta}$

(see figure). Find the least positive θ (in decimal degrees to three significant digits) such that Mercury is 3.78×10^7 miles from the sun.

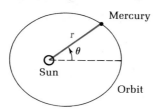

68.   *Astronomy*   In Problem 67, find the least positive θ (in decimal degrees to three significant digits) such that Mercury is 3.09×10^7 miles from the sun.

EXERCISE 5.4   CHAPTER REVIEW

A   *Evaluate exactly as real numbers.*

1. $\text{Cos}^{-1} \dfrac{\sqrt{3}}{2}$ 2. $\text{Arcsin} \dfrac{1}{2}$ 3. $\text{Arctan } 1$

4. $\text{Cot}^{-1} \dfrac{1}{\sqrt{3}}$ 5. $\text{Csc}^{-1} 2$ 6. $\text{Sec}^{-1} \dfrac{2}{\sqrt{3}}$

7. $\text{Arcsin}\left(-\dfrac{1}{2}\right)$ 8. $\text{Cos}^{-1}\left(-\dfrac{1}{2}\right)$ 9. $\text{Cot}^{-1}(-1)$

10. $\text{Arctan}(-1)$ 11. $\text{Sec}^{-1}(-2)$ 12. $\text{Csc}^{-1}(-2)$

Solve exactly for all real x, $0 \le x \le 2\pi$.

13. $2 \cos x - \sqrt{3} = 0$ 14. $2 \sin^2 x = \sin x$
15. $4 \cos^2 x - 3 = 0$ 16. $2 \cos^2 x + 3 \cos x + 1 = 0$

B   *Find exact values without a table or calculator.*

17. $\cos(\text{Cos}^{-1} 0.315)$ 18. $\sin\left[\text{Tan}^{-1}\left(-\dfrac{3}{4}\right)\right]$

19. $\sec\left[\text{Cot}^{-1}\left(-\dfrac{1}{2}\right)\right]$ 20. $\tan[\text{Csc}^{-1}(-\sqrt{5})]$

Find all solutions exactly for $0 \le x \le 360°$.

21. $2 \cos x + 2 = -\sin^2 x$ 22. $3 \sin x = -2 \cos^2 x$

23. $\sin^2 x = -\cos 2x$ 24. $\sin 2x = \dfrac{1}{2}$

25. $\sqrt{3} \cos x - \sin x = 0$

Evaluate exactly as real numbers and as degrees.

26. $\sin^{-1}\left(\frac{1}{2}\right)$

27. $\tan^{-1}(-1)$

Find all real solutions for each equation.

28. $\sqrt{3} - 2\cos x = 0$

29. $\sin 2x + \cos x = 0$

Use a calculator to evaluate as real numbers to two decimal places.

30. $\text{Cos}^{-1}(-0.304)$

31. $\text{Tan}^{-1} 4.319$

32. $\text{Cot}^{-1} 1.039$

33. $\text{Csc}^{-1}(-1.934)$

C   34. Solve to two decimal places in decimal degrees (use a calculator or table):

$$4\cos x - 3 = 0 \qquad 0 \le x \le 90°$$

35. Solve to two decimal places (use a calculator or table):

$$7 - 9\sin x = 0 \qquad 0 \le x \le \frac{\pi}{2}$$

36. Solve to two decimal places (use a calculator or table):

$$2(1 - \cos^2 x) = 1 - 2\sin x \qquad 0 \le x \le 2\pi$$

37. Find $\cos(\text{Sin}^{-1} x)$ in terms of x and free of trigonometric and inverse trigonometric functions.

ADDITIONAL TRIANGLE TOPICS; VECTORS

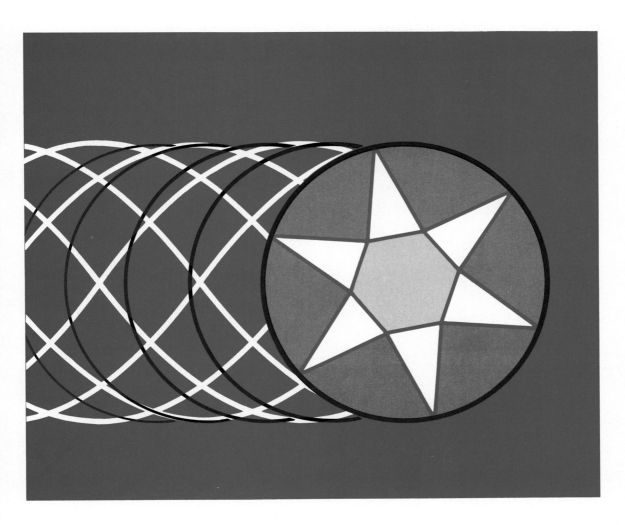

In this chapter we return to the problem of solving triangles—not just right triangles, but any triangle. We will use some of these ideas to introduce the concept of vector and to develop some useful area formulas for arbitrary triangles.

6.1   LAW OF SINES

The law of sines and the law of cosines (to be developed in Section 6.2) play fundamental roles in solving triangles that are not right triangles. Every triangle that is not a right triangle is either **acute** (all angles are between 0° and 90°) or **obtuse** (one angle is between 90° and 180°). Figure 1 illustrates the two cases.

The law of sines is relatively easy to prove using right triangle properties studied earlier. We will also use the fact that

$$\sin(180° - x) = \sin x$$

which can easily be obtained from the difference identity for sine. Referring to the triangles in Figure 1, we proceed as follows: For each triangle

$$\sin \alpha = \frac{h}{b} \qquad \text{and} \qquad \sin \beta = \frac{h}{a}$$

Therefore,

$$h = b \sin \alpha \qquad \text{and} \qquad h = a \sin \beta$$

Thus,

$$b \sin \alpha = a \sin \beta$$

and

$$\frac{\sin \alpha}{a} = \frac{\sin \beta}{b} \tag{1}$$

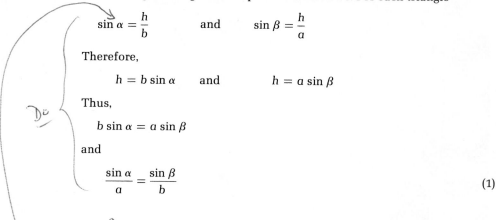

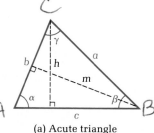

FIGURE 1   (a) Acute triangle   (b) Obtuse triangle

Similarly, for each triangle in Figure 1,

$$\sin \alpha = \frac{m}{c} \quad \text{and} \quad \sin \gamma = \sin(180° - \gamma)$$
$$= \frac{m}{a}$$

Therefore,

$$m = c \sin \alpha \quad \text{and} \quad m = a \sin \gamma$$

Thus,

$$c \sin \alpha = a \sin \gamma$$

and

$$\frac{\sin \alpha}{a} = \frac{\sin \gamma}{c} \tag{2}$$

If we combine (1) and (2), we obtain the law of sines.

LAW OF SINES

$$\frac{\sin \alpha}{a} = \frac{\sin \beta}{b} = \frac{\sin \gamma}{c}$$

This law is most useful when given:

Two sides and an angle opposite one of the sides
Two angles and a side opposite one of the angles

Need corresponding ∠ and side.

Before proceeding with examples, recall the rules governing significant digits that are listed in Table 1. (They are also listed inside the front cover for easy reference.)

TABLE 1

ANGLE TO NEAREST	SIGNIFICANT DIGITS FOR SIDE MEASURE
1°	2
10′ or 0.1°	3
1′ or 0.01°	4
10″ or 0.001°	5

EXAMPLE 1   Solve the triangle.

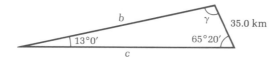

Solution   *Solve for* γ.

$$\alpha + \beta + \gamma = 180°$$
$$\gamma = 180° - (\alpha + \beta)$$
$$= 180° - (13°0' + 65°20')$$
$$= 101°40'$$

Solve for c.

$$\frac{\sin \alpha}{a} = \frac{\sin \gamma}{c}$$

$$c = \frac{a \sin \gamma}{\sin \alpha} = \frac{(35.0 \text{ km})(\sin 101°40')}{\sin 13°0'} = 152 \text{ km}$$

Calculator operations: First convert degree-minute quantities to decimal degrees:

A: $\boxed{35}$ $\boxed{\times}$ $\boxed{101.666\ldots}$ $\boxed{\sin}$ $\boxed{=}$ $\boxed{\div}$ $\boxed{13}$ $\boxed{\sin}$ $\boxed{=}$

P: $\boxed{35}$ $\boxed{\text{ENTER}}$ $\boxed{101.666\ldots}$ $\boxed{\sin}$ $\boxed{\times}$ $\boxed{13}$ $\boxed{\sin}$ $\boxed{\div}$

Solve for b.

$$\frac{\sin \alpha}{a} = \frac{\sin \beta}{b}$$

$$b = \frac{a \sin \beta}{\sin \alpha} = \frac{(35.0 \text{ km})(\sin 65°20')}{\sin 13°0'} = 141 \text{ km}$$

Dº

PROBLEM 1   Solve the triangle with $\alpha = 28°0'$, $\beta = 45°20'$, and $c = 120$ m.

AMBIGUOUS CASE   One can specify parts of a triangle so that more than one triangle is possible or no triangles are possible. Consider Example 2.

use 30°

EXAMPLE 2   Solve a triangle with $\alpha = \boxed{26°}$ $a = 10$ cm, and $b = 18$ cm.

Dº   Solution   If we try to draw a triangle with these values, we find that there are two possible triangles.

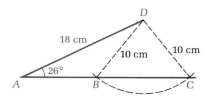
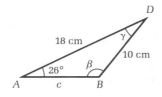
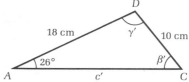

Solve for β and β′.

Here we will see that there are two possible values for β, hence, the two triangles.

$$\frac{\sin \beta}{b} = \frac{\sin \alpha}{a}$$

$$\sin \beta = \frac{b \sin \alpha}{a} = \frac{(18 \text{ cm})(\sin 26°)}{10 \text{ cm}} = 0.7891$$

The angle β can be either obtuse or acute:

$$\beta = 180° - \text{Sin}^{-1}\, 0.7891 \qquad \text{or} \qquad \beta' = \text{Sin}^{-1}\, 0.7891$$
$$= 180° - 52° = 128° \qquad\qquad\qquad = 52°$$

Solve for γ and γ′.

$$\gamma = 180° - (26° + 128°) = 26° \qquad \gamma' = 180° - (26° + 52°) = 102°$$

Solve for c and c′.

$$\frac{\sin \alpha}{a} = \frac{\sin \gamma}{c} \qquad\qquad \frac{\sin \alpha}{a} = \frac{\sin \gamma'}{c'}$$

$$c = \frac{a \sin \gamma}{\sin \alpha} \qquad\qquad c' = \frac{a \sin \gamma'}{\sin \alpha}$$

$$= \frac{(10 \text{ cm})(\sin 26°)}{\sin 26°} \qquad\qquad = \frac{(10 \text{ cm})(\sin 102°)}{\sin 26°}$$

$$= 10 \text{ cm} \qquad\qquad\qquad = 22 \text{ cm}$$

PROBLEM 2   Solve the triangle(s) with $a = 8.0$ mm, $b = 10$ mm, and $\alpha = 35°$.

If an angle and its opposite side are specified, as well as one side adjacent to the angle, then there may be the possibility (as indicated in Example 2) of more than one triangle, or even no triangle. Figure 2 (page 162) illustrates four possibilities. Since the altitude h of a triangle (if it exists) is $b \sin \alpha$, the possibilities illustrated in Figure 2 are summarized in Table 2. You do not have to memorize this table. Special cases become obvious in particular contexts.

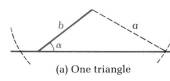

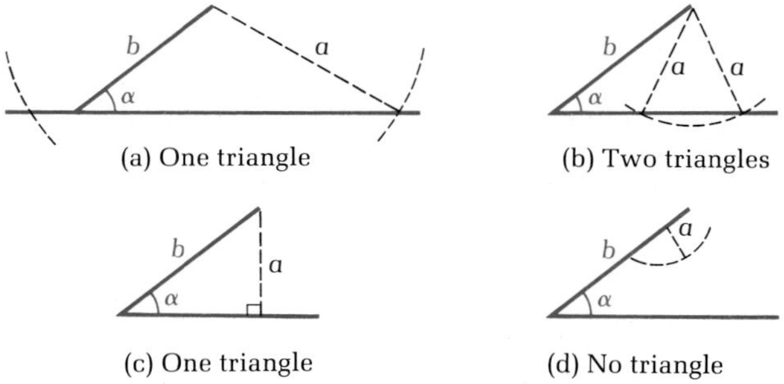

(a) One triangle

(b) Two triangles

(c) One triangle

(d) No triangle

FIGURE 2
Given α, a, and b

TABLE 2

TRIANGLE IN FIGURE 2	CASE	NUMBER OF TRIANGLES DETERMINED
(a)	$a \geq b$	1
(b)	$b \sin \alpha < a < b$	2
(c)	$a = b \sin \alpha$	1
(d)	$a < b \sin \alpha$	0

ANSWERS TO MATCHED PROBLEMS

1. $\gamma = 106°40'$, $a = 58.8$ m, $b = 89.1$ m
2. $\beta = 134°$, $\beta' = 46°$, $\gamma = 11°$, $\gamma' = 99°$, $c = 2.7$ mm, $c' = 14$ mm

EXERCISE 6.1

Solve each triangle.

A
1. $\beta = 12°40'$, $\gamma = 100°$, $b = 13.1$ km
2. $\beta = 77°$, $\gamma = 62°$, $a = 74$ cm
3. $\beta = 27°30'$, $\gamma = 54°30'$, $a = 9.27$ mm
4. $\alpha = 51°20'$, $\beta = 105°$, $c = 12.4$ ft

B
5. $\beta = 52°$, $a = 8.00$ cm, $b = 12.0$ cm, $0° < \alpha < 90°$
6. $\alpha = 54°20'$, $b = 30.0$ in., $a = 44.5$ in., $0° < \beta < 90°$
7. $\alpha = 18°55'$, $b = 105.0$ yd, $a = 48.35$ yd, $90° < \beta < 180°$
8. $\alpha = 12°10'$, $b = 2.43$ m, $a = 0.668$ m, $90° < \beta < 180°$
9. $a = 50$ m, $c = 40$ m, $\gamma = 30°$
10. $a = 23$ cm, $b = 20$ cm, $\beta = 37°$
11. $a = 14$ m, $b = 23$ m, $\alpha = 41°$
12. $\beta = 32°20'$, $a = 140$ cm, $b = 60$ cm

C
13. Mollweide's equation,

$$(a - b) \cos \frac{\gamma}{2} = c \sin \frac{\alpha - \beta}{2}$$

is often used to check the final solution of a triangle since all six parts of a triangle are involved in the equation. If, after substitution, the left side does not equal the right side, then an error has been made in solving a triangle. Use this equation to check Problem 1 (above).

14.   Use Mollweide's equation (see Problem 13) to check Problem 3.

15.   Use the law of sines and suitable identities to show that for any triangle

$$\frac{a - b}{a + b} = \frac{\tan \dfrac{\alpha - \beta}{2}}{\tan \dfrac{\alpha + \beta}{2}}$$

16.   Verify the formula in Problem 15 with values from Problem 1.

APPLICATIONS   **17.**   *Fire spotting*   A fire at F is spotted from two fire lookout stations A and B, which are located 10.3 miles apart. If station B reports the fire at angle $ABF = 52.6°$, and station A reports the fire at angle $BAF = 25.3°$, how far is the fire from station A? From station B?

18.   *Coast patrol*   Two lookout posts A and B, 12.4 miles apart, are established along a coast to watch for illegal foreign fishing boats coming within the 3 mile limit. If post A reports a ship S at angle $BAS = 37.5°$, and post B reports the same ship at angle $ABS = 19.7°$, how far is the ship from post A? How far is the ship from the shore (assuming the shore is along the line joining the two observation posts)?

19.   *Astronomy*   The orbits of the earth and Venus are approximately circular, with the sun at the center (see the figure). A sighting of Venus is made from earth and the angle α is found to be $18°40'$. If the diameter of the orbit of the earth is 2.99×10^8 km and the diameter of the orbit of Venus is 2.17×10^8 km, what are the possible distances from the earth to Venus?

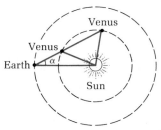

DISTANCE FROM THE EARTH TO VENUS

20.   *Astronomy*   In Problem 19 find the maximum value of α. [*Hint:* The value of α is maximum when a straight line joining the earth and Venus in the figure is tangent to Venus' orbit.]

*21.   *Coastal piloting*   A boat is traveling along a coast at night. A flashing buoy marks a reef. While proceeding on the same course, the navigator of the boat sights the buoy twice, 4.6 nautical miles apart, and forms triangles as in the accompanying figure. If the boat continues on course, by how far will it miss the reef?

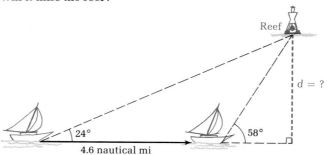

*22.   *Surveying*   Find the height of the mountain above the valley in the figure below:

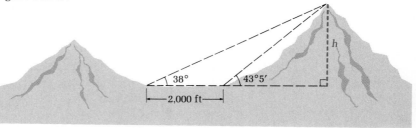

HOW HIGH IS THE MOUNTAIN?

*23.   *Surveying*   The scheme illustrated in the figure below is used to determine inaccessible heights when d, α, β, and γ can be measured. Show that

$$h = d \sin \alpha \csc(\alpha + \beta) \tan \gamma$$

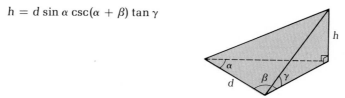

6.2   LAW OF COSINES

If three sides, or two sides and an included angle, are given in a triangle, then the law of sines is not particularly helpful in solving the triangle. How-

*Mention if
∠ = 90° ⟹
Pyth. thm*

ever, the Pythagorean theorem for right triangles can be generalized into another law, called the law of cosines, that will take care of these cases.

LAW OF COSINES

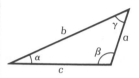

$a^2 = b^2 + c^2 - 2bc \cos \alpha$
$b^2 = a^2 + c^2 - 2ac \cos \beta$
$c^2 = a^2 + b^2 - 2ab \cos \gamma$
(All three equations say essentially the same thing.)

This law is most useful when given:

Three sides

Two sides and an included angle

We will derive this law for the first case only. (The other cases can then be obtained from the first case simply by relabeling the figure.) We start by locating a triangle in a rectangular coordinate system. Figure 3 indicates three typical triangles.

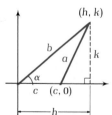 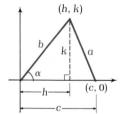 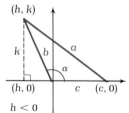

FIGURE 3

*Let st.
read*

For an arbitrary triangle located as in Figure 3, we obtain, using the distance-between-two-points formula,

$$a = \sqrt{(h - c)^2 + (k - 0)^2}$$

or, squaring both sides,

$$a^2 = (h - c)^2 + k^2 \qquad (1)$$
$$= h^2 - 2hc + c^2 + k^2$$

From Figure 3 we note that

$$b^2 = h^2 + k^2$$

Substituting b^2 for $h^2 + k^2$ in (1), we obtain

$$a^2 = b^2 + c^2 - 2hc \qquad (2)$$

But

$$\cos \alpha = \frac{h}{b}$$

$$h = b \cos \alpha$$

Thus, by replacing h in (2) with $b \cos \alpha$, we reach our objective,

$$a^2 = b^2 + c^2 - 2bc \cos \alpha$$

[*Note:*   If α is acute, then $\cos \alpha > 0$; if α is obtuse, then $\cos \alpha < 0$.]

EXAMPLE 3   Solve the triangle:

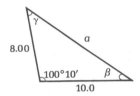

Solution   *Solve for a:*

$$a^2 = b^2 + c^2 - 2bc \cos \alpha$$
$$= (8.00)^2 + (10.0)^2 - 2(8.00)(10.0) \cos 100°10'$$
$$= 192.2419 \ldots$$
$$a = \sqrt{192.2419 \ldots} = 13.9 \qquad \text{To three significant digits}$$

Calculator operations: First set the calculator in degree mode and form decimal degrees:

A: ☐10 ☐÷ ☐60 ☐+ ☐100 ☐cos ☐× ☐10 ☐× ☐8 ☐× ☐2 ☐= ☐+/− ☐+

☐10 ☐x² ☐+ ☐8 ☐x² ☐= ☐√

P: ☐10 ☐ENTER ☐60 ☐÷ ☐100 ☐+ ☐cos ☐10 ☐× ☐8 ☐× ☐2 ☐× ☐+/−

☐10 ☐x² ☐+ ☐8 ☐x² ☐+ ☐√

Solve for β:
We now have a choice of using the law of sines or the law of cosines to find β. We choose the law of sines for simpler computation. We note first that both β and γ are acute angles. (Why?)

$$\frac{\sin \beta}{b} = \frac{\sin \alpha}{a} \qquad \text{Solve for } \sin \beta$$

$$\sin \beta = \frac{b \sin \alpha}{a} = \frac{(8.00) \sin 100°10'}{13.9} \qquad \text{Solve for } \beta;\ \text{remember } \beta \text{ is acute}$$

$$\beta = \text{Sin}^{-1} \left| \frac{(8.00) \sin 100°10'}{13.9} \right|$$

$$= 34.5° \qquad \text{or} \qquad 34°30' \qquad \text{To nearest } 0.1° \text{ or } 10'$$

Solve for γ:

$$\gamma = 180° - (100°10' + 34°30')$$

$$= 45°20'$$

PROBLEM 3   Solve a triangle with $\alpha = 78°0'$, $b = 10.0$, and $c = 18.0$.

EXAMPLE 4   Solve a triangle with $a = 9.23$, $b = 5.04$, and $c = 10.6$.

Solution   We first draw the triangle roughly to scale, then two acute angles will generally be obvious. [A triangle must always have two acute angles and an obtuse angle or three acute angles. (Why?)] We find the two acute angles first.

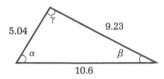

α and β are acute angles

Solve for α:

$$a^2 = b^2 + c^2 - 2bc \cos \alpha \qquad \text{Solve for } \cos \alpha$$

$$\cos \alpha = \frac{b^2 + c^2 - a^2}{2bc} \qquad \text{Solve for } \alpha$$

$$\alpha = \text{Cos}^{-1} \frac{b^2 + c^2 - a^2}{2bc}$$

$$= \text{Cos}^{-1} \frac{(5.04)^2 + (10.6)^2 - (5.04)^2}{2(5.04)^2(10.6)^2}$$

$$= 88.9° \qquad \text{or} \qquad 88°50' \qquad \text{To nearest } 0.1° \text{ or } 10'$$

Solve for β:
We can use either the law of cosines or the law of sines. We choose the latter because of simpler calculations.

$$\frac{\sin \alpha}{a} = \frac{\sin \beta}{b}$$ Solve for $\sin \beta$

$$\sin \beta = \frac{b \sin \alpha}{a}$$ Solve for β; remember β is acute

$$\beta = \text{Sin}^{-1} \frac{b \sin \alpha}{a}$$

$$= \text{Sin}^{-1} \frac{(5.04) \sin 88.9°}{9.23}$$

$$= 33.1° \quad \text{or} \quad 33°10' \qquad \text{To nearest } 0.1° \text{ or } 10'$$

Solve for γ:

$$\gamma = 180° - (88°50' + 33°10') = 58°0'$$

PROBLEM 4 Solve a triangle with $a = 1.20$, $b = 2.00$, and $c = 1.50$.

ANSWERS TO **3.** $a = 18.7$, $\beta = 31°30'$, $\gamma = 70°30'$
MATCHED PROBLEMS **4.** $\alpha = 36°40'$, $\beta = 95°0'$, $\gamma = 48°20'$

EXERCISE 6.2

A *Solve each triangle.*

1. $\alpha = 50°40'$, $b = 7.03$ mm, $c = 7.00$ mm
2. $\alpha = 71°$, $b = 5.32$ cm, $c = 5.00$ cm
3. $\gamma = 135.0°$, $a = 20.0$ m, $b = 8.00$ m
4. $\alpha = 120°$, $b = 5.00$ km, $c = 10.0$ km

B 5. $a = 9.00$ yd, $b = 6.00$ yd, $c = 10.0$ yd
6. $a = 5.00$ km, $b = 5.50$ km, $c = 6.00$ km
7. $a = 400$ km, $b = 700$ km, $c = 1,000$ km (answer to nearest 10′)
8. $a = 20.0$ cm, $b = 12.0$ cm, $c = 10.0$ cm (answer to nearest 10′)

C 9. Show, using the law of cosines, that if $\beta = 90°$, then $b^2 = c^2 + a^2$ (the Pythagorean theorem).
10. Show, using the law of cosines, that if $b^2 = c^2 + a^2$, then $\beta = 90°$.
11. Check Problem 1 above using Mollweide's equation,

$$(a - b) \cos \frac{\gamma}{2} = c \sin \frac{\alpha - \beta}{2}$$

12. Check Problem 3 using Mollweide's equation (see Problem 11).

APPLICATIONS 13. *Search and rescue* At midnight two search planes set out from New York to find a boat in distress. Plane A travels due east at 400 km/hr, and plane B travels northeast at 500 km/hr. At 2 AM plane A spots a flare from the boat and radios plane B to come and assist in the rescue. How far is plane B from plane A at this time? Compute the answer to two significant figures.

14.   *Navigation*   Los Angeles and San Francisco are approximately 600 km apart. A pilot flying from Los Angeles to San Francisco finds that after she is 200 km from Los Angeles she is 20° off course. How far is she from San Francisco at this time (to two significant figures)?

*15.   *Geometry—engineering*   A 58.3 cm chord subtends a central angle of 27.8°. Find the radius of the circle to three significant figures. [*Hint:*   Use law of cosines.]

*16.   *Geometry—engineering*   Find to the nearest centimeter the perimeter of a regular pentagon inscribed in a circle with radius 5 cm (see the figure).

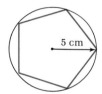

INSCRIBED PENTAGON

(*17.)   *Geometry—engineering*   Three circles of radius 2 cm, 5 cm, and 8 cm are tangent to each other (see the figure). Find to the nearest 10′ the three angles formed by the lines joining their centers.

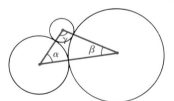

TANGENT CIRCLES

6.3   VECTORS

Physical quantities such as length, area, or volume can be completely specified by a single real number. Other quantities, such as directed distances, velocities, and forces, require both a magnitude and direction for their complete specification. The former are often called **scalar quantities** and the latter are referred to as **vector quantities.**

In this section we will limit our development to intuitive notions of geometric vectors in a plane. In a more rigorous treatment of the subject, geometric vectors become a special case of a more general type of vector. Vector forms are widely used in both pure and applied mathematics. They are extensively used in the social and life sciences as well as the physical sciences and engineering.

ADDITION OF GEOMETRIC VECTORS

We define a **geometric vector in a plane** to be a directed line segment from a fixed point O, called the **origin,** to a point P in the plane. We denote a vector by $\overrightarrow{OP}$ or by a boldface letter such as **v** (see Figure 4). If you use a single letter to represent a vector, then either underline it or place an arrow over it—it is difficult to write in boldface!

The magnitude of a vector $\overrightarrow{OP}$ is the length of the line segment from O to P and is denoted by $|\overrightarrow{OP}|$ or $|\mathbf{v}|$. The direction of a vector $\overrightarrow{OP}$ is the direction of the directed line segment from O to P. The zero vector is denoted by **0** and has magnitude zero and an arbitrary direction. Two vectors are equal if they have the same origin and the same magnitude and direction.

The **sum of two vectors u and v** with different directions is the diagonal of the parallelogram formed using **u** and **v** as adjacent sides (see Figure 5a). The diagonal **u** + **v** is also called the **resultant** of the two vectors **u** and **v**, and **u** and **v** are called **components** of **u** + **v**.

If the two vectors **u** and **v** have the same direction, then their sum is a vector with this direction and magnitude $|\mathbf{u}| + |\mathbf{v}|$. If **u** and **v** have opposite directions, then their sum is a vector with magnitude $\|\mathbf{u}\| - |\mathbf{v}\|$ and direction the same as the component with the largest magnitude. These last two cases are illustrated in Figure 5b.

FIGURE 4
Vector **v** or $\overrightarrow{OP}$

FIGURE 5
Vector addition

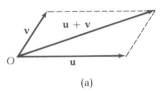

(a)

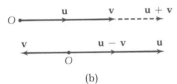

(b)

VELOCITY VECTORS

A vector that represents the direction and speed of an object in motion is called a **velocity vector.**

EXAMPLE 5

A power boat traveling at 24 km/hr relative to the water has a compass heading of 90°. A strong tidal current, with a heading of 30°, is flowing at 12 km/hr. What is the boat's actual heading and speed? [*Note:*   A navigational compass is marked clockwise in degrees starting at North.]

Solution

This problem can be solved using vector addition. We draw from the same point vectors representing the boat's heading and speed and the current's heading and speed. We then add these velocity vectors using the parallelogram definition of addition. The resultant (the diagonal vector) will represent the actual heading and speed of the boat (see Figure 6).

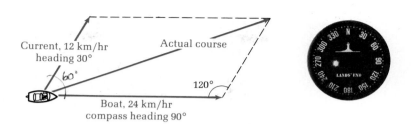

FIGURE 6
Adding velocity vectors

From Figure 6 we obtain the triangle below and then solve for α and c.

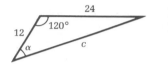

Boat's actual heading: $30° + \alpha$
Boat's actual speed: c

Solve for c.

$$c^2 = 12^2 + 24^2 - 2(12)(24)\cos 120°$$
$$= 1{,}008$$
$$c = \sqrt{1{,}008} \approx 32 \text{ km/hr} \qquad \text{Boat's actual speed}$$

Solve for α.
We now have the choice of using the law of sines or the law of cosines to find α. We choose the law of sines because of simpler computations. Note that α is acute. (Why?)

$$\frac{\sin \alpha}{24} = \frac{\sin 120°}{32}$$

$$\sin \alpha = \frac{24}{32} \sin 120°$$

$$\alpha = \text{Sin}^{-1}\left(\frac{24}{32} \sin 120°\right) = 41°$$

Actual heading $= 30° + 41° = 71°$.

Note: If the law of cosines had been used, we would obtain $\alpha = 40°$. The difference of 1° is due to round-off error.

PROBLEM 5   Repeat Example 5 with a current of 8 km/hr at 20° and the speedometer on the boat reading 30 km/hr with a compass heading of 90°.

FORCE VECTORS   A vector representing the direction and magnitude of an applied force is called a **force vector**.

EXAMPLE 6   Figure 7 shows a man and a horse pulling on a large piece of granite. If the sides of the indicated parallelogram are taken to be 200 lb and 1,000 lb, then the length of the main diagonal will be the actual magnitude of the force on the stone, and its direction will be the direction of the motion of the stone (if it moves). Find the magnitude and direction α of the resultant force.

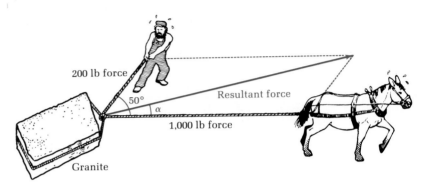

FIGURE 7
Adding force vectors

Solution   From Figure 7 we obtain the triangle below and then solve for α and b.

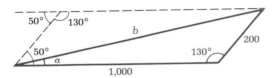

Solve for b.

$$b^2 = 1{,}000^2 + 200^2 - 2(1{,}000)(200) \cos 130°$$
$$= 1{,}297{,}115.04\ldots$$
$$b = \sqrt{1{,}297{,}115.04\ldots} \approx 1{,}100 \text{ lb}$$

Solve for α.

$$\frac{\sin \alpha}{200} = \frac{\sin 130°}{1{,}100}$$

$$\sin \alpha = \frac{200}{1{,}100} \sin 130°$$

$$\alpha = \text{Sin}^{-1}\left(\frac{200}{1{,}100} \sin 130°\right) \approx 8°$$

PROBLEM 6   Repeat Example 6 with the angle between the force vector from the horse and the force vector from the man 45° instead of 50°, and the magnitude of the force vector from the horse 800 lb instead of 1,000 lb.

RESOLUTION
OF A VECTOR
INTO COMPONENTS

Instead of adding vectors, many problems require the breaking down of vectors into components. Whenever a vector is expressed as a resultant of two vectors, these two vectors are called **components** of the given vector. For example, to find the horizontal and vertical components of the vector **v** in Figure 8, we find the magnitudes (the direction of the horizontal and vertical lines are already known) of these components using the sine and cosine functions.

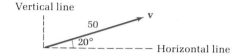

FIGURE 8

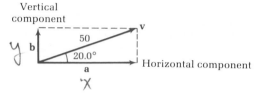

FIGURE 9

Magnitude of horizontal component:

$$\cos 20 = \frac{|\mathbf{a}|}{50}$$

$$|\mathbf{a}| = 50 \cos 20.0°$$

$$\approx 47.0$$

$x = r \cos \theta$

Horizontal projection of **v**

Magnitude of vertical component:

$$\sin 20.0° = \frac{|\mathbf{b}|}{50}$$

$$|\mathbf{b}| = 50 \sin 20.0°$$

$$\approx 17.1$$

$y = r \sin \theta$

Vertical projection of **v**

WORK

Intuitively, **work** is done when a force causes an object to be moved a certain distance. An example would be the work done by the force that the harness exerts on a dog sled in moving the sled a certain distance (see Figure 10).

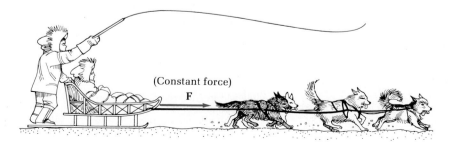

(Constant force)

F

FIGURE 10

The concept of work is very important in science studies dealing with energy and requires a definition that is quantitative: If the motion of an object takes place in a straight line in the direction of a constant force **F** causing the motion, then we define work W done by the force **F** to be the product of the magnitude of the force $|\mathbf{F}|$ and the distance d through which the object moves. Symbolically,

$$W = |\mathbf{F}|d$$

For example, suppose the dog sled in Figure 10 is moved a distance of 200 ft under a constant force of 100 lb, then the work done by the force is

$$W = (100 \text{ lb})(200 \text{ ft}) = 20{,}000 \text{ ft-lb}^\dagger$$

It is more usual to have a constant force responsible for the motion of an object acting in a direction other than along the line of motion. For example, a person pushing a lawn mower across a level lawn exerts a force downward along the handle (see Figure 11). In this case, only the component of force

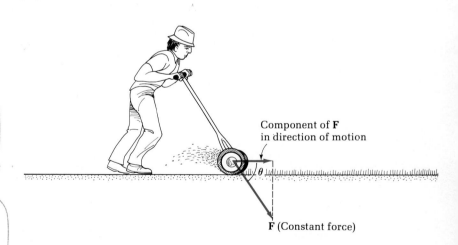

Component of **F** in direction of motion

θ

F (Constant force)

FIGURE 11

in the direction of the motion (the horizontal component) is responsible for the motion. This leads to the more general quantitative definition of work given in the box at the top of the next page.

---

†In scientific work, the basic unit of force is the **newton,** the basic unit of linear measure is the **meter,** and the basic unit of work is the **newton-meter** or **joule.** In the British-engineering system, which we will use, the basic unit of work is the **foot-pound (ft-lb).**

WORK DONE BY A CONSTANT FORCE

The work W done by a force $\mathbf{F}$ causing an object to move a distance d along a straight line is the product of the component of the force $\mathbf{F}$ in the direction of motion and the distance d. Symbolically,

$$W = (|\mathbf{F}| \cos \theta)d$$
$$= |\mathbf{F}|d \cos \theta$$

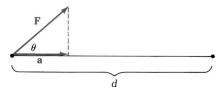

Magnitude of component in direction of motion: $|a| = |\mathbf{F}| \cos \theta$

EXAMPLE 7

$\mathcal{D}o$

A lawn mower is pushed 50 ft across a level lawn with a constant force of 60 lb directed down the handle. If the handle makes an angle of 40° relative to the horizontal, how much work is done?

Solution

$$W = (60 \text{ lb})(50 \text{ ft}) \cos 40°$$
$$= 2{,}300 \text{ ft-lb} \qquad \text{Two significant digits}$$

PROBLEM 7

Repeat Example 7 if the handle makes an angle of 45° relative to the horizontal and the distance is 75 ft.

Additional vector applications can be found in Chapter 8.

ANSWERS TO MATCHED PROBLEMS

5. True course: 34 km/hr at 76°
6. Resultant force: $b = 950$ lb, $\alpha = 9°$ 7.   3,200 ft-lb

___
EXERCISE 6.3

A

Express all angles in decimal degrees. Recall that in navigational problems a compass is divided clockwise into degrees starting at North.

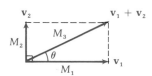

In Problems 1–6, find the magnitude M_3 of $\mathbf{v}_1 + \mathbf{v}_2$ and θ given the magnitude M_1 of $\mathbf{v}_1$ and M_2 of $\mathbf{v}_2$.

1. $M_1 = 62$ km/hr
    $M_2 = 34$ km/hr
2. $M_1 = 37$ km/hr
    $M_2 = 45$ km/hr
3. $M_1 = 48$ lb
    $M_2 = 31$ lb
4. $M_1 = 38$ lb
    $M_2 = 53$ lb
5. $M_1 = 143$ knots
    $M_2 = 57.4$ knots
6. $M_1 = 434$ knots
    $M_2 = 105$ knots

*In Problems 7–12, find the magnitudes H and V of the horizontal and vertical components, respectively, of the vector **v**, given |**v**| and θ.*

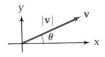

7. $|\mathbf{v}| = 40$ lb, $\theta = 34°$
8. $|\mathbf{v}| = 250$ lb, $\theta = 67°$
9. $|\mathbf{v}| = 23$ knots, $\theta = 62°$
10. $|\mathbf{v}| = 48$ knots, $\theta = 27°$
11. $|\mathbf{v}| = 244$ km/hr, $\theta = 43.2°$
12. $|\mathbf{v}| = 84.0$ km/hr, $\theta = 28.6°$

B   *In Problems 13–18, find M_3 and θ, given M_1, M_2, and α.*

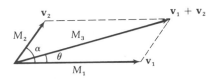

13. $M_1 = 120$ lb
    $M_2 = 80$ lb; $\alpha = 40°$
14. $M_1 = 60$ lb
    $M_2 = 20$ lb; $\alpha = 70°$
15. $M_1 = 8$ knots
    $M_2 = 2$ knots; $\alpha = 68°$
16. $M_1 = 20$ knots
    $M_2 = 3$ knots; $\alpha = 53°$

C   17. $M_1 = 655$ km/hr
    $M_2 = 97.3$ km/hr; $\alpha = 66.8°$
18. $M_1 = 487$ km/hr
    $M_2 = 74.2$ km/hr; $\alpha = 37.5°$

APPLICATIONS

19. *Navigation*   A river is flowing east (90°) at 3 km/hr. A boat crosses the river with a compass heading of 180° (south). If the speedometer on the boat reads 4 km/hr, what is the boat's actual speed and direction (resultant velocity) relative to the river bottom?

20. *Navigation*   A boat capable of traveling 12 knots on still water maintains a westward compass heading (270°) while crossing a river. If the river is flowing southward (180°) at 4 knots, what is the velocity (magnitude and direction) of the boat relative to the river bottom?

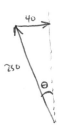

*21. *Navigation*   An airplane can cruise at 250 km/hr in still air. If a steady wind of 40 km/hr is blowing from the west, what compass heading should the pilot fly in order for his true course to be north (0°)? Compute the ground speed for this course.

*22. *Navigation*   Two docks are directly opposite each other on a southward flowing river. A boat pilot wishes to go in a straight line from the east dock to the west dock in a ferryboat with a cruising speed of 8 knots. If the river's current is 2.5 knots, what compass heading should be maintained while crossing the river? What is the actual speed of the boat relative to land?

23. *Resultant force*   Two tugs are trying to pull a barge off a shoal as indi-

cated in the figure. Find the magnitude of the resulting force and its direction relative to $\mathbf{F_1}$.

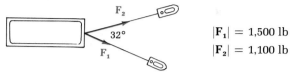

$$|\mathbf{F_1}| = 1{,}500 \text{ lb}$$
$$|\mathbf{F_2}| = 1{,}100 \text{ lb}$$

24.   *Resultant force*   Repeat Problem 23 with $\mathbf{F_1} = 1{,}300$ lb and the angle between the force vectors 45°.

25.   *Work*   A parent pulls a child in a wagon (see figure) for one block (440 ft). If a constant force of 15 lb is exerted along the handle, how much work is done?

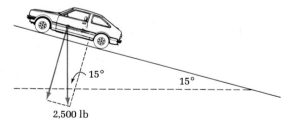

26.   *Work*   Repeat Problem 25 using a constant force of 12 lb, a distance of 5,300 ft (approximately a mile), and an angle of 35° relative to the horizontal.

27.   *Resolution of forces*   A car weighing 2,500 lb is parked on a hill inclined 15° to the horizontal. Neglecting friction, what magnitude of force parallel to the hill will keep the car from rolling down the hill? What is the force magnitude perpendicular to the hill?

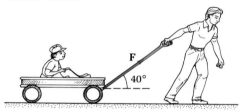

28.   *Resolution of forces*   Repeat Problem 27 with the car weighing 4,200 lb and the hill inclined 9°.

****29.**   *Resolution of forces*   If two weights are fastened together and placed on inclined planes as indicated in the figure, neglecting friction, which way will they slide?

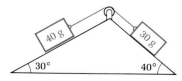

6.4   AREAS OF TRIANGLES (OPTIONAL)

We discuss below three frequently used methods of finding areas of triangles given the indicated information. The derivation of Heron's formula illustrates a significant use of identities.

Just discuss

BASE AND HEIGHT GIVEN   If the base b and the height h of a triangle are given, then the area is one-half the area of a parallelogram with the same base and height.

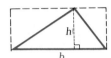

$$A = \frac{1}{2} bh$$

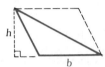

TWO SIDES AND INCLUDED ANGLE GIVEN   Given two sides, a and b, and an included angle θ, the formula above can be converted into the following form:

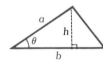

$$A = \frac{ab}{2} \sin \theta$$

$$A = \frac{1}{2} ab \sin C$$
$$= \frac{1}{2} ac \sin B$$
$$= \frac{1}{2} bc \sin A$$

This conversion is done by using the sine function to express h in terms of θ and a as follows:

$$h = a \sin \theta$$

which holds whether θ is acute or obtuse, since $\sin(180° - \theta) = \sin \theta$. Thus,

$$A = \frac{1}{2} bh = \frac{1}{2} ba \sin \theta = \frac{ab}{2} \sin \theta$$

EXAMPLE 8   Find the area of the triangle.

8 m

35.0°

5 m

Solution $A = \dfrac{ab}{2} \sin \theta$

$= \dfrac{1}{2}(8)(5) \sin 35.0°$

$= 11.5 \text{ m}^2$

PROBLEM 8   Find the area of a triangle with $a = 12$ cm, $b = 7$ cm, and $\theta = 125.0°$.

THREE SIDES GIVEN
(HERON'S FORMULA)

A very famous formula due to the Greek philosopher–mathematician Heron of Alexandria (75 AD) enables us to compute the area of a triangle directly, given only the lengths of the three sides of the triangle.

HERON'S FORMULA

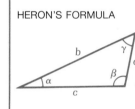

If the semiperimeter s is

$$s = \dfrac{a + b + c}{2}$$

then

$$A = \sqrt{s(s - a)(s - b)(s - c)}$$

Heron's formula is obtained from

$$A = \dfrac{bc}{2} \sin \alpha \tag{1}$$

by expressing $\sin \alpha$ in terms of the sides a, b, and c. Several identities and the law of cosines play a central role in the derivation of the formula. Since it is easier to first get $\sin(\alpha/2)$ and $\cos(\alpha/2)$ in terms of a, b, and c, we will do this; then we will use a double-angle identity in the form

$$\sin \alpha = 2 \sin \dfrac{\alpha}{2} \cos \dfrac{\alpha}{2} \tag{2}$$

to write $\sin \alpha$ in terms of a, b, and c.

We start with a half-angle identity for sine in the form

$$\sin^2 \dfrac{\alpha}{2} = \dfrac{1 - \cos \alpha}{2} \tag{3}$$

The following form of the law of cosines involves all three sides of the triangle a, b, and c and $\cos\alpha$:

$$a^2 = b^2 + c^2 - 2bc\cos\alpha$$

Solving for $\cos\alpha$, we obtain

$$\cos\alpha = \frac{b^2 + c^2 - a^2}{2bc} \tag{4}$$

Substituting (4) into (3), we can write $\sin^2(\alpha/2)$ in terms of a, b, and c:

$$\sin^2\frac{\alpha}{2} = \frac{1 - \dfrac{b^2 + c^2 - a^2}{2bc}}{2}$$

$$= \frac{2bc\left(1 - \dfrac{b^2 + c^2 - a^2}{2bc}\right)}{2bc(2)} \qquad \text{Convert to simple fraction}$$

$$= \frac{2bc - b^2 - c^2 + a^2}{4bc} \qquad \text{Numerator factors (not obvious)}$$

$$= \frac{(a + b - c)(a - b + c)}{4bc} \tag{5}$$

To bring the semiperimeter

$$s = \frac{a + b + c}{2} \tag{6}$$

into the picture, we write (5) in the form

$$\sin^2\frac{\alpha}{2} = \frac{(a + b + c - 2b)(a + b + c - 2c)}{4bc}$$

$$= \frac{2\left(\dfrac{a + b + c}{2} - b\right)2\left(\dfrac{a + b + c}{2} - c\right)}{4bc} \tag{7}$$

Substituting (6) into (7) and solving for $\sin(\alpha/2)$, we obtain

$$\sin\frac{\alpha}{2} = \sqrt{\frac{(s - b)(s - c)}{bc}} \tag{8}$$

If we repeat this reasoning starting with a half-angle identity for cosine in the form

$$\cos^2\frac{\alpha}{2} = \frac{1 + \cos\alpha}{2} \tag{9}$$

we obtain

$$\cos\frac{\alpha}{2} = \sqrt{\frac{s(s - a)}{bc}} \tag{10}$$

Substituting (8) and (10) into (2) produces

$$\sin \alpha = \frac{2}{bc} \sqrt{s(s - a)(s - b)(s - c)} \tag{11}$$

And we are almost there. We now substitute (11) into (1) to obtain Heron's formula:

$$A = \frac{bc}{2} \left[\frac{2}{bc} \sqrt{s(s - a)(s - b)(s - c)} \right]$$
$$= \sqrt{s(s - a)(s - b)(s - c)}$$

The derivation of Heron's formula provides a good illustration of the importance of identities. Try to imagine a derivation of this formula without identities.

EXAMPLE 9   Find the area to one decimal place of à triangle with sides $a = 12$ cm, $b = 8$ cm, and $c = 6$ cm.

Solution   First find the semiperimeter s:

$$s = \frac{a + b + c}{2} = \frac{12 + 8 + 6}{2} = 13 \text{ cm}$$

Then,

$$s - a = 13 - 12 = 1$$
$$s - b = 13 - 8 = 5$$
$$s - c = 13 - 6 = 7$$

Thus,

$$A = \sqrt{s(s - a)(s - b)(s - c)}$$
$$= \sqrt{13(1)(5)(7)}$$
$$= \sqrt{455}$$
$$= 21.3 \text{ cm}^2$$

PROBLEM 9   Find the area to one decimal place of a triangle with $a = 6$ m, $b = 10$ m, and $c = 8$ m.

ANSWERS TO MATCHED PROBLEMS   **8.**   34.4 cm²   **9.**   24.0 m²

Find the area of the triangle matching the information given below.

A

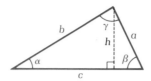 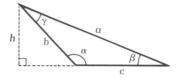

1. $h = 12$ m, $c = 17$ m
2. $h = 7$ ft, $c = 10$ ft
3. $\alpha = 30°$, $b = 6$ cm, $c = 8$ cm
4. $\alpha = 45°$, $b = 5$ m, $c = 6$ m
5. $a = 4$ in., $b = 6$ in., $c = 8$ in.
6. $a = 4$ ft, $b = 10$ ft, $c = 12$ ft

B

7. $\alpha = 23°20'$, $b = 403$ ft, $c = 512$ ft
8. $\alpha = 58°40'$, $b = 28.2$ in., $c = 6.4$ in.
9. $\alpha = 132.67°$, $b = 12.1$ cm, $c = 10.2$ cm
10. $\alpha = 147.5°$, $b = 125$ mm, $c = 67$ mm
11. $a = 12.7$ m, $b = 20.3$ m, $c = 24.4$ m
12. $a = 5.24$ cm, $b = 3.48$ cm, $c = 6.04$ cm

C

13. Obtain equation (10) following the same type of reasoning that was used to obtain equation (8) in the derivation of Heron's formula.

14. If $s = (a + b + c)/2$ is the semiperimeter of a triangle with sides a, b, and c, show that the radius r of an inscribed circle is given by

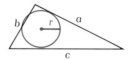

$$r = \sqrt{\frac{(s-a)(s-b)(s-c)}{s}}$$

15. Show that the diagonals of a parallelogram divide the figure into four triangles all having the same area.

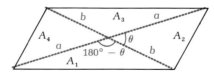

CHAPTER REVIEW

Where applicable, quantities in the problems refer to a triangle labeled as follows:

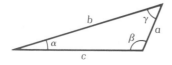

A   **1.** Solve the triangle with $\alpha = 66°$, $\beta = 32°$, and $b = 12$ m.

2. Solve the triangle with $\alpha = 25°$, $b = 22$ in., and $c = 27$ in.

3. Two vectors **u** and **v** are located in a coordinate system, as indicated in the figure. Find the direction and magnitude of **u** + **v** if $|\mathbf{u}| = 8$ and $|\mathbf{v}| = 5$.

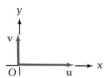

4. Find the magnitude of the horizontal and vertical components of the vector **v** located in a coordinate system as indicated in the figure.

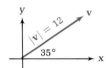

5. An airplane flies west at 300 km/hr and another plane flies northwest at 170 km/hr. If they leave the same place at the same time, how far apart will they be in 1 hr?

6. Find the area of the triangle with $\alpha = 20.0°$, $b = 12$ in., and $c = 5$ in.

B   **7.** Solve the triangle with $\alpha = 27.5°$, $b = 103$ m, and $c = 72.4$ m.

8. Solve the triangle with $\alpha = 35°20'$, $b = 15.7$ in., and $a = 13.2$ in. $(0° < \beta < 90°)$.

9. Repeat Problem 8 for $90° < \beta < 180°$.

10. Given the accompanying vector diagram, find $|\mathbf{u} + \mathbf{v}|$ and θ.

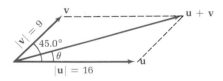

11. *Navigation*   An airplane flies with a speed of 230 km/hr and a compass heading of 68°. If a 55 km/hr wind is blowing in the direction of 5°, what is the plane's actual direction (relative to north) and ground speed?

12. Find the area of the triangle with $a = 14$ m, $b = 22$ m, and $c = 12$ m.

C   **13.** Find h in the triangle below.

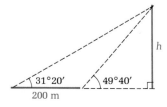

14.   Find the lengths of the sides of a parallelogram with diagonals 20 cm and 16 cm long intersecting at 36.4°.

15.   Find the area of the parallelogram in Problem 14.

16.   *Resultant force and work*   Two forces act on an object as indicated in the figure.

(A)   Find the magnitude of the resultant force and its direction relative to the horizontal force **F₁**.

(B)   How much work is done by the resultant force if the object is moved 22 ft?

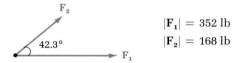

$|\mathbf{F_1}| = 352$ lb

$|\mathbf{F_2}| = 168$ lb

POLAR COORDINATES; COMPLEX NUMBERS 7

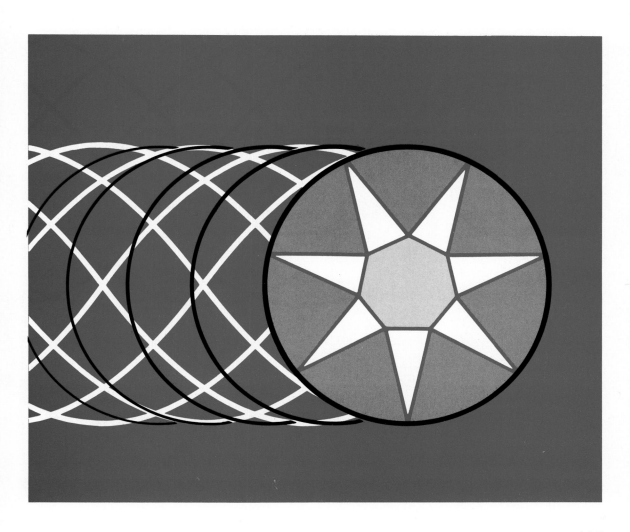

7.1   POLAR AND RECTANGULAR COORDINATES

How do we locate a point in a plane? Until now we have used a rectangular coordinate system in which two intersecting lines parallel to the coordinate axes determined the coordinates of the point (see Figure 1). Another type of coordinate system, called the **polar coordinate system,** is also widely used to locate points in a plane. In this system, a point P is located by specifying a **direction** θ and a **directed distance** r from the origin (see Figure 2). Each of these coordinate systems has desirable properties that make it particularly suited to special classes of applications. Let us now give a precise definition of the polar coordinate system and look at some of its interesting properties.

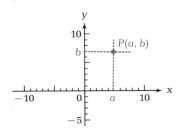

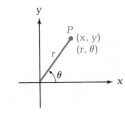

FIGURE 1

FIGURE 2

To form a polar coordinate system, we start with a fixed point in a plane called the **origin** or **pole** (see Figure 3). Attached to this point is a horizontal ray directed to the right, called the **polar axis.**

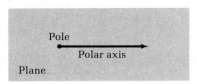

FIGURE 3

If P is an arbitrary point in the plane, then we associate polar coordinates (r, θ) with it as follows: Think of a number line, such as the one shown in Figure 4, fastened through its origin (0 point) to the pole in Figure 3 so that it is free to rotate like a spinner found in many games (see Figure 5). The second

FIGURE 4

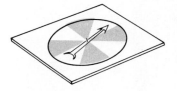

FIGURE 5

coordinate θ in (r, θ) is then the angle (in degree or radian measure) through which the number line is rotated so that it lies over P—counterclockwise for positive angle measures; clockwise for negative angle measures. The first coordinate r in (r, θ) is the number on the number line (directed distance

P
(3, 30°)

P
(−3, 210°)

P
(−3, −150°)

P
(3, $\frac{\pi}{6}$)

P
(−3, $\frac{7\pi}{6}$)

P
(−3, −$\frac{5\pi}{6}$)

FIGURE 6

Ex:

Cartesian (1,1)

Polar ($\sqrt{2}$, 45°) =

($\sqrt{2}$, 405°) =

($-\sqrt{2}$, 225°) =

⋮

from the pole) that is directly above P; r is positive if the positive end of the number line is over P and negative if the negative end is over P.

We can now see a major difference between the rectangular coordinate system and the polar coordinate system. In the former, each point has exactly one set of rectangular coordinates; in the latter, a point may have infinitely many polar coordinates. For example, Figure 6 illustrates a point P with six different sets of polar coordinates. Study this figure carefully.

The pole itself has polar coordinates of the form $(0, \theta)$, where θ is arbitrary. For example, (0, 37°) and (0, −π/4) are both polar coordinates of the pole, and there are infinitely many others.

Just as graph paper is readily available for work related to rectangular coordinate systems, polar graph paper is available for work related to polar coordinates. The examples below illustrate its use.

EXAMPLE 1

Do some →

Plot the following points in a polar coordinate system:

(A)   A(4, 45°),   B(−6, 45°),   C(7, 240°),   D(3, −75°)

(B)   A(8, π/6),   B(5, −3π/4),   C(−7, 2π/3),   D(−9, −π/6)

Solutions

(A)

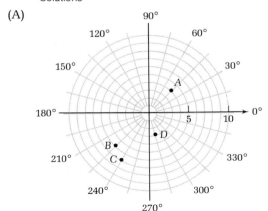

(B)

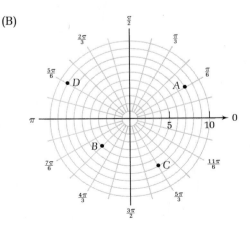

PROBLEM 1   Plot the following points in a polar coordinate system:
(A) $A(7, 30°)$, $B(-6, 165°)$, $C(-9, -90°)$
(B) $A(10, \pi/3)$, $B(8, -7\pi/6)$, $C(-5, -5\pi/4)$

EXAMPLE 2   For the point $(6, 60°)$, find three other sets of coordinates such that $-360° \leq \theta \leq 360°$.

Solution

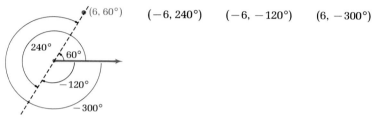

$(-6, 240°)$ $(-6, -120°)$ $(6, -300°)$

DO

PROBLEM 2   For the point $(7, \pi/6)$, find three other sets of coordinates such that $-2\pi \leq \theta \leq 2\pi$.

FROM POLAR FORM
TO RECTANGULAR
FORM AND VICE VERSA

It is frequently convenient to be able to transform coordinates or equations in rectangular form into polar form, or vice versa. The following relationships are useful in this regard:

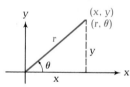

$$r^2 = x^2 + y^2$$

$$\sin \theta = \frac{y}{r} \qquad \cos \theta = \frac{x}{r} \qquad \tan \theta = \frac{y}{x}, x \neq 0$$

$$\theta = \sin^{-1} \frac{y}{r} \qquad \theta = \cos^{-1} \frac{x}{r} \qquad \theta = \tan^{-1} \frac{y}{x}, x \neq 0$$

$$y = r \sin \theta \qquad x = r \cos \theta$$

[*Note:*   The signs of x and y determine the quadrant for θ. The angle θ is usually chosen so that $|\theta|$ is minimum.]

EXAMPLE 3   Change $A(5, \pi/6)$, $B(-3, 3\pi/4)$, and $C(-2, -5\pi/6)$ to rectangular coordinates.

Solution   Use $x = r \cos \theta$ and $y = r \sin \theta$. For A:

Do 1,2

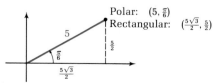

Polar: $(5, \frac{\pi}{6})$
Rectangular: $(\frac{5\sqrt{3}}{2}, \frac{5}{2})$

$$x = 5 \cos \frac{\pi}{6} = 5\left(\frac{\sqrt{3}}{2}\right) = \frac{5\sqrt{3}}{2}$$

$$y = 5 \sin \frac{\pi}{6} = 5\left(\frac{1}{2}\right) = \frac{5}{2}$$

Rectangular coordinates: $\left(\dfrac{5\sqrt{3}}{2},\dfrac{5}{2}\right)$

For B:

$$x = -3\cos\frac{3\pi}{4} = (-3)\left(\frac{-\sqrt{2}}{2}\right) = \frac{3\sqrt{2}}{2}$$

$$y = -3\sin\frac{3\pi}{4} = (-3)\left(\frac{\sqrt{2}}{2}\right) = \frac{-3\sqrt{2}}{2}$$

Rectangular coordinates: $\left(\dfrac{3\sqrt{2}}{2},\dfrac{-3\sqrt{2}}{2}\right)$

For C:

$$x = -2\cos\frac{-5\pi}{6} = (-2)\left(\frac{-\sqrt{3}}{2}\right) = \sqrt{3}$$

$$y = -2\sin\frac{-5\pi}{6} = (-2)\left(\frac{-1}{2}\right) = 1$$

Rectangular coordinates: $(\sqrt{3}, 1)$

PROBLEM 3 Change $A(8, \pi/3)$, $B(-6, 5\pi/4)$, and $C(-4, -7\pi/6)$ to rectangular coordinates.

EXAMPLE 4 Change $A(1, \sqrt{3})$ and $B(-\sqrt{3}, -1)$ into polar form with $r \geq 0$ and $0 \leq \theta < 2\pi$.

Solution Use $r^2 = x^2 + y^2$ and $\tan\theta = y/x$. For A:

$$r^2 = 1^2 + (\sqrt{3})^2 = 4$$
$$r = 2$$
$$\tan\theta = \frac{\sqrt{3}}{1}$$
$$\theta = \frac{\pi}{3} \qquad \text{Since } A \text{ is in the first quadrant}$$

Polar coordinates: $\left(2,\dfrac{\pi}{3}\right)$

For B:

$$r^2 = (-\sqrt{3})^2 + (-1)^2 = 4$$
$$r = 2$$
$$\tan\theta = \frac{-1}{-\sqrt{3}}$$
$$\theta = \frac{7\pi}{6} \qquad \text{Since } B \text{ is in the third quadrant}$$

Polar coordinates: $\left(2,\dfrac{7\pi}{6}\right)$

PROBLEM 4   Change $A(\sqrt{3}, 1)$ and $B(1, -\sqrt{3})$ into polar form with $r \geq 0$ and $0 \leq \theta < 2\pi$.

EXAMPLE 5   Change $x^2 + y^2 - 2x = 0$ to polar form.

Solution   Use $r^2 = x^2 + y^2$ and $x = r \cos \theta$:

$$x^2 + y^2 - 2x = 0$$
$$r^2 - 2r \cos \theta = 0$$
$$r(r - 2 \cos \theta) = 0$$
$$r = 0 \quad \text{or} \quad r - 2 \cos \theta = 0$$

The graph of $r = 0$ is the pole, and since the pole is included as a solution of $r - 2 \cos \theta = 0$ (let $\theta = \pi/2$), we can discard $r = 0$ and keep only

$$r - 2 \cos \theta = 0$$

or

$$r = 2 \cos \theta$$

Do

PROBLEM 5   Change $x^2 + y^2 - 2y = 0$ to polar form.

EXAMPLE 6   Change $r + 3 \sin \theta = 0$ to rectangular form.

Solution   The conversion of this equation, as it stands, to rectangular form gets messy. A simple trick, however, makes the conversion easy. We multiply both sides by r, which simply adds the pole to the graph. But the pole is already included as a solution of $r + 3 \sin \theta = 0$ (let $\theta = 0$), so we have not actually changed anything by doing this. Thus,

$$\begin{cases} r + 3 \sin \theta = 0 \\ r^2 + 3r \sin \theta = 0 \\ x^2 + y^2 + 3y = 0 \end{cases} \qquad r^2 = x^2 + y^2, \quad y = r \sin \theta$$

Do

$r + 3 \sin \theta = 0$

$r + 3 \left(\frac{y}{r} \right) = 0$ OR

$r^2 + 3y = 0$

$x^2 + y^2 + 3y = 0$

PROBLEM 6   Change $r = -8 \cos \theta$ to rectangular form.

ANSWERS TO MATCHED PROBLEMS

1. (A)

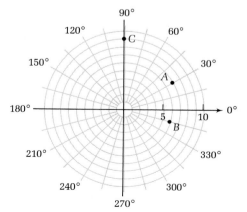

(B)

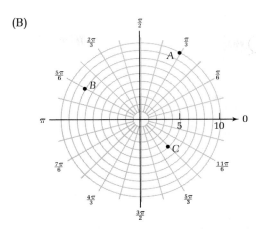

2. $(-7, 7\pi/6)$, $(-7, -5\pi/6)$, $(7, -11\pi/6)$
3. $A(4, 4\sqrt{3})$, $B(3\sqrt{2}, 3\sqrt{2})$, $C(2\sqrt{3}, -2)$
4. $A(2, \pi/6)$, $B(2, 5\pi/3)$ **5.** $r = 2\sin\theta$ **6.** $x^2 + y^2 + 8x = 0$

EXERCISE 7.1

A   *Plot in a polar coordinate system.*

1. $A(8, 0°)$, $B(5, 90°)$, $C(6, 30°)$
2. $A(4, 0°)$, $B(7, 180°)$, $C(9, 45°)$
3. $A(-8, 0°)$, $B(-5, 90°)$, $C(-6, 30°)$
4. $A(-4, 0°)$, $B(-7, 180°)$, $C(-9, 45°)$
5. $A(5, -30°)$, $B(4, -45°)$, $C(9, -90°)$
6. $A(8, -45°)$, $B(6, -60°)$, $C(4, -30°)$
7. $A(-5, -30°)$, $B(-4, -45°)$, $C(-9, -90°)$
8. $A(-8, -45°)$, $B(-6, -60°)$, $C(-4, -30°)$
9. $A(6, \pi/6)$, $B(5, \pi/2)$, $C(8, \pi/4)$
10. $A(8, \pi/3)$, $B(4, \pi/4)$, $C(10, 0)$
11. $A(-6, \pi/6)$, $B(-5, \pi/2)$, $C(-8, \pi/4)$
12. $A(-8, \pi/3)$, $B(-4, \pi/4)$, $C(-10, 0)$
13. $A(6, -\pi/6)$, $B(5, -\pi/2)$, $C(8, -\pi/4)$
14. $A(8, -\pi/3)$, $B(4, -\pi/4)$, $C(10, -\pi/6)$
15. $A(-6, -\pi/2)$, $B(-5, -\pi/3)$, $C(-8, -\pi/4)$
16. $A(-6, -\pi/6)$, $B(-5, -\pi/2)$, $C(-8, -\pi/4)$

Change to rectangular coordinates.

17. $(8, \pi/3)$ **18.** $(4, \pi/4)$ **19.** $(-9, \pi/2)$
20. $(-8, \pi)$ **21.** $(-4, \pi/4)$ **22.** $(-6, \pi/6)$
23. $(10, 5\pi/6)$ **24.** $(8, 7\pi/6)$ **25.** $(6, -7\pi/6)$
26. $(4, -7\pi/4)$ **27.** $(-4, -\pi/6)$ **28.** $(-5, -\pi/3)$

Change to polar coordinates with r ≥ 0 and 0 ≤ θ < 2π.

29. $(2\sqrt{3}, 2)$ 30. $(3, 3\sqrt{3})$ 31. $(-4\sqrt{2}, 4\sqrt{2})$
32. $(-6\sqrt{3}, 6)$ 33. $(-4, -4\sqrt{3})$ 34. $(5\sqrt{2}, -5\sqrt{2})$
35. $(0, -7)$ 36. $(-10, 0)$

B   *Plot in a polar coordinate system.*

37. $A(5, 210°)$, $B(-5, 210°)$, $C(-5, -210°)$
38. $A(9, 120°)$, $B(-9, 120°)$, $C(-9, -120°)$
39. $A(7, 7\pi/4)$, $B(-7, 7\pi/4)$, $C(-7, -7\pi/4)$
40. $A(6, 4\pi/3)$, $B(-6, 4\pi/3)$, $C(-6, -4\pi/3)$

Change to polar form.

41. $6x - x^2 = y^2$ 42. $y^2 = 5y - x^2$ 43. $2x + 3y = 5$
44. $3x - 5y = -2$ 45. $x^2 + y^2 = 9$ 46. $y = x$

Change to rectangular form.

47. $r(2\cos\theta + \sin\theta) = 4$ 48. $r(3\cos\theta - 4\sin\theta) = -1$
49. $r = 8\cos\theta$ 50. $r = -2\sin\theta$
51. $r = 4$ 52. $\theta = \pi/4$

C   53. Change $r = 3/(\sin\theta - 2)$ into rectangular form.
     54. Change $(y - 3)^2 = 4(x^2 + y^2)$ into polar form.

7.2   SKETCHING POLAR EQUATIONS

To graph an equation such as

$$r = 4\cos\theta$$

in a polar coordinate system, we locate all points with coordinates that satisfy the equation. An approximation of the graph is found (as in rectangular coordinates) by making a table of values that satisfy the equation, plotting these, and then joining the points with a smooth curve. We can use special angles, a table, or a calculator. In Example 7 we use special angles.

EXAMPLE 7   Graph $r = 4\cos\theta$.

Solution   We form a table by using special angles, selecting values until the graph starts to repeat.

θ	0	$\pi/6$	$\pi/4$	$\pi/3$	$\pi/2$	$2\pi/3$	$3\pi/4$	$5\pi/6$	π
r	4	$2\sqrt{3}$	$2\sqrt{2}$	2	0	-2	$-2\sqrt{2}$	$-2\sqrt{3}$	-4

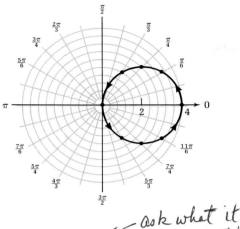

PROBLEM 7   Graph r = 4 sin θ. ← *ask what it would look like.*

EXAMPLE 8   Repeat Example 7, using a calculator and θ in degrees.

Solution

θ	0°	30°	45°	60°	90°	120°	135°	150°	180°
r	4	3.5	2.8	2	0	−2	−2.8	−3.5	−4

The graph is the same as that found in Example 7, except degrees are marked around the polar coordinate system instead of radians. For accurate graphing, a calculator is very helpful—it is easy to intersperse additional points until the accuracy of the graph is assured.

PROBLEM 8   Use a calculator and construct a table for r = 4 sin θ, using the same values for θ as in Example 8.

RAPID SKETCHING
OF POLAR EQUATIONS

If only a rough sketch of a polar graph involving sin θ or cos θ is desired, then we can speed up the process described above as follows: For reference, we draw (or mentally visualize) one period of y = sin x and y = cos x in rectangular coordinate systems (Figure 7). At a glance, we can tell how each function behaves in each of the four quadrants, and we can use this information to sketch some polar equations without the need of tedious point-by-point plotting. The examples below should make the process clear.

FIGURE 7

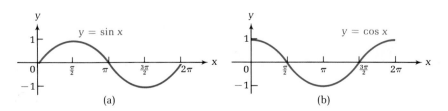

EXAMPLE 9   Sketch $r = 5 + 5 \cos \theta$.

Solution   We start by setting up a table that indicates how r changes as we allow θ to vary through each set of quadrant values (see Table 1). This procedure is effective because $\cos \theta$ behaves in a uniform way through each of the four quadrants. Thus, as θ increases from 0 to $\pi/2$, r decreases from 10 to 5, and so on. Using the information in the table, and, perhaps, one or two particular points for parts of the curve that are in doubt, we can quickly sketch the graph of the equation, as shown in Figure 8.

TABLE 1

θ	$\cos \theta$	$5 \cos \theta$	$r = 5 + 5 \cos \theta$
0 to $\pi/2$	1 to 0	5 to 0	10 to 5
$\pi/2$ to π	0 to -1	0 to -5	5 to 0
π to $3\pi/2$	-1 to 0	-5 to 0	0 to 5
$3\pi/2$ to 2π	0 to 1	0 to 5	5 to 10

Draw

① $r = 2 + 2 \cos \theta$

② $r = 3 + 2 \cos \theta$

③ $r = 1 + 2 \cos \theta$

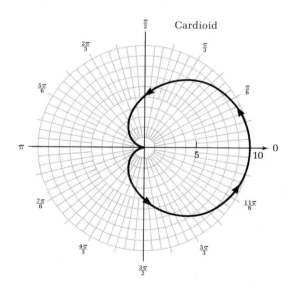

Cardioid

FIGURE 8
$r = 5 + 5 \cos \theta$

PROBLEM 9   Sketch $r = 4 + 4 \sin \theta$.

EXAMPLE 10   Sketch $r = 6 \sin 2\theta$.

Do

Solution   In this case, we let 2θ range through each set of quadrant values. That is, we start with values for 2θ (the second column in Table 2), fill in the table to the right, and then add values for θ (the first column), since we are ultimately interested in how θ and r are related. Thus, as 2θ increases from 0 to $\pi/2$, θ increases from 0 to $\pi/4$, and r increases from 0 to 6. As 2θ increases

TABLE 2

θ	2θ	$\sin 2\theta$	$r = 6 \sin 2\theta$
0 to $\pi/4$	0 to $\pi/2$	0 to 1	0 to 6
$\pi/4$ to $\pi/2$	$\pi/2$ to π	1 to 0	6 to 0
$\pi/2$ to $3\pi/4$	π to $3\pi/2$	0 to -1	0 to -6
$3\pi/4$ to π	$3\pi/2$ to 2π	-1 to 0	-6 to 0
π to $5\pi/4$	2π to $5\pi/2$	0 to 1	0 to 6
$5\pi/4$ to $3\pi/2$	$5\pi/2$ to 3π	1 to 0	6 to 0
$3\pi/2$ to $7\pi/4$	3π to $7\pi/2$	0 to -1	0 to -6
$7\pi/4$ to 2π	$7\pi/2$ to 4π	-1 to 0	-6 to 0

from $\pi/2$ to π, θ increases from $\pi/4$ to $\pi/2$, and r decreases to 0. As 2θ increases from π to $3\pi/2$, θ increases from $\pi/2$ to $3\pi/4$, and r decreases from 0 to -6. And so on. Continuing in the same way (until the table starts to repeat) and plotting the results in a polar coordinate system, we finally obtain the complete graph of the equation, as shown in Figure 9.

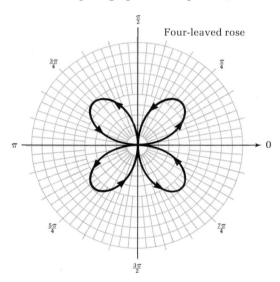

Four-leaved rose

FIGURE 9
$r = 6 \sin 2\theta$

PROBLEM 10   Sketch $r = 8 \cos 2\theta$.

The simplest type of equations to graph in a polar coordinate system are $\theta = $ Constant and $r = $ Constant. Figure 10 illustrates two particular cases.

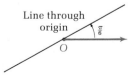

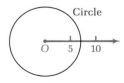

FIGURE 10

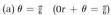 (a) $\theta = \frac{\pi}{6}$ $(0r + \theta = \frac{\pi}{6})$

(b) $r = 7$ $(0\theta + r = 7)$

APPLICATIONS Polar coordinate systems are very useful in many types of applications. Two applications from astronomy are included in Exercise 7.2. Figure 11, which was supplied by the United States Coast and Geodetic Survey, illustrates an application from oceanography. Each arrow represents the direction and magnitude of the tide at a particular time of the day at the San Francisco Light Station (a navigational light platform in the Pacific Ocean about 12 nautical mi west of the Golden Gate Bridge).

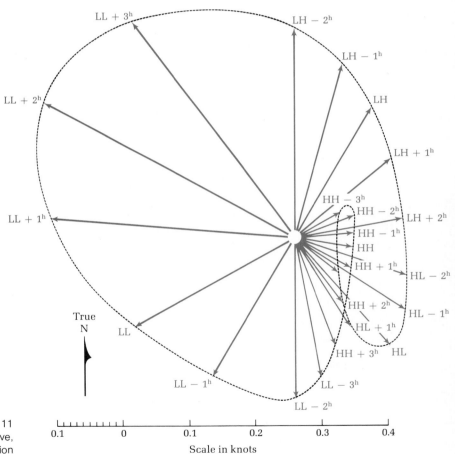

FIGURE 11
Tidal current curve,
San Francisco Light Station

7.

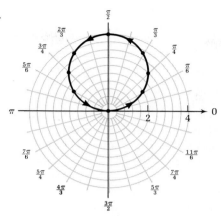

8.

θ	0°	30°	45°	60°	90°	120°	135°	150°	180°
r	0	2	2.8	3.5	4	3.5	2.8	2	0

9.

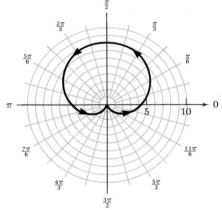

10.

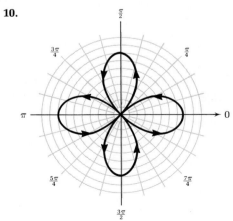

EXERCISE 7.2

A

1. Graph $r = 10 \cos \theta$ by assigning θ the values 0, $\pi/6$, $\pi/4$, $\pi/3$, $\pi/2$, $2\pi/3$, $3\pi/4$, $5\pi/6$, and π. Then join the resulting points with a smooth curve.
2. Repeat Problem 1 (above) for $r = 8 \sin \theta$, using the same set of values for θ.
3. Graph $r = 3 + 3 \cos \theta$, $0 \le \theta \le 360°$, using multiples of $30°$, starting at 0. Use a calculator or properties of special angles.
4. Graph $r = 4 + 4 \sin \theta$, $0 \le \theta \le 360°$, using multiples of $30°$, starting at 0. Use a calculator or properties of special angles.
5. Graph $r = \theta$, $0 \le \theta \le 2\pi$, using multiples of $\pi/6$ for θ starting at $\theta = 0$.
6. Graph $r = \theta/2$, $0 \le \theta \le 2\pi$, using multiples of $\pi/2$ for θ starting at $\theta = 0$.

Graph each polar equation.

7. $r = 5$ 8. $r = 8$ 9. $\theta = \pi/4$ 10. $\theta = \pi/3$

B *Sketch each polar equation using rapid sketching techniques.*

11. $r = 4 \cos \theta$	12. $r = 4 \sin \theta$
13. $r = 8 \cos 2\theta$	14. $r = 10 \sin 2\theta$
15. $r = 6 \sin 3\theta$	16. $r = 5 \cos 3\theta$
17. $r = 3 + 3 \cos \theta$	18. $r = 2 + 2 \sin \theta$
19. $r = 2 + 4 \cos \theta$	20. $r = 2 + 4 \sin \theta$
21. $r = 4 - 2 \sin \theta$	22. $r = 4 - 2 \cos \theta$
23. $r = \dfrac{\theta}{\pi}$, $\theta \ge 0$	24. $r\theta = \pi$, $\theta \ge 0$

C 25. $r = 5 + 5 \cos \dfrac{\theta}{2}$ 26. $r = 5 + 5 \sin \dfrac{\theta}{2}$

27. $r^2 = 64 \cos 2\theta$ 28. $r^2 = 64 \sin 2\theta$

29. Find all ordered pairs of numbers (r, θ), $0 \le \theta \le \pi$, that satisfy the following system, and interpret geometrically:

$$r = 2 \cos \theta \quad \text{and} \quad r = 2 \sin \theta$$

[*Note:* (r_1, θ_1) must satisfy both equations; some points of intersection of the two graphs may have coordinates that do not satisfy both equations.]

30. Using a calculator, graph the equation

$$r = \frac{8}{1 - e \cos \theta}$$

for the values of e given below.

(A) $e = \frac{1}{2}$ (B) $e = 1$ (C) $e = 2$

Can you identify each curve?

31. *Astronomy* **(A) The planet Mercury travels around the sun in an elliptical orbit given approximately by

$$r = \frac{3.442 \times 10^7}{1 - 0.206 \cos \theta}$$

where r is in miles. Graph the orbit with the sun at the pole. Find the distance from Mercury to the sun at **aphelion** (greatest distance from the sun) and at **perihelion** (shortest distance from the sun).

(B) Kepler (1571–1630) showed that a line joining a planet to the sun swept out equal areas in space in equal intervals in time (see the figure). Use this information to determine whether a planet travels faster or slower at aphelion than at perihelion.

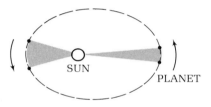

7.3   COMPLEX NUMBERS IN RECTANGULAR AND POLAR FORMS

Since a complex number is any number that can be written in the form

$a + bi$

where a and b are real and i is the imaginary unit (see Appendix A, Section A.2), each complex number can be associated with a unique ordered pair of real numbers, and vice versa. For example,

$2 - 3i$ corresponds to $(2, -3)$

and, in general,

$a + bi$ corresponds to (a, b)

Therefore, each complex number can be associated with a unique point in a rectangular coordinate system, and each point in a rectangular coordinate system can be associated with a unique complex number (see Figure 12).

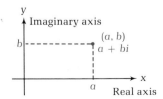

FIGURE 12
Complex plane

The cartesian plane, with points associated with ordered pairs (a, b), is often called the **complex plane** when points (a, b) are associated with complex numbers $a + bi$. The x axis is then called the **real axis** and the y axis is called the **imaginary axis.**

EXAMPLE 11   Plot each number in a complex plane:
(A) $3 + 4i$ (B) $-3 - 2i$ (C) -5 (D) $-3i$

Solutions

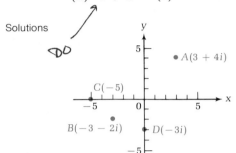

PROBLEM 11   Plot each number in a complex plane:
(A) $4 + 2i$ (B) $3 - 3i$ (C) -4 (D) $3i$

Complex numbers can be changed from rectangular form to **polar (or trigonometric) form** by using the relationships developed in Section 7.1. Thus, using

$$x = r \cos \theta \qquad \text{and} \qquad y = r \sin \theta$$

we can write (see Figure 13)

$$x + iy = r \cos \theta + ir \sin \theta$$
$$= r(\cos \theta + i \sin \theta)$$

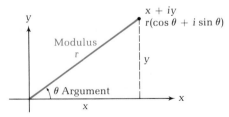

FIGURE 13
Polar and rectangular forms

Since sine and cosine each have periods of 2π, we can write

$$\sin(\theta + 2k\pi) = \sin \theta \qquad k \text{ any integer}$$
$$\cos(\theta + 2k\pi) = \cos \theta \qquad k \text{ any integer}$$

Hence, we have the more general polar forms:

FROM RECTANGULAR TO POLAR FORM

$$z = x + iy = r[\cos(\theta + 2k\pi) + i\sin(\theta + 2k\pi)] \qquad k \text{ any integer}$$
$$= r \, \text{cis}(\theta + 2k\pi)$$

where

$$\text{cis } \theta = \cos \theta + i \sin \theta$$

and where the quadrant for θ is determined by x and y.

The number r is called the **modulus** or **absolute value** of z, denoted by **mod z** or $|z|$. The polar angle that the line joining z to the origin makes with the positive x axis is called the **argument** of z, denoted **arg z**. From Figure 13 we can see the following relationships:

MODULUS AND ARGUMENT FOR $z = x + iy$

$$\text{mod } z = r = |z| = \sqrt{x^2 + y^2}$$

$$\arg z = \theta + 2k\pi = \sin^{-1}\frac{y}{r} = \cos^{-1}\frac{x}{r} = \tan^{-1}\frac{y}{x}$$

To write a complex number in polar form, we usually take the smallest positive angle for arg z.

EXAMPLE 12   Write (A) $z_1 = 1 + i\sqrt{3}$,   (B) $z_2 = -1 - i$, and (C) $z_3 = 2i$ in polar form.

Solutions   It helps to graph these numbers in a rectangular coordinate system first; then if x and y are associated with special angles, we can often determine r and θ by inspection.

(A) $z_1 = 2\left(\cos\dfrac{\pi}{3} + i \sin\dfrac{\pi}{3}\right)$

$= 2 \, \text{cis} \dfrac{\pi}{3}$

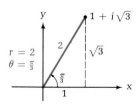

(B) $z_2 = \sqrt{2}\left(\cos\dfrac{5\pi}{4} + i\sin\dfrac{5\pi}{4}\right)$ (C) $z_3 = 2\left(\cos\dfrac{\pi}{2} + i\sin\dfrac{\pi}{2}\right)$

$\quad\quad = \sqrt{2}\,\text{cis}\,\dfrac{5\pi}{4}$ $= 2\,\text{cis}\,\dfrac{\pi}{2}$

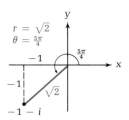

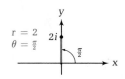

PROBLEM 12   Write   (A) $z_1 = \sqrt{3} + i$ (B) $z_2 = -1 + i$,   and   (C) $z_3 = -5i$ in polar form.

EXAMPLE 13   Write   (A) $z_1 = \sqrt{2}\,\text{cis}\,(\pi/4)$ (B) $z_2 = 2\,\text{cis}\,150°$,   and
(C) $z_3 = 5\,\text{cis}\,\pi$ in rectangular form.

Do ↗

Solutions   (A) $z_1 = \sqrt{2}\,\text{cis}\,\dfrac{\pi}{4} = \sqrt{2}\left(\cos\dfrac{\pi}{4} + i\sin\dfrac{\pi}{4}\right)$

$\quad\quad\quad\quad\quad = \sqrt{2}\left(\dfrac{1}{\sqrt{2}} + i\,\dfrac{1}{\sqrt{2}}\right)$

$\quad\quad\quad\quad\quad = 1 + i$

(B) $z_2 = 2\,\text{cis}\,150° = 2(\cos 150° + i\sin 150°)$

$\quad\quad\quad\quad = 2\left(-\dfrac{\sqrt{3}}{2} + i\,\dfrac{1}{2}\right)$

$\quad\quad\quad\quad = -\sqrt{3} + i$

(C) $z_3 = 5\,\text{cis}\,\pi = 5(\cos\pi + i\sin\pi)$

$\quad\quad\quad\quad = 5(-1 + 0i)$

$\quad\quad\quad\quad = -5$

PROBLEM 13   Write   (A) $z_1 = 4\,\text{cis}\,(\pi/6)$ (B) $z_2 = 3\sqrt{2}\,\text{cis}\,315°$,   and
(C) $z_3 = 6\,\text{cis}\,(3\pi/2)$ in rectangular form.

MULTIPLICATION AND
DIVISION IN POLAR
FORM

We will now see a particular advantage in representing complex numbers in polar form: multiplication and division become easy. Let us start with multiplication.

$z_1 z_2 = (r_1\,\text{cis}\,\theta_1)(r_2\,\text{cis}\,\theta_2)$

$\quad\quad = r_1 r_2(\cos\theta_1 + i\sin\theta_1)(\cos\theta_2 + i\sin\theta_2)$

$\quad\quad = r_1 r_2(\cos\theta_1\cos\theta_2 + i\cos\theta_1\sin\theta_2 + i\sin\theta_1\cos\theta_2 - \sin\theta_1\sin\theta_2)$

$$z_1 z_2 = r_1 r_2 [(\cos \theta_1 \cos \theta_2 - \sin \theta_1 \sin \theta_2) + i(\cos \theta_1 \sin \theta_2 + \sin \theta_1 \cos \theta_2)]$$
$$= r_1 r_2 [\cos(\theta_1 + \theta_2) + i \sin(\theta_1 + \theta_2)] \qquad \text{Sum identities}$$
$$= r_1 r_2 \operatorname{cis}(\theta_1 + \theta_2)$$

Thus, to multiply two complex numbers in polar form, we multiply r_1 and r_2 and add θ_1 to θ_2. Similarly, one can show that

$$\frac{z_1}{z_2} = \frac{r_1 \operatorname{cis} \theta_1}{r_2 \operatorname{cis} \theta_2} = \frac{r_1}{r_2} \operatorname{cis}(\theta_1 - \theta_2)$$

The proof of this quotient form is left to Problem 57, Exercise 7.3. We summarize both forms as follows:

PRODUCTS AND QUOTIENTS IN POLAR FORM

$$z_1 z_2 = (r_1 \operatorname{cis} \theta_1)(r_2 \operatorname{cis} \theta_2) = r_1 r_2 \operatorname{cis}(\theta_1 + \theta_2)$$

$$\frac{z_1}{z_2} = \frac{r_1 \operatorname{cis} \theta_1}{r_2 \operatorname{cis} \theta_2} = \frac{r_1}{r_2} \operatorname{cis}(\theta_1 - \theta_2)$$

EXAMPLE 14   If $z_1 = 8 \operatorname{cis} 50°$ and $z_2 = 4 \operatorname{cis} 30°$, find $z_1 z_2$ and z_1/z_2.

Solution

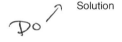

$$z_1 z_2 = (8 \operatorname{cis} 50°)(4 \operatorname{cis} 30°) \qquad \frac{z_1}{z_2} = \frac{8 \operatorname{cis} 50°}{4 \operatorname{cis} 30°}$$

$$= (8 \cdot 4) \operatorname{cis}(50° + 30°) \qquad = \frac{8}{4} \operatorname{cis}(50° - 30°)$$

$$= 32 \operatorname{cis} 80° \qquad = 2 \operatorname{cis} 20°$$

PROBLEM 14   If $z_1 = 21 \operatorname{cis} 140°$ and $z_2 = 3 \operatorname{cis} 105°$, find $z_1 z_2$ and z_1/z_2.

ANSWERS TO
MATCHED PROBLEMS

11.

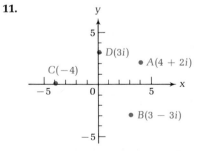

12.   (A) $z_1 = 2\left(\cos \dfrac{\pi}{6} + i \sin \dfrac{\pi}{6}\right) = 2 \operatorname{cis} \dfrac{\pi}{6}$

(B) $z_2 = \sqrt{2}\left(\cos \dfrac{3\pi}{4} + i \sin \dfrac{3\pi}{4}\right) = \sqrt{2} \operatorname{cis} \dfrac{3\pi}{4}$

(C) $z_3 = 5\left(\cos \dfrac{3\pi}{2} + i \sin \dfrac{3\pi}{2}\right) = 5 \operatorname{cis} \dfrac{3\pi}{2}$

13. (A) $z_1 = 2\sqrt{3} + 2i$ (B) $z_2 = 3 - 3i$ (C) $z_3 = -6i$

14. $z_1 z_2 = 63 \text{ cis } 245°,$ $z_1/z_2 = 7 \text{ cis } 35°$

EXERCISE 7.3

A   *Plot each set of complex numbers in a complex plane.*

1. $A = 4 + 5i,$ $B = -3 + 4i,$ $C = -3i$
2. $A = 3 - 2i,$ $B = -4 - 2i,$ $C = 5i$
3. $A = 4 + i,$ $B = -2 + 3i,$ $C = -4$
4. $A = -3 - i,$ $B = 5 - 4i,$ $C = 4$
5. $A = 8 \text{ cis}(\pi/4),$ $B = 6 \text{ cis}(\pi/2),$ $C = 3 \text{ cis}(\pi/6)$
6. $A = 5 \text{ cis}(5\pi/6),$ $B = 3 \text{ cis}(3\pi/2),$ $C = 4 \text{ cis}(7\pi/4)$
7. $A = 5 \text{ cis } 270°,$ $B = 4 \text{ cis } 60°,$ $C = 8 \text{ cis } 150°$
8. $A = 3 \text{ cis } 310°,$ $B = 4 \text{ cis } 180°,$ $C = 5 \text{ cis } 210°$

Find $z_1 z_2$ and z_1/z_2. Write answer in r cis θ form.

9. $z_1 = 10 \text{ cis } 45°,$ $z_2 = 5 \text{ cis } 32°$
10. $z_1 = 4 \text{ cis } 62°,$ $z_2 = 2 \text{ cis } 51°$
11. $z_1 = 7 \text{ cis } 163°,$ $z_2 = 3 \text{ cis } 102°$
12. $z_1 = 5 \text{ cis } 204°,$ $z_2 = 2 \text{ cis } 34°$

B   *Change to polar form.*

13. $1 + i$	**14.** $-1 - i$	**15.** $-\sqrt{3} + i$
16. $1 + i\sqrt{3}$	**17.** $4i$	**18.** -4
19. $-1 - i\sqrt{3}$	**20.** $\sqrt{3} - i$	**21.** $2 - i2\sqrt{3}$
22. $-3 + 3i$	**23.** $-8i$	**24.** 10

Change to rectangular form.

25. $\sqrt{2} \text{ cis } 45°$	**26.** $4 \text{ cis}(\pi/3)$	**27.** $\sqrt{2} \text{ cis } 135°$
28. $6 \text{ cis } 150°$	**29.** $8 \text{ cis } \pi$	**30.** $7 \text{ cis } 0$
31. $12 \text{ cis } 90°$	**32.** $11 \text{ cis } 270°$	**33.** $6 \text{ cis}(4\pi/3)$
34. $2\sqrt{2} \text{ cis}(5\pi/4)$	**35.** $4 \text{ cis } 330°$	**36.** $8 \text{ cis } 300°$

Find each of the following directly and by using polar forms. Write answers in a + bi and in r cis θ forms.

37. $(1 + i)^2$	**38.** $(-1 + i)^2$
39. $(1 + i\sqrt{3})(\sqrt{3} + i)$	**40.** $(-1 + i)(1 + i)$

Change to polar form, with $r \geq 0$ and $0° \leq \theta < 360°$. Compute θ to one decimal place and leave r in exact radical form.

C

41. $3 + 5i$	**42.** $5 + 6i$	**43.** $-7 + 3i$
44. $-4 - 9i$	**45.** $6 - 5i$	**46.** $3 - 7i$

Change to rectangular form a + bi. Compute a and b to three significant figures.

47. 9 cis 37°20′   48. 7 cis 23°40′
49. 5 cis 197.2°   50. 3 cis 133.8°
51. 11 cis 321°20′   52. 10 cis 305°30′

53. If $z = r$ cis θ, show that $z^2 = r^2$ cis 2θ and $z^3 = r^3$ cis 3θ. What do you think z^n is for n a natural number?
54. Show that $r^{1/2}$ cis$(\theta/2)$ is a square root of r cis θ. [*Hint:* Square this and see what happens.]
55. Show that $r^{1/n}$ cis(θ/n), for n a natural number, is an nth root of r cis θ, assuming $z^n = r^n$ cis $n\theta$.
56. Show that $r^{1/n}$ cis$[(\theta + 2k\pi)/n]$, for n a natural number and k any integer, is an nth root of r cis θ, assuming $z^n = r^n$ cis $n\theta$.
57. Prove that

$$\frac{r_1 \text{ cis } \theta_1}{r_2 \text{ cis } \theta_2} = \frac{r_1}{r_2} \text{ cis}(\theta_1 - \theta_2)$$

7.4   DE MOIVRE'S THEOREM

We now apply the product formula for polar forms to powers z^2, z^3, . . . :

$$z^2 = (x + iy)^2 = (r \text{ cis } \theta)(r \text{ cis } \theta) = r^2 \text{ cis } 2\theta$$
$$z^3 = z^2 z = (r \text{ cis } \theta)^2(r \text{ cis } \theta) = (r^2 \text{ cis } 2\theta)(r \text{ cis } \theta) = r^3 \text{ cis } 3\theta$$

You can probably guess what z^4 would be. If you guessed r^4 cis 4θ, you are right! Now you are probably brave enough to jump to the general case for z^n, n any natural number. If so, you have discovered De Moivre's famous theorem:

DE MOIVRE'S THEOREM

$$z^n = (x + iy)^n = (r \text{ cis } \theta)^n = r^n \text{ cis } n\theta \qquad n \text{ a natural number}$$

A general proof of De Moivre's theorem requires a technique called **mathematical induction,** a topic usually considered in an advanced algebra course.

EXAMPLE 15   Find $(\sqrt{3} + i)^{13}$ and write the answer in the form $a + bi$.

Solution   Write $\sqrt{3} + i$ in polar form, use De Moivre's theorem, and then convert back to rectangular form.

$$(\sqrt{3} + i)^{13} = (2 \text{ cis } 30°)^{13}$$
$$= 2^{13} \text{ cis } 390°$$
$$= 2^{13} \text{ cis } 30° \qquad \text{Why?}$$
$$= 8{,}192(\cos 30° + i \sin 30°)$$
$$= 8{,}192\left(\frac{\sqrt{3}}{2} + i\frac{1}{2}\right)$$
$$= 4{,}096\sqrt{3} + 4{,}096i$$

$Do \longrightarrow$ **PROBLEM 15**   Find $(-1 + i)^6$ and write the answer in the form $a + bi$.

Now let us take a look at roots of complex numbers. We say **w is an nth root of z,** n a natural number, if

$$w^n = z$$

For example, if $w^2 = z$, then w is a square root of z; if $w^3 = z$, then w is a cube root of z; and so on. Let us show that if

$$z = r \text{ cis } \theta \tag{1}$$

then

$$r^{1/2} \text{ cis } \frac{\theta}{2} \tag{2}$$

is a square root of z. We simply square (2), using De Moivre's theorem, to obtain (1):

$$\left(r^{1/2} \text{ cis } \frac{\theta}{2}\right)^2 = (r^{1/2})^2 \text{ cis } 2\left(\frac{\theta}{2}\right)$$
$$= r \text{ cis } \theta$$

We can proceed in the same way to show that $r^{1/n} \text{ cis}(\theta/n)$ is an nth root of $r \text{ cis } \theta$, n a natural number:

$$\left(r^{1/n} \text{ cis } \frac{\theta}{n}\right)^n = (r^{1/n})^n \text{ cis } n\left(\frac{\theta}{n}\right)$$
$$= r \text{ cis } \theta$$

But we can do even better than this. The following theorem shows us how to find *all* the nth roots of a complex number:

---

*n*TH ROOT THEOREM

$$r^{1/n} \text{ cis } \frac{\theta + k360°}{n} \qquad k = 0, 1, \ldots, (n-1)$$

are n distinct nth roots of $z = r \text{ cis } \theta$, and there are no others.

The proof of this theorem is left to Problems 21 and 22, Exercise 7.4.

EXAMPLE 16 Find the six distinct sixth roots of $1 + i\sqrt{3}$ and graph them.

Solution First, write $1 + i\sqrt{3}$ in polar form:

Do

$$1 + i\sqrt{3} = 2 \text{ cis } 60°$$

All six roots are given by

$$2^{1/6} \text{ cis } \frac{60° + k360°}{6} \qquad k = 0, 1, 2, 3, 4, 5$$

Thus,

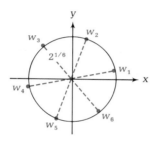

Notice how the six roots
are evenly spaced around
the circle.

$$w_1 = 2^{1/6} \text{ cis } \frac{60° + 0 \cdot 360°}{6} = 2^{1/6} \text{ cis } 10°$$

$$w_2 = 2^{1/6} \text{ cis } \frac{60° + 1 \cdot 360°}{6} = 2^{1/6} \text{ cis } 70°$$

$$w_3 = 2^{1/6} \text{ cis } \frac{60° + 2 \cdot 360°}{6} = 2^{1/6} \text{ cis } 130°$$

$$w_4 = 2^{1/6} \text{ cis } \frac{60° + 3 \cdot 360°}{6} = 2^{1/6} \text{ cis } 190°$$

$$w_5 = 2^{1/6} \text{ cis } \frac{60° + 4 \cdot 360°}{6} = 2^{1/6} \text{ cis } 250°$$

$$w_6 = 2^{1/6} \text{ cis } \frac{60° + 5 \cdot 360°}{6} = 2^{1/6} \text{ cis } 310°$$

PROBLEM 16 Find three distinct cube roots of $1 + i$.

EXAMPLE 17 Solve $x^3 - 1 = 0$. Write final answers in rectangular form.

Solution $x^3 - 1 = 0$

$$x^3 = 1$$

DO ⟶

*Could also do by
using $x = 1$ as one root,
then factoring to find
the others.*

Therefore, x is a cube root of 1, and there are three of them. First, we
write 1 in polar form:

$$1 = 1 + 0i = 1 \text{ cis } 0°$$

All three cube roots of 1 are given by

$$1^{1/3} \text{ cis } \frac{0° + k360°}{3} \qquad k = 0, 1, 2$$

Thus,

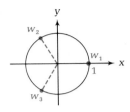

$$w_1 = 1^{1/3} \text{ cis } \frac{0° + 0 \cdot 360°}{3} = 1$$

$$w_2 = 1^{1/3} \text{ cis } \frac{0° + 1 \cdot 360°}{3} = \text{cis } 120° = -\frac{1}{2} + \frac{\sqrt{3}}{2} i$$

$$w_3 = 1^{1/3} \text{ cis } \frac{0° + 2 \cdot 360°}{3} = \text{cis } 240° = -\frac{1}{2} - \frac{\sqrt{3}}{2} i$$

PROBLEM 17   Solve $x^3 + 1 = 0$. Write final answers in rectangular form.

We have only touched on a subject that has far-reaching consequences. The theory of functions of a complex variable provides a powerful tool for engineers, scientists, and mathematicians.

ANSWERS TO   15. $8i$ 16. $w_1 = 2^{1/6} \text{ cis } 15°$, $w_2 = 2^{1/6} \text{ cis } 135°$, $w_3 = 2^{1/6} \text{ cis } 255°$
MATCHED PROBLEMS
17. $w_1 = \frac{1}{2} + \frac{\sqrt{3}}{2}i$, $w_2 = -1$, $w_3 = \frac{1}{2} - \frac{\sqrt{3}}{2}i$

EXERCISE 7.4

A   *Find the value of each using De Moivre's theorem. Leave your answer in polar form.*

1. $(3 \text{ cis } 15°)^3$ 2. $(2 \text{ cis } 30°)^8$ 3. $(\sqrt{2} \text{ cis } 45°)^{10}$
4. $(\sqrt{2} \text{ cis } 60°)^8$ 5. $(\sqrt{3} + i)^6$ 6. $(1 + i\sqrt{3})^3$

B   *Find the value of each using De Moivre's theorem, and write the result in the form $a + bi$.*

7. $(-1 + i)^4$ 8. $(-\sqrt{3} - i)^4$ 9. $(-\sqrt{3} + i)^5$

10. $(1 - i)^8$ 11. $\left(-\frac{1}{2} - \frac{\sqrt{3}}{2}i\right)^3$ 12. $\left(-\frac{1}{2} + \frac{\sqrt{3}}{2}i\right)^3$

For n and z as indicated, find all nth roots of z. Leave answer in polar form.

13. $z = 4 \text{ cis } 30°$, $n = 2$ 14. $z = 16 \text{ cis } 60°$, $n = 2$
15. $z = 8 \text{ cis } 90°$, $n = 3$ 16. $z = 27 \text{ cis } 120°$, $n = 3$
17. $z = -1 + i$, $n = 5$ 18. $z = 1 - i$, $n = 5$
19. $z = 1$, $n = 6$ 20. $z = i$, $n = 3$

C   21.   Show that

$$\left(r^{1/n} \operatorname{cis}\frac{\theta + k360°}{n}\right)^n = r \operatorname{cis} \theta$$

for n any natural number and k any integer.

22.   Show that

$$r^{1/n} \operatorname{cis}\frac{\theta + k360°}{n}$$

is the same number for $k = 0$ and $k = n$.

Solve each equation for all roots. Write final answers in rectangular form, $a + bi$, where a and b are computed to three decimal places. A calculator will be helpful.

23. $x^5 - 1 = 0$ 24. $x^4 + 1 = 0$

25. $x^3 + 5 = 0$ 26. $x^5 - 6 = 0$

EXERCISE 7.5   CHAPTER REVIEW

A   *In Problems 1–3 plot in a polar coordinate system.*

1. $A(5, 210°)$, $B(-7, 180°)$, $C(-5, -45°)$
2. $r = 4 + 4\cos\theta$
3. $r = 8$
4.   Change $(2\sqrt{2}, \pi/4)$ to rectangular coordinates.
5.   Change $(-\sqrt{3}, 1)$ to polar coordinates $(r \geq 0, \quad 0 \leq \theta < 2\pi)$.
6.   Graph $-3 - 2i$ in a rectangular coordinate system.
7.   Graph $z = 5\operatorname{cis}60°$ in a polar coordinate system.
8.   Find $z_1 z_2$ and z_1/z_2 for $z_1 = 9\operatorname{cis}42°$ and $z_2 = 3\operatorname{cis}37°$. Leave answer in polar form.
9.   Find $(2\operatorname{cis}10°)^4$ using De Moivre's theorem.

B   *In Problems 10–12 plot in a polar coordinate system.*

10. $A(-5, \pi/4)$, $B(5, -\pi/3)$, $C(-8, \quad 4\pi/3)$
11. $r = 8\sin 3\theta$
12. $\theta = \pi/6$
13.   Change $8x - y^2 = x^2$ to polar form.
14.   Change $r(3\cos\theta - 2\sin\theta) = -2$ to rectangular form.
15.   Change $r = -3\cos\theta$ to rectangular form.
16.   Convert $-\sqrt{3} - i$ into polar form $(r \geq 0, \quad 0° \leq \theta < 360°)$.
17.   Convert $3\sqrt{2}\operatorname{cis}(3\pi/4)$ to rectangular form.
18.   Convert $(2 + i2\sqrt{3})(-\sqrt{2} + i\sqrt{2})$ to polar form and evaluate.
19.   Divide $(-\sqrt{2} + i\sqrt{2})/(2 + i2\sqrt{3})$ by converting to polar form first.

20. Find $(-1 - i)^4$ using De Moivre's theorem and write the result in the form $a + bi$.

21. Find all cube roots of $-4\sqrt{3} - 4i$.

22. Show that $2 \text{ cis } 30°$ is a square root of $2 + i2\sqrt{3}$

C 23. Graph $r = 6/(2 - \cos\theta), 0 \le \theta \le 2\pi$, using multiples of $\pi/6$ starting at 0.

24. Plot $r = 4 + 4 \cos(\theta/2)$ in a polar coordinate system.

25. Change $r(\sin\theta - 2) = 3$ to rectangular form.

26. Find all roots of $x^3 - 12 = 0$. Write answers in rectangular form, $a + bi$, where a and b are computed to three decimal places (use a calculator).

27. Show that

$$\left(r^{1/3} \text{ cis } \frac{\theta + k360°}{3}\right)^3 = r \text{ cis } \theta \qquad k = 0, 1, 2$$

ADDITIONAL APPLICATIONS

INTRODUCTORY REMARKS

This chapter is a sampler of applications of trigonometry from a variety of fields. You are not expected to become an expert in any of the areas discussed. The purpose of the chapter is to let you see the variety of applications so that you will gain some appreciation of the power and widespread use of trigonometry. The chapter can be covered in several ways. The following are a few suggestions:

SUGGESTIONS FOR USING THIS CHAPTER

1. A brief survey of the material will give you some indication of the variety of applications in which trigonometric functions play a central role.

2. In addition to a brief survey, an occasional pause to work a problem or two will give you a deeper understanding of the trigonometric functions and their uses.

3. A topic of particular interest can serve as a springboard for an expository paper on that subject.

Most of the areas of applications to be considered make use of the following important properties of the sine and cosine functions:

FOR $y = A \sin(Bt + C)$ OR $y = A \cos(Bt + C)$, $B > 0$

$$\text{Amplitude} = |A| \qquad \text{Period} = \frac{2\pi}{B}$$

$$\text{Frequency} = \frac{B}{2\pi} \qquad \text{Period} = \frac{1}{\text{Frequency}}$$

$$\text{Phase shift} = \begin{cases} \left|\dfrac{C}{B}\right| \text{ units to the right} & \text{if } \dfrac{C}{B} < 0 \\[2ex] \dfrac{C}{B} \text{ units to the left} & \text{if } \dfrac{C}{B} > 0 \end{cases}$$

Phenomena that can be described by either of the above equations are said to be **simple harmonic.** All the material in the box should appear familiar. If t is time, then the reciprocal of the period is **frequency.** For example, if a particular sine wave has a period of 1/50 sec, then 1 cycle is completed in 1/50 sec. Thus, 50 cycles are completed in 1 sec, and the wave has a frequency of 50 Hz (cycles per second). The unit Hz is named after the German physicist Heinrich Rudolph Hertz (1857–1894), who was the first to confirm the existence of electromagnetic radiation.

8.2   SOUND WAVES

Sound is produced by a vibrating object, which in turn excites air molecules into motion. The vibrating air molecules cause a periodic change in air pressure that travels through air at about 1,100 ft/sec. (Sound is not transmitted in a vacuum.) When this periodic change in air pressure reaches your eardrum, the drum vibrates at the same frequency as the source, and the vibration is transmitted to the brain as sound.

If in place of an eardrum we use a microphone, then the air disturbance can be changed into a pulsating electrical signal that can be visually displayed on an oscilloscope (Figure 1). A pure tone from a tuning fork will look like a sine curve (also called a **sine wave**). The sound from a tuning fork can be accurately described by either the sine function or cosine function.

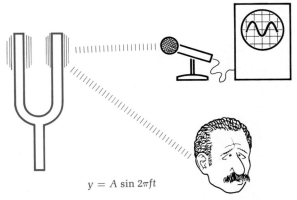

FIGURE 1
A simple sound wave

$$y = A \sin 2\pi ft$$

For example, a tuning fork vibrating at 264 Hz ($f = 264$) with an amplitude of 0.002 in., produces C on the musical scale, and the wave on an oscilloscope can be described by an equation of the form

$$
\begin{aligned}
y &= A \sin 2\pi ft \\
&= 0.002 \sin 2\pi(264)t
\end{aligned}
$$

Most sounds are more complex than that produced by a tuning fork. Figure 2 (page 214) represents a note produced by a guitar. You may be surprised to learn that even these more complex sound forms can be described in terms of simple sine waves by means of a Fourier series (see Section 3.5). Theoretically, any sound can be reproduced by an appropriate combination of pure tones from tuning forks.

EXAMPLE 1   For a sound wave given by $y = 2.7 \sin 1600\pi t$, find the amplitude, period, and frequency if t is time in seconds and y is in centimeters.

Solution $\text{Amplitude} = |2.7| = 2.7 \text{ cm}$ $\text{Period} = \dfrac{2\pi}{B} = \dfrac{2\pi}{1{,}600\pi} = \dfrac{1}{800} \text{ sec}$

$\text{Frequency} = \dfrac{B}{2\pi} = 800 \text{ Hz}$

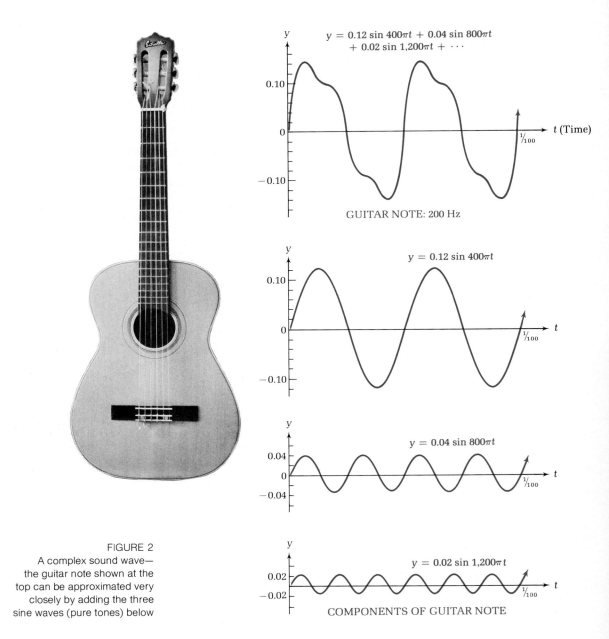

FIGURE 2
A complex sound wave—
the guitar note shown at the
top can be approximated very
closely by adding the three
sine waves (pure tones) below

PROBLEM 1 A sound wave has an amplitude of 1.4 cm and a frequency of 400 Hz. Assuming the wave has an equation of the form $y = A \sin Bt$, find A and B.

ANSWERS TO   1. $A = 1.4$, $B = 800\pi$
MATCHED PROBLEMS

EXERCISE 8.2

1.   If a sound wave is given by $y = 0.004 \sin 1{,}000\pi t$, what is its amplitude, period, and frequency if t is time in seconds and y is in meters?

2.   A sound wave has an amplitude of 0.2 cm and a frequency of 800 Hz. Assuming the wave has an equation of the form $y = A \sin Bt$, find A and B.

3.   Graph the equation in Problem 1 (above) for $0 \le t \le \frac{1}{250}$.

4.   Graph the equation in Problem 2 (above) for $0 \le t \le \frac{1}{400}$.

5.   Graph the sound wave given by $y = 0.06 \sin 400\pi t + 0.03 \sin 800\pi t$ for $0 \le t \le \frac{1}{200}$.

8.3   CREATING ELECTRIC CURRENT

Michael Faraday, a research scientist at the Royal Institute in London, and Joseph Henry, a professor at the Albany Academy in New York, independently discovered that a flow of electricity could be created by moving a wire in a magnetic field. Suppose we bend a wire in the form of a rectangle, and we locate this wire between the south and north poles of magnets, as indicated in Figure 3. We now rotate the wire at a constant counterclockwise speed, starting with BC in its lowest position. As BC turns toward the horizontal, an electrical current will flow from C to B. The strength of the current (measured in amperes) will be 0 at the lowest position, and will increase to a maximum value at the horizontal position. As BC continues to turn from the horizontal to the top position, the current flow decreases to 0. As BC starts down from the top position in a counterclockwise direction, the current flow reverses, going from B to C, and again reaches a maximum at the left horizontal position. When BC moves from this horizontal position

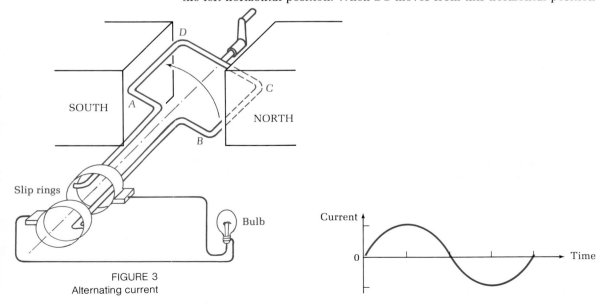

FIGURE 3
Alternating current

back to the original bottom position, the current flow again decreases to 0. This pattern repeats itself for each revolution, and hence is periodic. Is it too much to expect that a trigonometric function can describe the relationship between current and time? If we measure and graph the strength of the current I in the wire relative to time t, we will find that we can indeed find an equation of the form

$$I = A \sin(Bt + C)$$

that will give us the same curve. For example,

$$I = 30 \sin 120\pi t$$

represents a 60 Hz alternating current flow with a maximum value of 30 amperes.

EXAMPLE 2   An alternating current generator produces an electrical current (measured in amperes) that is described by the equation

$$I = 35 \sin(40\pi t - 10\pi)$$

where t is time in seconds. What is the amplitude, frequency, and phase shift for the current?

Solution

$$\overset{A}{} \quad \overset{B}{} \quad \overset{C}{}$$
$$y = 35 \sin(40\pi t - 10\pi)$$

Amplitude $= |35| = 35$ amperes   Frequency $= \dfrac{B}{2\pi} = \dfrac{40\pi}{2\pi} = 20$ Hz

Phase shift $= \left|\dfrac{C}{B}\right| = \dfrac{10\pi}{40\pi} = \dfrac{1}{4}$ sec   (right)

PROBLEM 2   Repeat Example 2 for $I = -25 \sin(30\pi t + 5\pi)$.

**ANSWERS TO
MATCHED PROBLEMS**

2.   Amplitude $= 25$ amperes,   Frequency $= 15$ Hz, Phase shift $= \frac{1}{6}$ sec (left)

EXERCISE 8.3

1.   An alternating current generator produces a current given by

$$I = 10 \sin(120\pi t - 60\pi)$$

where t is in seconds and I is in amperes. What is the amplitude, frequency, and phase shift for this current?

2.   An alternating current generator produces a 60 Hz current flow with a maximum value of 10 amperes. Write an equation of the form $I = A \sin Bt$ for this current.

8.4   SEASONS

What causes seasons? Why is it hotter in the summer and colder in the winter? If the axis of rotation of the earth were exactly perpendicular to the plane of the orbit of the earth around the sun, then there would be no seasons. We obviously have seasons, so there must be a reason. The earth's axis of rotation actually tilts 23½° away from the perpendicular (see Figure 4). It is because of this that sunrise and sunset times change, and we have summer, fall, winter, and spring. Sun rays strike more perpendicularly to the surface of the earth in the northern hemisphere in the summer and at greater angles in the winter. Figure 5 is a plot of sunrise times over a 2 year period. Note that the data are periodic and the graph is approximately a sine curve. Graphs of other seasonal phenomena such as sunset times, average weekly temperatures, and smog levels are also approximately sine curves, and will look very much like Figure 5.

FIGURE 4
Seasons

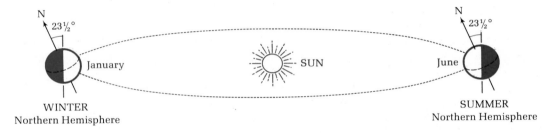

FIGURE 5
Sunrise times

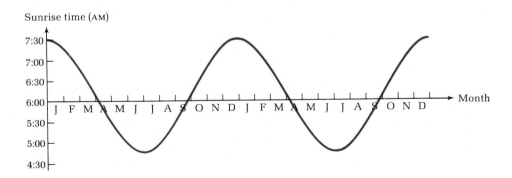

EXERCISE 8.4   **1.**   The table on page 218 gives sunset times for a 1 year interval. These times repeat with a period of *approximately* 1 year. Graph sunset times for a 2 year interval. Indicate the period in months and approximate the amplitude in minutes.

DATE[†]	SUNSET	DATE[†]	SUNSET
1/1	5:02	7/20	7:31
1/21	5:21	8/9	7:13
2/10	5:43	8/29	6:47
3/2	6:05	9/18	6:16
3/22	6:24	10/8	5:45
4/11	6:42	10/28	5:15
5/1	7:01	11/17	4:58
5/21	7:19	12/13	4:51
6/10	7:33	12/27	4:58
6/30	7:38		

[†]From United States Coast and Geodetic Survey sources. Time interval is 20 days.

8.5 WATER WAVES

Water waves are perhaps the most familiar wave form. You can actually see the wave in motion! Water waves are formed by particles of water rotating in circles (Figure 6). A particle actually only moves a short distance as the wave passes through. These wave forms are moving waves, and it can be shown that in their simplest form, they are sine curves. Hence, the waves can be represented by an equation of the form

$$y = A \sin 2\pi \left(ft - \frac{r}{\lambda}\right) \qquad \text{Moving wave equation*} \qquad (1)$$

where f is frequency, t is time, r is distance from the source, and λ is wavelength. Thus, for a wave of a given frequency and wavelength, y is a function of the two variables, t and r.

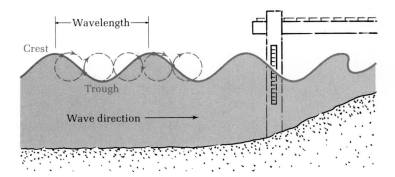

FIGURE 6
Water waves

*Actually, this general wave equation can be used to describe *any* longitudinal (compressional) or any transverse wave motion of one frequency in a nondispersive medium. A sound wave is an example of a longitudinal wave; water waves and electromagnetic waves (to be studied later) are examples of transverse waves.

Equation (1) is typical for moving waves. If a wave passes a pier piling with a vertical scale attached (Figure 6), then the graph of the water level on the scale relative to time would look something like Figure 7a, since r would be fixed. On the other hand, if we actually photograph a wave, freezing the motion in time, then the profile of the wave would look something like Figure 7b.

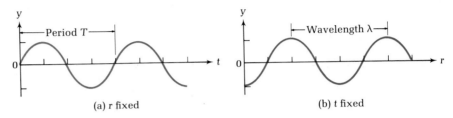

FIGURE 7
Water wave

(a) r fixed

(b) t fixed

Experiments have shown that the **wavelength λ** for water waves is given approximately by

$$\lambda = 5.12T^2 \quad \text{(in feet)}$$

where T is the period of the wave in seconds (see Figure 7a), and the **speed S of the wave** is given approximately by

$$S = \sqrt{\frac{g\lambda}{2\pi}} \quad \text{(in feet per second)}$$

where $g = 32$ ft/sec².

Thus, a wave with a period of 8 sec and an amplitude of 3 ft would have an equation of the form (holding r fixed)

$$y = 3 \sin 2\pi \left(\frac{1}{8}t - \frac{r}{\lambda}\right)$$

$$= 3 \sin \left(\frac{\pi}{4}t + c\right)$$

where c is a constant. Its wavelength would be

$$\lambda = 5.12(8^2) = 328 \text{ ft}$$

and it would move at a speed of approximately

$$S = \sqrt{\frac{32(328)}{2(3.14)}} \approx 41 \text{ ft/sec}$$

or, in miles per hour (mph),

$$(41 \text{ ft/sec}) \frac{3{,}600 \text{ sec/hr}}{5{,}280 \text{ ft/mi}} \approx 28 \text{ mph}$$

EXERCISE 8.5

1. A water wave at a fixed position has an equation of the form

$$y = 15 \sin \frac{\pi}{4}t$$

where t is time in seconds and y is in feet. How high is the wave from trough to crest? (See Figure 6.) What is its wavelength in feet? How fast is it traveling in feet per second?

2. A water wave has an amplitude of 30 ft and a period of 14 sec. If its equation is given by

$$y = A \sin Bt$$

at a fixed position, find A and B. What is the wavelength in feet? How fast is it traveling in feet per second? How high would the wave be from a trough to a crest?

3. Graph the equation in Problem 1 (above) for $0 \le t \le 16$.
4. Graph the equation in Problem 2 (above) for $0 \le t \le 28$.
5. A water wave has an equation of the form

$$y = 25 \sin 2\pi \left(\frac{t}{10} - \frac{r}{512} \right)$$

(A)   Graph the equation for a fixed position, say $r = 1{,}024$ ft, for $0 \le t \le 20$.
(B)   Graph the equation for a fixed time, say $t = 0$, for $0 \le r \le 1{,}024$. [Use $\sin(-x) = -\sin x$.]

8.6   FLOATING OBJECTS

Did you know that you can determine the mass of a floating object (Figure 8) simply by making it bob up and down in the water and timing its period of oscillation? If the rest position of the floating object is taken to be 0 and we start counting time as the object passes up through 0 when it is made to oscillate, then its equation of motion can be shown to be (neglecting water and air resistance)

$$y = D \sin \sqrt{\frac{1{,}000gA}{M}}\, t$$

where y and D are in meters, t is in seconds, $g = 9.75$ m/sec^2, A is horizontal cross-sectional area in square meters, and M is mass in kilograms. The equation indicates that the motion is simple harmonic with

$$\text{Amplitude} = |D| \text{ meters} \qquad \text{Period} = \frac{2\pi}{\sqrt{1{,}000gA/M}}$$

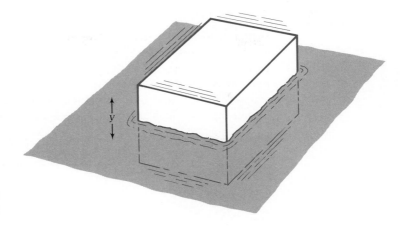

FIGURE 8
Floating object

EXAMPLE 3   A cylindrical buoy with cross-sectional area 1.25 m² is observed (after being pushed) to bob up and down with a period of 0.5 sec. Approximately what is the mass of the buoy in kilograms?

Solution   We use the formula

$$\text{Period} = \frac{2\pi}{\sqrt{1{,}000gA/M}}$$

with $\pi = 3.14$, $g = 9.75$ m/sec², $A = 1.25$ m², and Period $= 0.5$ second, and solve for M. Thus,

$$0.5 = \frac{2(3.14)}{\sqrt{(1{,}000)(9.75)(1.25)/M}}$$

$$M \approx 77 \text{ kg}$$

PROBLEM 3   Write the equation of motion of the buoy in Example 3 in the form $y = D \sin Bt$, assuming its amplitude is 0.4 m.

ANSWERS TO MATCHED PROBLEMS

3. $y = 0.4 \sin 4\pi t$

EXERCISE 8.6

1.   A 3 m × 3 m × 1 m swimming float in the shape of a rectangular solid is observed to bob up and down with a period of 1 sec. Approximately what is the mass of the float in kilograms?

2.   Write an equation of motion for the float in Problem 1 (above) in the form $y = D \sin Bt$, assuming the amplitude of the motion is 0.2 m.

3.   Graph the equation found in Problem 2 (above) for $0 \le t \le 2$.

8.7 LIGHT AND OTHER ELECTROMAGNETIC WAVES

Visible light is a transverse wave form with a frequency range between 4×10^{14} Hz (red) and 7×10^{14} Hz (violet). The retina of the eye responds to these vibrations, and through a complicated chemical process, the vibrations are eventually perceived by the brain as light in various colors. Light is actually a small part of a continuous spectrum of electromagnetic wave forms—most of which are not visible (see Figure 9). Included in the spectrum

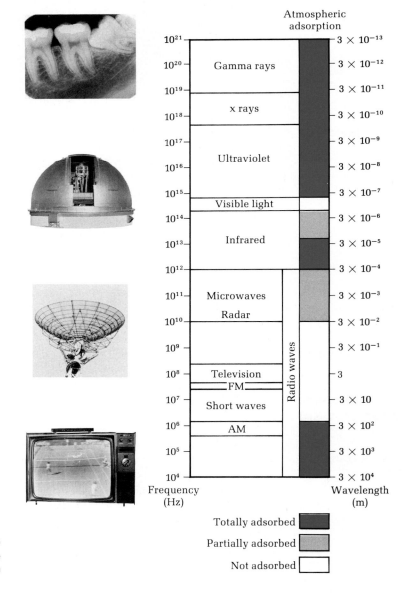

FIGURE 9
Electromagnetic wave
spectrum

are radio waves (AM and FM), microwaves, x rays, and gamma rays. All these waves travel at the speed of light, 3×10^{10} cm/sec (186,000 mph) and many of them are either partially or totally adsorbed in the atmosphere of the earth, as indicated in Figure 9. Their distinguishing characteristics are wavelength and frequency, which are related by the formula

$$\lambda \nu = c \tag{1}$$

where c is speed of light, λ is wavelength, and ν is frequency.

Electromagnetic waves are traveling waves that can be described by an equation of the form

$$E = A \sin 2\pi \left(\nu t - \frac{r}{\lambda} \right) \qquad \text{Traveling wave equation} \tag{2}$$

where t is time and r is the distance from the source. This is a function of two variables, t and r. If we freeze time, then the graph of (2) looks something like Figure 10a. If we look at the electromagnetic field at a single point in space, that is, if we hold r fixed, then the graph of (2) looks something like Figure 10b.

FIGURE 10
Electromagnetic wave

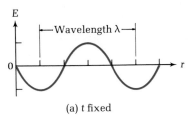

(a) t fixed

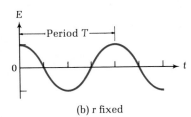

(b) r fixed

Electromagnetic waves are produced by exciting electrons into oscillatory motion. This creates a combination of electric and magnetic fields that moves through space at the speed of light. The frequency of the oscillation of the electron determines the nature of the wave (see Figure 9), and a receiver responds to the wave through induced oscillation of the same frequency (see Figure 11).

FIGURE 11
Electromagnetic field

Electron   Electron

Figures 12 and 13 on page 224 illustrate FM, AM, and radar waves at fixed points in space.

Intensity

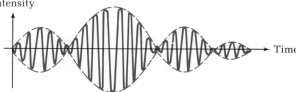

(a) FM—a frequency-modulated carrier wave
with constant amplitude and variable frequency

Intensity

FIGURE 12

(b) AM—an amplitude-modulated carrier wave
with constant frequency and variable amplitude

Intensity

FIGURE 13
Radar pulses

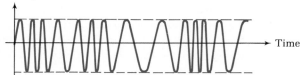

EXAMPLE 4   If an electromagnetic wave has a frequency of $\nu = 10^{12}$ Hz, what is its period? What is its wavelength in meters?

Solution $$\text{Period} = \frac{1}{\nu} = \frac{1}{10^{12}} = 10^{-12} \text{ sec}$$

To find the wavelength λ, we use the formula

$$\lambda\nu = c$$

with the speed of light, c, in metric units, given by

$$c \approx 3 \times 10^8 \text{ m/sec}$$

$$\lambda = \frac{3 \times 10^8 \text{ m/sec}}{10^{12} \text{ Hz}} = 3 \times 10^{-4} \text{ m}$$

PROBLEM 4   Repeat Example 4 for $\nu = 10^6$ Hz.

EXAMPLE 5   An ultraviolet wave has an equation of the form

$$y = A \sin Bt$$

Find B if the wavelength is $\lambda = 3 \times 10^{-9}$ m.

Solution   We first use $\lambda \nu = c$ ($c \approx 3 \times 10^8$ m/sec) to find the frequency and then use $\nu = B/2\pi$ to find B.

$$\nu = \frac{c}{\lambda} = \frac{3 \times 10^8 \text{ m/sec}}{3 \times 10^{-9} \text{ m/sec}} = 10^{17} \text{ Hz}$$

$$B = 2\pi\nu = 2\pi \times 10^{17}$$

PROBLEM 5   A gamma ray has an equation of the form $y = A \sin Bt$. Find B if the wavelength is $\lambda = 3 \times 10^{-12}$ m.

ANSWERS TO   **4.**   Period $= 10^{-6}$ sec; $\lambda = 300$ m   **5.** $B = 2\pi \times 10^{20}$
MATCHED PROBLEMS

EXERCISE 8.7   **1.**   Suppose an electron oscillates at 10^8 Hz ($\nu = 10^8$ Hz) creating an electromagnetic wave. What is its period? What is its wavelength in meters?
2.   Repeat Problem 1 (above) with $\nu = 10^{18}$ Hz.
3.   An x ray has an equation of the form

$$y = A \sin Bt$$

Find B if the wavelength of the x ray, λ, is 3×10^{-10} m.   [*Note:* $c \approx 3 \times 10^8$ m/sec.]
4.   A microwave has an equation of the form

$$y = A \sin Bt$$

Find B if the wavelength is $\lambda = 0.003$ m.
5.   An AM radio wave for a given station has an equation of the form

$$y = A(1 + 0.02 \sin 2\pi \cdot 1{,}200t) \sin 2\pi \cdot 10^5 t$$

for a given 1,200 Hz tone. The expression in parentheses modulates the amplitude A of the carrier wave $y = A \sin 2\pi \cdot 10^5 t$ (see Figure 12b). What is the period and frequency of the carrier wave for time t in seconds? Can a wave with this frequency pass through the atmosphere? (See Figure 9.)

6.   An FM radio wave for a given station has an equation of the form

$$y = A \sin(2\pi \cdot 10^8 t + 0.02 \sin 2\pi \cdot 1{,}200t)$$

for a given tone of 1,200 Hz. The second term within the parentheses modulates the frequency of the carrier wave $y = A \sin 2\pi \cdot 10^8 t$ (see Figure 12a). What is the period and frequency of the carrier wave? Can a wave with this frequency pass through the atmosphere? (See Figure 9.)

8.8   BOW WAVES

BOATS   A boat moving at a constant rate, faster than the water waves it produces, generates a **bow wave** that extends back from the bow of the boat at a given angle (Figure 14). If we know the speed of the boat and the speed of the waves produced by the boat, then we can determine the angle of the bow wave. Actually, if we know any two of these quantities, we can always find the third. Surprisingly, the solution to this problem can be applied to sonic booms and high-energy particle physics, as we will see below.

Referring to Figure 15, we reason as follows: When the boat is at P_1, the water wave it produces will radiate out in a circle, and by the time the boat reaches P_2, the wave will have moved a distance of r_1—which is less than the distance between P_1 and P_2, since the boat is assumed to be traveling faster than the wave. By the time the boat reaches the apex position B in Figure 15, the wave motion at P_2 will have moved r_2 units, and the wave motion at P_1 will have continued on out to r_3. Because of the constant speed of the boat and the constant speed of the wave motion, these circles of wave radiation will all have a common tangent that passes through the boat. Of

FIGURE 14
Bow waves of boats, sonic booms, and high-energy physics are related in a curious way; this section explains how·

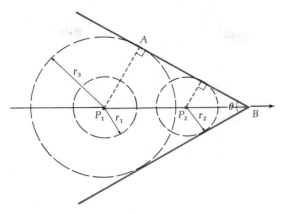

FIGURE 15

course, the motion of the boat is continuous, and what we have said about P_1 and P_2 applies to all points along the path of the boat. The result of this phenomenon is the clearly visible wave fronts produced by the bow of a boat. Refer again to Figure 15, and you will see that the boat travels from P_1 to B in the same time t that the bow wave travels from P_1 to A; hence, if S_b is the speed of the boat and S_w is the speed of the bow wave, then (using $d = rt$)

Distance from P_1 to $A = S_w t$ Distance from P_1 to $B = S_b t$

and, since triangle P_1BA is a right triangle,

$$\sin \frac{\theta}{2} = \frac{S_w t}{S_b t} = \frac{S_w}{S_b}$$

Thus,

$$\sin \frac{\theta}{2} = \frac{S_w}{S_b}$$

where S_w is the speed of the bow wave, S_b is the speed of the boat, and $S_b > S_w$.

SONIC BOOMS

Following exactly the same line of reasoning as above, an aircraft flying faster than the speed of sound produces sound waves that pile up behind the aircraft in the form of a cone (see Figure 16 on page 228). The cone intersects the ground in the form of a hyperbola, and along this curve we experience a phenomenon called a **sonic boom.**

As in the case of the boat, we have

$$\sin \frac{\theta}{2} = \frac{S_s}{S_a}$$

where S_s is the speed of the sound wave, S_a is the speed of the aircraft, and $S_a > S_s$.

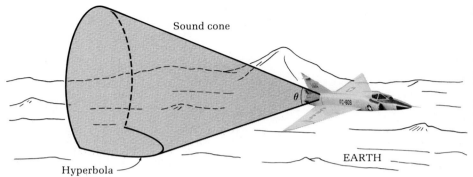

FIGURE 16
Sound cones and sonic
booms

HIGH-ENERGY PHYSICS

In certain materials, such as glass, nuclear particles can be made to move faster than the velocity of light in glass. In 1958 three physicists (Cerenkov, Frank, and Tamm) jointly received a Nobel prize for the work they did based on this fact. Interestingly, the bow wave analysis for boats applies equally well here. Instead of a sound cone as in Figure 16, we obtain a light cone as shown in Figure 17. By measuring the cone angle θ, Cerenkov, Frank, and Tamm were able to determine the speed of the particle (because the speed of light in glass is readily determined). They used the formula

$$\sin \frac{\theta}{2} = \frac{S_\ell}{S_p}$$

where S_ℓ is the speed of light in glass, S_p is the speed of the particle, and $S_p > S_\ell$. By determining the speed of the particle, they were then able to determine its energy by routine procedures.

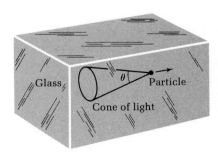

FIGURE 17
Particle energy

EXAMPLE 6   If a speed boat travels at 40 km/hr and the angle between the bow waves is 70°, how fast is the bow wave traveling?

Solution   We use

$$\sin\frac{\theta}{2} = \frac{S_w}{S_b}$$

where $\theta = 70°$ and $S_b = 40$ km/hr. Then we solve for S_w:

$$S_w = (40\ \text{km/hr})(\sin 35°) \approx 23\ \text{km/hr}$$

PROBLEM 6   If a sound wave produced by a supersonic aircraft travels at 1,210 km/hr and the angle of the sound cone is 74.5°, how fast is the aircraft flying?

ANSWERS TO   **6.** 2,000 km/hr
MATCHED PROBLEMS

EXERCISE 8.8   **1.** If the bow waves of a boat travel at 20 km/hr and create an angle of 60°, how fast is the boat traveling?
2. A boat traveling at 55 km/hr produces bow waves that separate at an angle of 54°. How fast is the bow wave traveling?
3. If a supersonic aircraft flies at twice the speed of sound in air, what will the cone angle be?
4. If a supersonic aircraft flies at three times the speed of sound in air, what will the cone angle be?
5. In crown glass, light travels at 2×10^{10} cm/sec. If a high-energy particle passing through this glass creates a light cone of 30°, how fast is it traveling?

8.9   ANGULAR VELOCITY

We will put our notion of radian measure to use in defining two special kinds of velocity that involve rotating motion. These velocities, called **angular velocity** and **linear velocity on a circle,** are used extensively in engineering and physics.

Starting with θ in radian measure, recall (Section 2.2) that

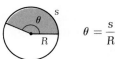 $\theta = \dfrac{s}{R}$

We rewrite the formula in the equivalent form

$$s = R\theta \tag{1}$$

If we think of a point moving on the circumference of the circle with uniform speed, then a radial line from the center of the circle to this point will sweep out an angle θ at a uniform rate. Thus, if we divide both sides of equation (1) by time t, we obtain

$$\frac{s}{t} = R\frac{\theta}{t} \qquad (2)$$

Now,

$$\frac{s}{t} = \text{Change in arc length per unit change in time}$$

$$= \text{Linear velocity of point on the circle}$$

$$= V$$

and

$$\frac{\theta}{t} = \text{Change in angle in radian measure per unit change in time}$$

$$= \text{Angular velocity}$$

$$= \omega$$

Making these substitutions in (2), we obtain the following important formulas:

LINEAR AND ANGULAR VELOCITY ON A CIRCLE
$V = R\omega$ $\qquad$ V is linear velocity on a circle
$\omega = \dfrac{V}{R}$ $\qquad$ ω is angular velocity (radians per unit time)

EXAMPLE 7 If a 6 m diameter wheel turns at 4 rad/sec, what is the velocity of a point on the wheel in meters per second?

Solution $\qquad V = R\omega$

$\qquad\qquad = 3(4) = 12$ m/sec

PROBLEM 7 If a 3 ft diameter wheel turns at 10 rad/min, what is the velocity of a point on the wheel in feet per minute?

EXAMPLE 8 A point on the rim of a 6 in. diameter wheel is traveling at 75 ft/sec. What is the angular velocity of the wheel in radians per second?

Solution

$$\omega = \frac{V}{R}$$

$$= \frac{75}{0.25} = 300 \text{ rad/sec}$$

[*Note:*   3 in. = 0.25 ft.]

PROBLEM 8   A point on the rim of a 4 in. diameter wheel is traveling at 88 ft/sec. What is the angular velocity of the wheel in radians per second?

EXAMPLE 9   If a 6 cm diameter shaft is rotating at 4,000 rpm (revolutions per minute), what is the speed of a particle on its surface in centimeters per minute?

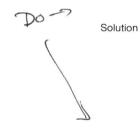

Solution   Since 1 revolution is equivalent to 2π rad, we multiply 4,000 by 2π to obtain the angular velocity of the shaft in radians per minute:

$$\omega = 8{,}000\pi \text{ rad/min}$$

Now we use $V = R\omega$ to complete the solution:

$$V = 3(8{,}000\pi) = 24{,}000\pi \text{ cm/min}$$

PROBLEM 9   If an 8 cm diameter drive shaft in a boat is rotating at 350 rpm, what is the speed of a particle on its surface in centimeters per second?

ANSWERS TO   **7.**   15 ft/min   **8.**   528 rad/sec   **9.**   147 cm/sec
MATCHED PROBLEMS

EXERCISE 8.9   *Find the velocity V of a point on the rim of a wheel given the indicated information.*

1. $R = 6$ mm, $\omega = 0.5$ rad/sec   **2.** $R = 4{,}000$ km, $\omega = 0.04$ rad/hr

Find the angular velocity ω given the following information:

3. $R = 6$ cm, $V = 102$ cm/sec   **4.** $R = 250$ km, $V = 500$ km/hr

5.   A 16 mm diameter shaft rotates at 1,500 rps (revolutions per second). What is the speed of a particle on its surface in meters per second?

6.   A 6 cm diameter shaft rotates at 500 rps. What is the speed of a particle on its surface in meters per second?

7.   An earth satellite travels in a circular orbit at 20,000 mph. If the radius of the orbit is 4,300 miles, what angular velocity (in radians per hour) is generated?

8.   A bicycle is ridden at a speed of 25.2 km/hr (7 m/sec). If the wheel diameter is 70 cm, what is the angular velocity in rad/sec?

9. The velocity of sound in air is approximately 335.3 m/sec. If an airplane has a 3 m diameter propeller, at what angular velocity (in radians per second) will its tip pass through the sound barrier? Transform your answer into revolutions per second.

10. If an electron in an atom travels around the nucleus at 8.11×10^6 cm/sec (see the figure), what angular velocity (in radians per second) does it generate, assuming the radius of the orbit is 5×10^{-9} cm?

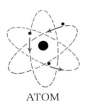

ATOM

11. The earth revolves about the sun in an orbit that is approximately circular with a radius of 9.3×10^7 mi (see the figure). The radius of the orbit sweeps out an angle with what angular velocity in radians per hour? How fast (in miles per hour) is the earth traveling along its orbit?

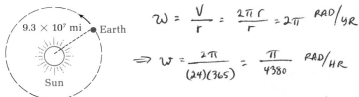

$$\omega = \frac{V}{r} = \frac{2\pi r}{r} = 2\pi \text{ RAD}/\text{YR}$$

$$\Rightarrow \omega = \frac{2\pi}{(24)(365)} = \frac{\pi}{4380} \text{ RAD}/\text{HR}$$

VELOCITY OF THE EARTH $\Rightarrow V = \left(\frac{\pi}{4380}\right)(9.3 \times 10^7) \text{ mi}/\text{HR}$

12. Take into consideration only the daily rotation of the earth to find out how fast (in miles per hour) a person would be moving on the equator. The radius of the earth is approximately 4,000 mi. [*Hint:* Find the angular velocity first.]

8.10   LIGHT WAVES AND REFRACTION

Did you ever look at a pencil in a glass of water or poke a straight pole into a clear pool of water? The objects appear to bend at the surface (Figure 18). This bending phenomenon is caused by refracted light: When light waves pass from one medium to another of less or greater density, they bend. The reason for this is that light waves behave according to Fermat's least-time principle, that is, they follow the path from A to B that requires the least amount of time. This can be shown to be a bent path as shown in Figure 19.

In physics it is shown that

$$\frac{c_1}{c_2} = \frac{\sin \alpha}{\sin \beta} \tag{1}$$

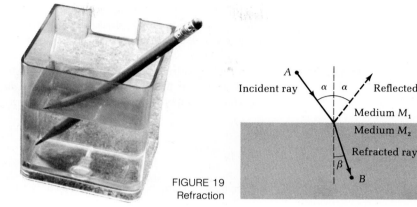

FIGURE 18
One or two pencils? There is
actually only one.

FIGURE 19
Refraction

where c_1 is the speed of light in medium M_1, c_2 is the speed of light in medium M_2, and α and β are as indicated in Figure 19.

A more convenient form of equation (1) uses the notion of **index of refraction,** which is the ratio of the speed of light in a vacuum to the speed of light in a given substance:

$$\text{Index of refraction, } n = \frac{\text{Speed of light in a vacuum}}{\text{Speed of light in a substance}}$$

If we let c represent the speed of light in a vacuum, then

$$\frac{c_1}{c_2} = \frac{c_1/c}{c_2/c} = \frac{1/n_1}{1/n_2} = \frac{n_2}{n_1} \qquad (2)$$

where n_1 is the index of refraction in medium M_1, and n_2 is the index of refraction in medium M_2. (The index of refraction of a substance is the property that is generally tabulated, rather than the velocity of light in that substance.) Substituting (2) into (1), we obtain **Snell's law:**

$$\frac{n_2}{n_1} = \frac{\sin \alpha}{\sin \beta} \qquad (3)$$

[*Note:* n_2 is on top and n_1 is on the bottom in (3), while c_1 is on the top and c_2 is on the bottom in (1).]

For any two given substances the ratio n_2/n_1 is a constant. Thus, equation (3) is equivalent to

$$\frac{\sin \alpha}{\sin \beta} = \text{Constant}$$

The discovery that the sines of the angles of incidence and refraction are in a constant ratio to each other has been attributed to Willegrord Snell, a Dutch astronomer and mathematician (1591–1626), though there is now some doubt that he actually made the discovery. In any case, we will yield to common usage and will continue to refer to the law as "Snell's law."

EXAMPLE 10   A spotlight shining on a pond strikes the water so that the angle of incidence α in Figure 19 is 23.5°. Find the refracted angle β. Use Snell's law and the fact that $n = 1.33$ for water and $n = 1.00$ for air.

Solution   Use

$$\frac{n_2}{n_1} = \frac{\sin \alpha}{\sin \beta}$$

where $n_2 = 1.33$, $n_1 = 1.00$, and $\alpha = 23.5°$, and solve for β.

$$\frac{1.33}{1.00} = \frac{\sin 23.5°}{\sin \beta}$$

$$\sin \beta = \frac{\sin 23.5°}{1.33}$$

$$\beta = \text{Sin}^{-1}\left(\frac{\sin 23.5°}{1.33}\right) = 17.4°$$

PROBLEM 10   Repeat Example 10 with $\alpha = 18.4°$

The fact that light bends when passing from one medium into another is what makes telescopes, microscopes, and cameras possible. It is the carefully controlled bending of light rays that produces the useful results in optical instruments. Figures 20–22 illustrate a variety of phenomena connected with refracted light.

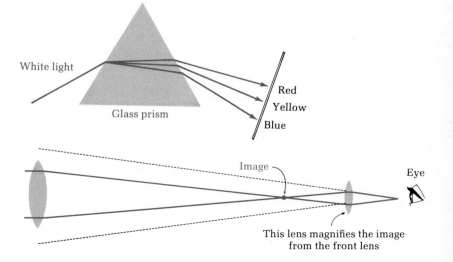

FIGURE 20
Light spectrum—different wavelengths, different index of refraction, different colors

White light

Glass prism

Red
Yellow
Blue

FIGURE 21
Telescope optics

Image

Eye

This lens magnifies the image from the front lens

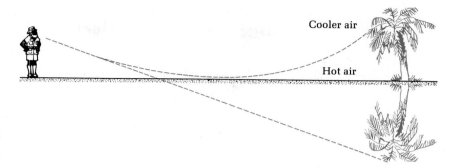

FIGURE 22
Mirage—light is refracted and
bent to appear as if reflected
off water

Table 1 includes refractive indexes for several common materials.

TABLE 1   REFRACTIVE INDEXES

Material	Refractive Index
Air	1.0003
Crown glass	1.52
Diamond	2.42
Flint glass	1.66
Ice	1.31
Water	1.33

EXAMPLE 11   If an underwater flashlight is directed toward the surface of a swimming pool, at what angle of incidence α will the light beam be totally reflected? (See figure.)

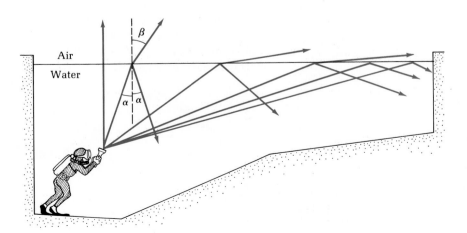

Solution   The index of refraction for water is $n_1 = 1.33$ and that for air is $n_2 = 1.00$. Find the angle of incidence α in the figure such that the angle of refraction β is 90°.

$$\frac{\sin \alpha}{\sin \beta} = \frac{n_2}{n_1}$$

$$\sin \alpha = \frac{1.00}{1.33} \sin 90°$$

$$\sin \alpha = \frac{1.00}{1.33} \quad (1)$$

$$\alpha = \operatorname{Sin}^{-1} \frac{1.00}{1.33} = 48.8°$$

Thus, the light will be totally reflected if $\alpha \geq 48.8°$ (no light will be transmitted through the surface).

PROBLEM 11   Show that the light beam passing through the flint glass prism is totally reflected off of the slanted surface.

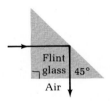

In general, the angle of incidence α such that the angle of refraction β is 90° is called the **critical angle.** For any angle of incidence larger than the critical angle, a light ray will be totally reflected. This critical angle (for total reflection) is important in the design of many optical instruments (such as binoculars) and is at the heart of the emerging science of **fiber optics.** A small diameter glass fiber bent in a curve as shown in Figure 23 will "trap" a light ray entering one end and the ray will be totally reflected and will emerge out of the other end.

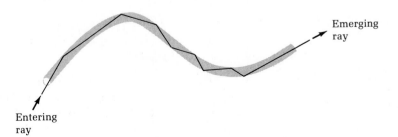

FIGURE 23
Fiber optics

Emerging ray

Entering ray

Important uses of fiber optics are found in medicine and communications. Physicians use fiber optic instruments to see inside of functioning organs. Communication companies are now using fiber optic cables to transmit signals. And this is just a beginning!

ANSWERS TO MATCHED PROBLEMS

10. 13.7°   **11.**

$$\frac{\sin 45°}{\sin \beta} = \frac{1.00}{1.66}$$
$$\sin \beta = 1.66 \sin 45°$$
$$\sin \beta = 1.17$$

Since the maximum value of sine is 1, the condition $\sin \beta = 1.17$ cannot physically happen! In other words, the light must be totally reflected as indicated in the figure.

EXERCISE 8.10

1. A light ray passing through air strikes the surface of a pool of water so that the angle of incidence $\alpha = 40.6°$. Find the angle of refraction β.

2. Repeat Problem 1 above with $\alpha = 34.2°$.

3. A light ray from an underwater spotlight passes through a porthole (of flint glass) of a sunken ocean liner. If the angle of incidence α is 32.0°, what is the angle of refraction β?

4. Repeat Problem 3 above with $\alpha = 45.0°$.

5. If light inside a diamond strikes one of its facets (flat surfaces), what is the critical angle of incidence α for total reflection? (The diamond is surrounded by air.)

6. If light inside a triangular flint glass prism strikes one of the flat surfaces, what is the critical angle of incidence α for total reflection? (The prism is surrounded by air.)

7. An Eskimo hunter standing on an ice floe is about to spear a seal under the water. Should he aim high or low? Study Figure 19 to determine where the seal will appear to be relative to its actual position.

*8. In an attempt to measure the speed of light in a sample of glass to be used for a lens, we send a light ray from a vacuum into the glass and measure angles α and β (Figure 19) to be 30.00° and 18.22°, respectively. If light travels 3.00×10^8 m/sec in a vacuum, how fast will it travel in the glass?

8.11   SPRING–MASS SYSTEMS

An object of mass M hanging on a spring will produce simple harmonic motion when pulled down and released (if we neglect friction and air resistance; see Figure 24).

FIGURE 24

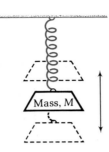

FIGURE 25

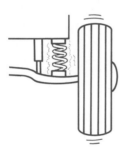

Spring only
Simple harmonic motion (neglect friction and air resistance)

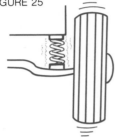
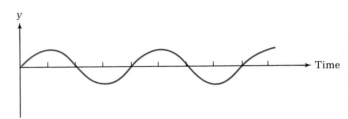

Spring and shock absorber
Damped harmonic motion

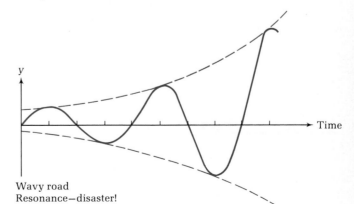

Wavy road
Resonance—disaster!

Figure 25 illustrates simple harmonic motion, damped harmonic motion, and resonance. **Damped harmonic motion** occurs when amplitude decreases to 0 as time increases. **Resonance** occurs when amplitude increases without bound as time increases. Damped harmonic motion is essential in the design of suspension systems for cars, buses, trains, and motorcycles, as well as in the design of buildings, bridges, and aircraft. Resonance is useful in some electric circuits, but it is disastrous in bridges, buildings, and aircraft. Commercial jets have lost wings in flight, and large bridges have collapsed because of resonance. Marching soldiers must break step while crossing a bridge in order to avoid the creation of resonance—and the collapse of the bridge!

EXAMPLE 12   Graph

$$y = \frac{1}{t} \sin \frac{\pi}{2} t \qquad 1 \le t \le 8$$

Solution   The $1/t$ factor in front of $\sin(\pi/2)t$ affects the amplitude. Since the maximum and minimum that $\sin(\pi/2)t$ can assume are 1 and -1, respectively, if we graph $y = 1/t$ first ($1 \le t \le 8$) and reflect the graph across the t axis, we will have upper and lower bounds for the graph of $y = (1/t)\sin(\pi/2)t$. Of course, $(1/t)\sin(\pi/2)t$ will still be zero when $\sin(\pi/2)t$ is zero.

Step 1   Graph $y = 1/t$ and its reflection [called the **envelope** for the graph of $y = (1/t)\sin(\pi/2)t$].

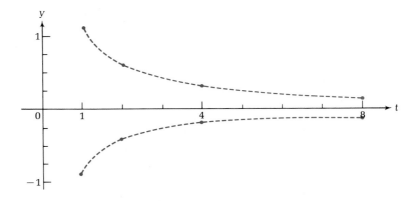

Step 2   Now sketch the graph of $y = \sin(\pi/2)t$, but keep high and low points within the envelope, as shown on page 240.

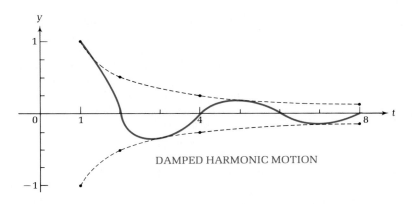

DAMPED HARMONIC MOTION

PROBLEM 12   Graph

$$y = t \sin \frac{\pi}{2} t \qquad 1 \le t \le 8$$

ANSWERS TO MATCHED PROBLEMS   12.

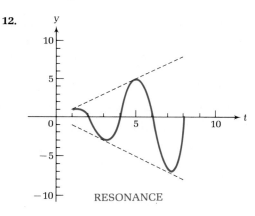

RESONANCE

EXERCISE 8.11

1. Graph $y = \sin 2\pi t, \quad 0 \le t \le 2$.

2. Graph $y = \frac{1}{t} \sin 2\pi t, \quad \frac{1}{4} \le t \le 2$.

3. Graph $y = t \sin 2\pi t, \quad 0 \le t \le 2$.

4. Which form of oscillatory motion (simple harmonic motion, damped harmonic motion, or resonance) is illustrated by each equation in Problems 1, 2, and 3?

8.12   EQUILIBRIUM AND VECTORS

STATIC EQUILIBRIUM

We will now see how horizontal and vertical components of force vectors (see Section 6.3) can be used to solve certain types of physics–engineering problems. We start with two basic ideas regarding forces and objects subjected to the forces:

STATIC EQUILIBRIUM

1. A body at rest is said to be in static equilibrium.
2. For a body to remain in static equilibrium (to remain at rest) in a plane, it is necessary that the sum of the horizontal components and the sum of the vertical components of all forces acting on the body both be zero.

EXAMPLE 13

Two skiers on a chair lift deflect the cable relative to the horizontal as indicated in the figure below. If the skiers and chair weigh 320 lb (neglect cable weight), what is the tension in each cable running to each tower?

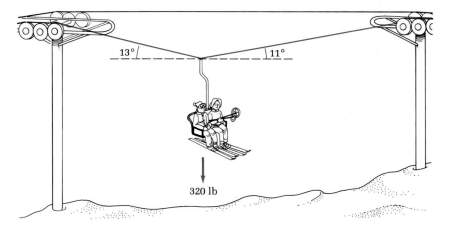

Solution

We introduce geometric vectors for each force involved and draw a force diagram.

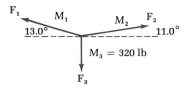

Let M_1 be the magnitude of the tension in the cable on the left, M_2 the magnitude of the tension in the cable on the right, and M_3 the magnitude of the downward force. Then:

Magnitude of:

Horizontal component of $\mathbf{F_1} = M_1 \cos 13°$
Horizontal component of $\mathbf{F_2} = M_2 \cos 11°$

$$\text{Vertical component of } \mathbf{F_1} = M_1 \sin 13°$$
$$\text{Vertical component of } \mathbf{F_2} = M_2 \sin 11°$$

For the system to be in static equilibrium, we must have

$$-M_1 \cos 13° + M_2 \cos 11° = 0$$
$$M_1 \sin 13° + M_2 \sin 11° = 320$$

Solving this system of two equations and two unknowns, we find that

$$M_1 = 772 \text{ lb} \qquad \text{and} \qquad M_2 = 767 \text{ lb}$$

PROBLEM 13   Repeat Example 13 with 13.0° replaced with 14.5°, 11.0° replaced with 9.5°, and 320 lb replaced with 435 lb.

CENTRIFUGAL FORCE   Ideally, a racing car driving around a curve will not slide sideways if the track is banked an appropriate amount. How much should a track be banked for a given speed v and a given curve of radius r to eliminate sideway forces? Figure 26 shows the relevant forces. In order for there to be no sideway forces, the components of the forces parallel to the track surface due to the mass of the car and its movement around the curve (centrifugal force) must be zero. That is,

$$mg \sin \theta = \frac{mv^2}{r} \cos \theta$$

$$\frac{\sin \theta}{\cos \theta} = \frac{mv^2}{mgr}$$

$$\tan \theta = \frac{v^2}{gr}$$

$$\theta = \text{Tan}^{-1} \frac{v^2}{gr} \qquad \begin{array}{l} v \text{ in m/sec} \\ r \text{ in m} \\ g = 9.81 \text{ m/sec}^2 \end{array}$$

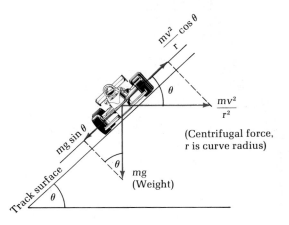

FIGURE 26

EXAMPLE 14   At what angle must a track be banked at a curve to eliminate sideway forces parallel to the track, given that the curve has a radius of 525 m and the racing car is moving at 85 m/sec (about 190 mi/hr)?

Solution
$$\theta = \text{Tan}^{-1}\frac{v^2}{gr}$$

$$= \text{Tan}^{-1}\frac{(85 \text{ m/sec})^2}{(9.81 \text{ m/sec}^2)(525 \text{ m})}$$

$$= 54.5°$$

PROBLEM 14   Repeat Example 14 with a car moving at 25 m/sec (about 56 mi/hr) around a curve with a radius of 125 m.

ANSWERS TO MATCHED PROBLEMS   **13.** $M_1 = 1{,}050$ lb; $M_2 = 1{,}040$ lb   **14.** $\theta = 27.0°$

EXERCISE 8.12   **1.** A tightrope walker weighing 112 lb deflects a rope as indicated in the figure. How much tension is in each part of the rope?

4.2°   5.3°

2. Repeat Problem 1 (above) with the left angle 5.5°, the right angle 6.2°, and the weight of the person 155 lb.

8.13   PSYCHOLOGY–PERCEPTION

An important field of study in psychology has to do with sensory perception—hearing, seeing, smelling, feeling, and tasting. It is well-known that individuals see certain objects differently in different surroundings. Lines that appear to be parallel in one setting may appear to be curved in another. Lines of the same length may appear to have different lengths in two different settings. Is a square always a square? (See Figure 27 on page 244.)

Figure 28 illustrates a perspective illusion in which the people farthest away appear to be larger than those closest—they are actually all the same size.

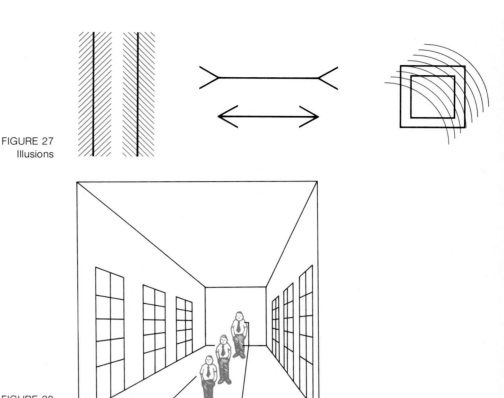

FIGURE 27
Illusions

FIGURE 28
Which person is tallest?

An interesting experiment in visual perception was conducted by psychologists Berliner and Berliner. A tilted field of parallel lines was presented to several subjects, who were then asked to estimate the position of a horizontal line in the field. Berliner and Berliner (*American Journal of Psychology,* vol. 65, pp. 271–277, 1952) reported that most subjects were consistently off, and that the difference in degrees d between their estimates and the actual horizontal could be approximated by the equation

$$d = a + b \sin 4\theta$$

where a and b were constants associated with a particular individual and θ was the angle of tilt of the visual field in degrees (see Figure 29).

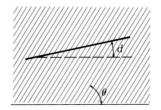

FIGURE 29
Visual perception

EXERCISE 8.13

1. Find d to the nearest degree for $a = -2.2$, $b = -4.5$, and $\theta = 30°$.

2. Set up an experiment and try to verify the formula for yourself or a friend. Make the field of parallel lines large and hang it on a wall.

APPENDIX: COMMENTS ON NUMBERS

A.1   SCIENTIFIC NOTATION AND SIGNIFICANT FIGURES

SCIENTIFIC NOTATION   Work in science and engineering often involves the use of very, very large numbers: The distance that light travels in 1 year is called a **light-year.** It is approximately

9,440,000,000,000 kilometers

Very, very small numbers are also used: The mass of a water molecule is approximately

0.000 000 000 000 000 000 000 03 gram

Writing and calculating with numbers of this type in standard decimal notation is troublesome. It is generally better to convert them to **scientific notation;** that is, as the product of a number between 1 and 10 and a power of 10.

EXAMPLE 1   Decimal fractions and scientific notation

$$4 = 4 \times 10^0 \qquad\qquad 0.36 = 3.6 \times 10^{-1}$$
$$63 = 6.3 \times 10 \qquad\qquad 0.0702 = 7.02 \times 10^{-2}$$
$$805 = 8.05 \times 10^2 \qquad\qquad 0.00532 = 5.32 \times 10^{-3}$$
$$3,143 = 3.143 \times 10^3 \qquad\qquad 0.00067 = 6.7 \times 10^{-4}$$
$$43,160,000 = 4.316 \times 10^7 \qquad 0.000\ 000\ 006\ 43 = 6.43 \times 10^{-9}$$

Can you discover a simple mechanical rule that relates the number of decimal places the decimal is moved with the power of 10 that is used?

PROBLEM 1   Write in scientific notation:

(A)   450   (B)   360,000   (C)   0.0372   (D)   0.000 001 43

SIGNIFICANT FIGURES   Most calculations involving problems of the real world deal with figures that are only approximate. It would therefore seem reasonable to assume that a final answer could not be any more accurate than the least accurate figure used in the calculation. This is an important point, since calculators tend to give the impression that greater accuracy is achieved than is warranted.

Suppose we wish to compute the diagonal of a rectangular field from measurements of its sides of 118.2 m and 73.4 m. Using the Pythagorean theorem and a calculator, we find

$$d^2 = 118.2^2 + 73.4^2$$
$$= 19,358.8$$
$$d = \sqrt{19,358.8}$$
$$= 139.1359048$$

The calculator answer suggests an accuracy that is not justified. What accuracy is justified? To answer this question, we introduce the idea of **significant figures.**

The measurement 73.4 m indicates that the measurement was made to the nearest tenth of a meter; that is, the actual width is between 73.35 m and 73.45 m. The number 73.4 has three significant figures. If we had written, instead, 73.40 m as the width, then the actual width would be between 73.395 m and 73.405 m, and our measurement, 73.40 m, would have four significant figures. In general:

The number of significant figures in a number is found by counting the digits from left to right, starting with the first nonzero digit and ending with the last digit present.

We underline the significant digits in the following numbers:

<u>203.15</u>   <u>4,035</u>   <u>73.40</u>   0.00<u>630</u>   0.00<u>5</u>

Note that the number of significant figures does not depend on the position of the decimal point. The definition above takes care of all cases except one. Consider, for example, the number

3,500

It is not clear whether the number has been rounded to the hundreds place, the tens place, or the units place. This ambiguity can be resolved by writing this type of number in scientific notation. Thus,

3.5×10^3 would have two significant figures

3.50×10^3 would have three significant figures

3.500×10^3 would have four significant figures

All three are equal to 3,500 when written without powers of 10.

In calculations involving multiplication, division, powers, and roots we will adopt the following convention:

We will round off the answer to match the number of significant figures in the number with the least number of significant figures used in the calculation.

Thus, in computing the diagonal of the field above, we would write

$d = 139$ m   Three significant figures

since the two numbers used in the calculation have three and four significant figures, respectively.

One final note: In rounding a number that is exactly halfway between a larger and smaller number, we will use the convention of making the final result even.

EXAMPLE 2　　Round each number to three significant figures:

(A)　3.1495　　　(B)　0.004135　　　(C)　32,450　　　(D)　4.31476409 × 10¹²

Solutions　　(A)　3.15
(B)　0.00414 ⎫　Convention of making the digit before the 5 even, if it is odd,
(C)　32,400 ⎭　or leaving it alone, if it is even
(D)　4.31 × 10¹²

PROBLEM 2　　Repeat Example 2 with:
(A)　43.0690　　　(B)　48.05　　　(C)　48.15　　　(D)　8.017632 × 10⁻³

EXAMPLE 3　　How many significant figures should be used in the answer for the following?

$$a = \sqrt{21.3^2 + 14.0^2 - 2(21.3)(14.0)(0.6039)}$$
↑
2 is exact

Solution　　Three, since there are three significant figures in the numbers with the least number of significant figures in the calculation.

PROBLEM 3　　How many significant figures should be in the answer for the following?

$$c = \frac{12(0.3815)}{(0.7605)}$$

ANSWERS TO　　**1.** (A)　4.5 × 10²　　(B)　3.6 × 10⁵　　(C)　3.72 × 10⁻²　　(D)　1.43 × 10⁻⁶
MATCHED PROBLEMS　　**2.** (A)　43.1　　(B)　48.0　　(C)　48.2　　(D)　8.02 × 10⁻³
3. Two

EXERCISE A.1　　*Write in scientific notation.*

1. 640	2. 384	3. 5,460,000,000
4. 38,400,000	5. 0.73	6. 0.00493
7. 0.000 000 32	8. 0.0836	9. 0.000 049 1
10. 435,640	11. 67,000,000,000	12. 0.000 000 043 2

Write as a decimal fraction.

13. 5.6 × 10⁴	14. 3.65 × 10⁶	15. 9.7 × 10⁻³
16. 6.39 × 10⁻⁶	17. 1.08 × 10⁻¹	18. 4.61 × 10¹²

Round off to three significant figures.

19. 635,431	20. 4,089,100	21. 86.85
22. 7.075	23. 0.004 652 3	24. 0.000 380 0

Write in scientific notation, rounding to two significant figures.

25. 734   26. 908   27. 0.040
28. 700   29. 0.000435   30. 635.46813

Indicate how many significant figures should be used in the final answer.

31. 32.8(0.2035)   32. $\dfrac{103.8}{0.4137}$

33. $\dfrac{(23.5)(0.0135)}{2.0}$ 34. $\sqrt{(43.08)^2 + (1.5)^2}$

35. $\dfrac{(5.03 \times 10^{-3})(6 \times 10^4)}{8.0}$ 36. $\dfrac{(0.035)^2 + (0.104)^2}{(2.0)(0.5917)}$

A.2   COMPLEX NUMBERS

The Pythagoreans (500–275 BC) found that the simple equation

$$x^2 = 2 \tag{1}$$

had no rational number solutions. (A rational number is any number that can be expressed as P/Q where P and Q are integers and $Q \neq 0$.) If (1) were to have a solution, then a new kind of number had to be invented—the **irrational number.**

The irrational numbers $\sqrt{2}$ and $-\sqrt{2}$ are both solutions to (1). The invention of these irrational numbers evolved over a period of about 2,000 years, and it was not until the nineteenth century that they were finally put on a firm foundation. The rational numbers and irrational numbers together constitute the real number system.

Is there any need to extend the real number system further? Yes, since we find that another simple equation

$$x^2 = -1$$

has no real solutions. (What does $\sqrt{-1}$ mean?) Once again we are forced to invent a new kind of number, a number that has the possibility of being negative when it is squared. These new numbers are called **complex numbers.** The complex numbers, like the irrational numbers, evolved over a long period of time, dating mainly back to Cardono (1545). But it was not until the nineteenth century that they were firmly established as numbers in their own right.

THE COMPLEX
NUMBER SYSTEM

A complex number is defined to be a number of the form

$$a + bi$$

where a and b are real numbers, and i is called the imaginary unit. Thus,

$$5 + 3i \qquad \frac{1}{3} - 6i \qquad \sqrt{7} + \frac{1}{4}i \qquad 0 + 6i \qquad \frac{1}{2} + 0i \qquad 0 + 0i$$

are all complex numbers. Particular kinds of complex numbers are given special names, as follows:

$a + 0i = a$	Real number
$0 + bi = bi$	Pure imaginary number
$0 + 0i = 0$	Zero
$1i = i$	Imaginary unit
$a - bi$	Conjugate of $a + bi$

Thus, we see that just as every integer is a rational number, every real number is a complex number.

To use complex numbers we must know how to add, subtract, multiply, and divide them. We start by defining equality, addition, and multiplication.

$$
\begin{aligned}
\textbf{Equality:} \quad & a + bi = c + di \quad \text{if and only if} \quad a = c \text{ and } b = d \\
\textbf{Addition:} \quad & (a + bi) + (c + di) = (a + c) + (b + d)i \\
\textbf{Multiplication:} \quad & (a + bi)(c + di) = (ac - bd) + (ad + bc)i
\end{aligned}
$$

These definitions, particularly the one for multiplication, may seem a little strange to you. But it turns out that if we want many of the same basic properties that hold for real numbers to hold for complex numbers, and if we also want the possibility of having the square of a number negative, then we must define addition and multiplication as above. Let us use the definition of multiplication to see what happens to i when it is squared:

$$
\begin{aligned}
i^2 &= (0 + 1i)(0 + 1i) \\
&= (0 \cdot 0 - 1 \cdot 1) + (0 \cdot 1 + 1 \cdot 0)i \\
&= -1 + 0i \\
&= -1
\end{aligned}
$$

Thus:

$$i^2 = -1$$

This is an important result. We also can write:

$$i = \sqrt{-1} \qquad \text{and} \qquad -i = -\sqrt{-1}$$

Fortunately, you do not have to memorize the definitions of addition and multiplication above. We can show that the complex numbers, under these definitions, are closed, associative, and commutative, and multiplication distributes over addition. As a consequence, we can manipulate complex numbers as if they were binomial forms in real number algebra, with the exception that i^2 is to be replaced with -1. Example 4 illustrates the mechanics of carrying out addition, subtraction, multiplication, and division.

EXAMPLE 4   Write each of the following in the form $a + bi$:

(A) $(2 + 3i) + (3 - i)$ (B) $(2 + 3i) - (3 - i)$ (C) $(2 + 3i)(3 - i)$
(D) $(2 + 3i)/(3 - i)$

Solutions   We treat these as we would ordinary binomials in elementary algebra, with one exception: whenever i^2 turns up, we replace it with -1.

(A) $(2 + 3i) + (3 - i) = 2 + 3i + 3 - i$
$= 2 + 3 + 3i - i$
$= 5 + 2i$

(B) $(2 + 3i) - (3 - i) = 2 + 3i - 3 + i$
$= 2 - 3 + 3i + i$
$= -1 + 4i$

(C) $(2 + 3i)(3 - i) = 6 + 7i - 3i^2$
$= 6 + 7i - 3(-1)$
$= 6 + 7i + 3$
$= 9 + 7i$

(D) To eliminate i from the denominator, we multiply the numerator and denominator by the complex conjugate of $3 - i$, namely, $3 + i$. [Recall from elementary algebra that $(a - b)(a + b) = a^2 - b^2$.]

$$\frac{2 + 3i}{3 - i} \cdot \frac{3 + i}{3 + i} = \frac{6 + 11i + 3i^2}{9 - i^2}$$
$$= \frac{6 + 11i - 3}{9 + 1}$$
$$= \frac{3 + 11i}{10} = \frac{3}{10} + \frac{11}{10}i$$

PROBLEM 4   Write each of the following in the form $a + bi$:

(A) $(3 - 2i) + (2 + i)$ (B) $(3 - 2i) - (2 - i)$ (C) $(3 - 2i)(2 - i)$
(D) $(3 - 2i)/(2 - i)$

At this time your experience with complex numbers has likely been limited to solutions of equations, particularly quadratic equations. Recall that if $b^2 - 4ac$ is negative in

$$x = \frac{-b \pm \sqrt{b^2 - 4ac}}{2a}$$

then the solutions to the quadratic equation $ax^2 + bx + c = 0$ are complex. This is easy to verify, since a square root of a negative number can be written in the form

$$\sqrt{-k} = i\sqrt{k} \qquad \text{for } k > 0$$

To check this last equation, we square $i\sqrt{k}$ to see if we get $-k$:

$$(i\sqrt{k})^2 = i^2(\sqrt{k})^2 = -k$$

[*Note:* We write $i\sqrt{k}$ instead of $\sqrt{k}i$ so that i will not mistakenly be included under the radical.]

EXAMPLE 5   Write in the form $a + bi$:

(A) $\sqrt{-4}$ (B) $4 + \sqrt{-4}$ (C) $(-3 - \sqrt{-7})/2$
(D) $1/(1 - \sqrt{-9})$

Solutions   (A) $\sqrt{-4} = i\sqrt{4} = 2i$ (B) $4 + \sqrt{-4} = 4 + i\sqrt{4} = 4 + 2i$

(C) $\dfrac{-3 - \sqrt{-7}}{2} = \dfrac{-3 - i\sqrt{7}}{2} = -\dfrac{3}{2} - \dfrac{\sqrt{7}}{2}i$

(D) $\dfrac{1}{1 - \sqrt{-9}} = \dfrac{1}{1 - 3i} = \dfrac{1}{1 - 3i} \cdot \dfrac{1 + 3i}{1 + 3i} = \dfrac{1 + 3i}{1 - 9i^2}$

$= \dfrac{1 + 3i}{10} = \dfrac{1}{10} + \dfrac{3}{10}i$

PROBLEM 5   Write in the form $a + bi$:

(A) $\sqrt{-16}$ (B) $5 + \sqrt{-16}$ (C) $(-5 - \sqrt{-2})/2$
(D) $1/(3 - \sqrt{-4})$

**ANSWERS TO
MATCHED PROBLEMS**

4.   (A) $5 - i$ (B) $1 - i$ (C) $4 - 7i$ (D) $\frac{8}{5} - \frac{1}{5}i$
5.   (A) $4i$ (B) $5 + 4i$ (C) $-\frac{5}{2} - (\sqrt{2}/2)i$ (D) $\frac{3}{13} + \frac{2}{13}i$

EXERCISE A.2   *Perform the indicated operations and write each answer in the standard form $a + bi$.*

1. $(3 - 2i) + (4 + 7i)$ 2. $(4 + 6i) + (2 - 3i)$
3. $(3 - 2i) - (4 + 7i)$ 4. $(4 + 6i) - (2 - 3i)$
5. $(6i)(3i)$ 6. $(5i)(4i)$
7. $2i(3 - 4i)$ 8. $4i(2 - 3i)$
9. $(3 - 4i)(1 - 2i)$ 10. $(5 - i)(2 - 3i)$
11. $(3 + 5i)(3 - 5i)$ 12. $(7 - 3i)(7 + 3i)$

13. $\dfrac{1}{2 + i}$

14. $\dfrac{1}{3 - i}$

15. $\dfrac{2 - i}{3 + 2i}$

16. $\dfrac{3 + i}{2 - 3i}$

17. $\dfrac{-1 + 2i}{4 + 3i}$

18. $\dfrac{-2 - i}{3 - 4i}$

Convert square roots of negative numbers to complex forms, perform the indicated operations, and express answers in the standard form a + bi.

19. $(3 + \sqrt{-4}) + (2 - \sqrt{-16})$

20. $(2 + \sqrt{-9}) + (3 - \sqrt{-25})$

21. $(5 - \sqrt{-1}) - (2 - \sqrt{-36})$

22. $(2 + \sqrt{-9}) - (3 - \sqrt{-25})$

23. $(-3 - \sqrt{-1})(-2 + \sqrt{-49})$

24. $(3 - \sqrt{-9})(-2 - \sqrt{-1})$

25. $\dfrac{5 - \sqrt{-1}}{2 + \sqrt{-4}}$

26. $\dfrac{-2 + \sqrt{-16}}{3 - \sqrt{-25}}$

Evaluate.

27. $(1 - i)^2 - 2(1 - i) + 2$

28. $(1 + i)^2 - 2(1 + i) + 2$

29. $\left(-\dfrac{1}{2} + \dfrac{\sqrt{3}}{2}i\right)^3$

30. $\left(-\dfrac{1}{2} - \dfrac{\sqrt{3}}{2}i\right)^3$

APPENDIX: FUNCTIONS AND INVERSE FUNCTIONS

B

B.1 RELATIONS AND FUNCTIONS

Seeking relationships among various types of phenomena is undoubtedly one of the most important aspects of science. In physics, one attempts to find a relationship between the current in an electrical circuit and time; in chemistry, a relationship between the speed of a chemical reaction and the concentration of a given substance; in economics, a relationship between the price of an object and the demand for the object; in psychology, a relationship between IQ and school performance; . . . . The list could be continued without end.

Establishing and working with relationships or correspondences among various types of phenomena—whether through tables, graphs, or equations—is so fundamental to pure and applied science that it has become necessary to describe this activity in the precise language of mathematics.

RELATIONS
AND FUNCTIONS
AS CORRESPONDENCES

A **relation** R is a correspondence that assigns to each element x from a set of elements X one or more values y from a set Y. A **function** f is a relation that assigns to each x from a set X one and only one y from a set Y. Note that all functions are relations, but some relations are not functions. The set X is called the **domain** of the relation or function and the set Y is called the **range.** Each element in the range Y corresponds to one or more elements in the domain X.

Most equations in two variables that you have encountered establish relations. If in an equation in two variables, say, x and y, there corresponds exactly one range value y for each domain value x, then the correspondence established by the equation is a function.

EXAMPLE 1 Given the relations

$$R: \quad x^2 + y^2 = 4$$
$$G: \quad x^2 + y = 4$$

with domains X the set of all real numbers that produce real y, which relation is a function and why?

Solution

To test which is a function we ask how many y's correspond to each choice of x out of the domain X? If only one y corresponds to each x, then the relation is a function. The relation R is not a function, since, for example, $(0, 2)$ and $(0, -2)$ both satisfy the equation; that is, if $x = 0$, then y can either be 2 or -2. On the other hand, G is a function, since for each value x there exists exactly one value y, and no more. For example, if $x = 0$, then $y = 4$; if $x = 1$, then $y = 3$; if $x = 2$, then $y = 0$; and so on.

PROBLEM 1 Repeat Example 1 for

$$R: \quad y = x^2 - 3$$
$$G: \quad y^2 = x - 3$$

RELATIONS AND FUNCTIONS AS SETS OF ORDERED PAIRS

Since relations and functions involve correspondences between two sets of elements (domains and ranges), we can match domain values with range values and form sets of ordered pairs of elements. This observation leads to alternate definitions of relations and functions that are useful in certain situations.

We define a **relation** as any set of ordered pairs of elements (may be produced by a table, graph, equation, etc.), and a **function** is a relation with the added restriction that no two distinct ordered pairs can have the same first component. The set of first components in the set of ordered pairs is called the **domain,** and the set of second components is called the **range.**

EXAMPLE 2

Given the relations

$$H = \{(1, 1), (2, 1), (3, 2), (3, 4)\}$$
$$F = \{(2, 4), (3, -1), (4, 4)\}$$

(A) Which relation is a function?

(B) Give its domain and range.

Solutions

(A) H is not a function since (3, 2) and (3, 4) both have the same first components (the domain value 3 corresponds to more than one range value); F is a function (each domain value corresponds to exactly one range value)

(B) Domain of $F = X = \{2, 3, 4\}$
Range of $F = Y = \{-1, 4\}$

PROBLEM 2

Repeat Example 2 for the relations

$$M = \{(3, 4), (5, 4), (6, -1)\}$$
$$N = \{(-1, 2), (0, 4), (1, 2), (0, 1)\}$$

FUNCTION NOTATION

If x represents an element in the domain of a function f, then we will often use the symbol $f(x)$ in place of y to designate the number in the range of f to which x is paired. It is important not to confuse this new symbol and think of it as a product of f and x. The symbol is read "f of x" or "the value of f at x." The correct use of this function symbol should be mastered early. Thus, if

$$f(x) = 2x + 3$$

then

$$f(5) = 2(5) + 3 = 13$$

That is, the function f assigns the range value 13 to the domain value 5. Thus, (5, 13) belongs to f. Can you find another ordered pair that belongs to f?

The correspondence between domain values and range values is usually illustrated in one of the two ways shown in Figure 1 on page 260.

$f: x \rightarrow 2x + 3$

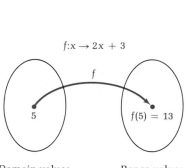

Domain values   Range values

OR

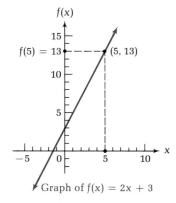

FIGURE 1   (a) Map   (b) Graph

EXAMPLE 3   If $f(x) = (x/2) - 1$ and $g(x) = 1 - x^2$, find:

(A) $f(4)$ (B) $g(-3)$ (C) $f(2) - g(0)$ (D) $f(2 + h)$
(E) $[f(3 + h) - f(3)]/h$ (F) $f[g(3)]$

Solutions   (A) $f(4) = \dfrac{4}{2} - 1 = 2 - 1 = 1$

(B) $g(-3) = 1 - (-3)^2 = 1 - 9 = -8$

(C) $f(2) - g(0) = \left(\dfrac{2}{2} - 1\right) - (1 - 0^2) = 0 - 1 = -1$

(D) $f(2 + h) = \dfrac{2 + h}{2} - 1 = \dfrac{2 + h - 2}{2} = \dfrac{h}{2}$

(E) $\dfrac{f(3 + h) - f(3)}{h} = \dfrac{\left(\dfrac{3 + h}{2} - 1\right) - \left(\dfrac{3}{2} - 1\right)}{h} = \dfrac{\dfrac{1 + h}{2} - \dfrac{1}{2}}{h} = \dfrac{1}{2}$

(F) $f[g(3)] = f[1 - 3^2] = f(-8) = \dfrac{-8}{2} - 1 = -5$

PROBLEM 3   If $f(x) = 2x - 3$ and $g(x) = x^2 - 2$, find:

(A) $f(3)$ (B) $g(-2)$ (C) $f(0) + g(1)$ (D) $f(3 + h)$
(E) $[f(2 + h) - f(2)]/h$ (F) $f[g(2)]$

EXAMPLE 4   If the function f defined by $f(x) = x^2 - x$ has domain $X = \{-2, -1, 0, 1, 2\}$, what is the range of f?

Solution $f(-2) = (-2)^2 - (-2) = 6$ $f(1) = 1^2 - 1 = 0$
$f(-1) = (-1)^2 - (-1) = 2$ $f(2) = 2^2 - 2 = 2$
$f(0) = 0^2 - 0 = 0$

Thus, the range of $f = Y = \{0, 2, 6\}$.

PROBLEM 4   Repeat Example 4 for $g(x) = x^2 - 4$, $X = \{-2, -1, 0, 1, 2\}$.

FUNCTION
CLASSIFICATION

Functions are classified in special categories for more efficient study. You have already had some experience with many algebraic functions (defined by means of the algebraic operations addition, subtraction, multiplication, division, powers, and roots), exponential functions, and logarithmic functions. In this book, we add two more classes of functions to this list, namely, the trigonometric functions and the inverse trigonometric functions. These five classes of functions are called the **elementary functions.** They are related to each other as follows:

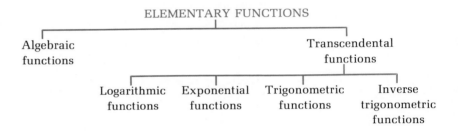

Since most of the applications in this book require the use of one or more of these five classes of functions, it is important that you become familiar with their basic properties.

ANSWERS TO
MATCHED PROBLEMS

1. R is a function, since for each domain value there corresponds exactly one range value; G is not a function, since, for example, both 1 and -1 correspond to the domain value 4.
2. (A) M is a function   (B) $X = \{3, 5, 6\}$, $Y = \{-1, 4\}$
3. (A) 3   (B) 2   (C) -4 (D) $3 + 2h$ (E) 2   (F) 1
4. $Y = \{-4, -3, 0\}$

EXERCISE B.1   For $f(x) = 4x - 1$, find each of the following:

1. $f(1)$ 2. $f(2)$ 3. $f(-1)$
4. $f(-2)$ 5. $f(0)$ 6. $f(5)$

If $g(x) = x - x^2$, find each of the following:

7. $g(1)$ 8. $g(3)$ 9. $g(5)$
10. $g(4)$ 11. $g(-2)$ 12. $g(-3)$

For $f(x) = 1 - 2x$ and $g(x) = 4 - x^2$, find each of the following:

13. $f(0) + g(0)$ 14. $g(0) - f(0)$ 15. $\dfrac{f(3)}{g(1)}$

16. $[g(-2)][f(-1)]$ 17. $2f(-1)$ 18. $\dfrac{1}{5}g(-3)$

19. $f(2 + h)$ 20. $g(2 + h)$ 21. $\dfrac{f(2 + h) - f(2)}{h}$

22. $\dfrac{g(2 + h) - g(2)}{h}$ 23. $g[f(2)]$ 24. $f[g(2)]$

Which of the following equations define functions, assuming a solution is an ordered pair of numbers of the form (x, y)?

25. $x^2 + y^2 = 25$ 26. $y = 3x + 1$ 27. $2x - 3y = 6$
28. $y = x^2 - 2$ 29. $y^2 = x$ 30. $y = x^2$
31. $y = |x|$ 32. $|y| = x$

33. If the function f, defined by $f(x) = x^2 - x + 1$, has domain $X = \{-2, -1, 0, 1, 2\}$, find the range of f.
34. If the function g, defined by $g(x) = 1 + x - x^2$, has domain $X = \{-2, -1, 0, 1, 2\}$, find the range of g.
35. Indicate which relation is a function and write down its domain and range:

$F = \{(-2, 1), (-1, 1), (0, 0)\}$
$G = \{(-4, 3), (0, 3), (-4, 0)\}$

36. Repeat Problem 35 for

$H = \{(-1, 3), (2, 3), (-1, -1)\}$
$L = \{(-1, 1), (0, 1), (1, 1)\}$

Problems 37–39 pertain to the following relationship: The distance d (in meters) that an object falls in a vacuum in t seconds is given by

$d = s(t) = 4.88t^2$

37. Find $s(0)$, $s(1)$, $s(2)$, and $s(3)$.
38. *Calculator exercise* $[s(2 + h) - s(2)]/h$ represents the average speed of the falling object over the time interval from $t = 2$ to $t = 2 + h$. Use a calculator to compute each of the following; then guess at the speed of a free-falling object at the end of 2 sec.

(A) $\dfrac{s(3) - s(2)}{1}$ (B) $\dfrac{s(2.1) - s(2)}{0.1}$ (C) $\dfrac{s(2.01) - s(2)}{0.01}$

(D) $\dfrac{s(2.001) - s(2)}{0.001}$ (E) $\dfrac{s(2.0001) - s(2)}{0.0001}$

39. Find $[s(2 + h) - s(2)]/h$ and simplify. What happens as h gets closer and closer to zero? Interpret physically.

B.2   INVERSE RELATIONS AND FUNCTIONS

If we start with a relation, we can form a new relation simply by reversing the order of the elements in each ordered pair of the set. The new relation is called the **inverse** of the original relation. Thus, if R is a relation, then we define its inverse, denoted by R^{-1}, to be[†]

$$R^{-1} = \{(b, a)|(a, b) \in R\}$$

For example, if

$$R = \{(1, 5), (-1, 3), (1, 2)\}$$

then

$$R^{-1} = \{(5, 1), (3, -1), (2, 1)\}$$

It follows from the definition that the **domain of R^{-1}** is the range of R and the **range of R^{-1}** is the domain of R.

If a relation is specified by an equation, say,

$$R: \quad y = 2x - 2 \tag{1}$$

then how do we find R^{-1}? The answer is easy: We interchange the variables in (1). Thus,

$$R^{-1}: \quad x = 2y - 2 \tag{2}$$

Or, solving for y in terms of x,

$$R^{-1}: \quad y = \frac{x + 2}{2} \tag{3}$$

Any ordered pair of numbers that satisfies (1), when reversed in order, will satisfy (2) and (3). For example, (4, 6) satisfies (1) and (6, 4) satisfies (2) and (3). (Check this.) The graphs of R and R^{-1} are given in Figure 2.

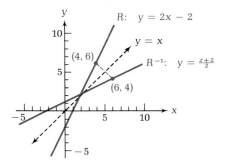

FIGURE 2

[†]R^{-1} is a function symbol and does not mean $1/R$.

It is useful to sketch in the line $y = x$ and to observe that if we fold the paper along this line, then R and R^{-1} will match. Actually, we can graph R^{-1} by drawing R with wet ink and folding the paper along $y = x$ before the ink dries; R will then print R^{-1}. To prove this, one has to show that the line $y = x$ is the perpendicular bisector of the line segment joining (a, b) to (b, a). Knowing that the graphs of R and R^{-1} are symmetric relative to the line $y = x$ makes it easy to graph R^{-1} if R is known, and vice versa.

Another way of looking at the relation R and its inverse R^{-1} is indicated in Figure 3. Note that R sends 4 into 6 and R^{-1} sends 6 back into 4.

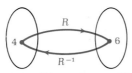

FIGURE 3

Domain of R Range of R
Range of R^{-1} Domain of R^{-1}

In Figure 2 we observe that both R and R^{-1} are functions. However, this is not always the case. Inverses of some functions may not be functions. Consider Example 5.

EXAMPLE 5   If R is given by $y = x^2$:

(A)   Find R^{-1} (B)   Graph R and R^{-1}
(C)   Indicate which are functions

Solutions   (A) R: $y = x^2$
                   R^{-1}: $x = y^2$ or R^{-1}: $y = \pm\sqrt{x}$

(B)

(C) R is a function; R^{-1} is not a function, since for each domain value, more than one range value is possible.

PROBLEM 5   Repeat Example 6 for R given by $|y| = x$.

Given a function, how can we tell before finding its inverse, whether its inverse is also a function? Theorem 1 provides us with an easily applied test.

THEOREM 1   A function has an inverse that is a function if and only if f is one-to-one (that is, if and only if each domain value is associated with exactly one range value and each range value is associated with exactly one domain value).

EXAMPLE 6   Which of the following functions are one-to-one; that is, have inverses that are functions?

(A) $f(x) = x^2$ (B) $g(x) = x^2, \quad x \geq 0$

Solutions   (A)   Write

$$f: \quad y = x^2$$

and note that each x corresponds to exactly one y; however, each value of y may correspond to more than one value of x (e.g., if $y = 4$, then $x = \pm 2$). Thus, f is not one-to-one, and the inverse of f is not a function.

(B)   Write

$$g: \quad y = x^2 \quad x \geq 0$$

Again, for each x there corresponds exactly one y. But now, having restricted the domain of f to produce g, we see that for each y there corresponds exactly one x (e.g., if $y = 4$, then $x = 2$). Thus, g is one-to-one, and hence has an inverse that is a function. Compare the graphs of f and g in Figure 4.

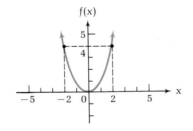

FIGURE 4   (a) f is not one-to-one   (b) g is one-to-one

PROBLEM 6   Which of the following functions are one-to-one and thus have an inverse that is a function?

(A) $f(x) = (x - 1)^2$ (B) $g(x) = (x - 1)^2, \quad x \geq 1$

Example 6 and Problem 6 illustrate an important idea: A function that is not one-to-one can, in order to obtain an inverse that is a function, have its domain restricted so that it becomes one-to-one. The trigonometric functions must be restricted in this way in order to obtain the inverse trigonometric functions.

We conclude this section with Theorem 2.

THEOREM 2   If f is one-to-one, then:

1. $f[f^{-1}(x)] = x$
2. $f^{-1}[f(x)] = x$

EXAMPLE 7   For $f(x) = 2x - 3$, find $f^{-1}(x)$ and show that $f[f^{-1}(x)] = x$.

Solution   We write

$$f: \quad y = 2x - 3$$

and obtain

$$f^{-1}: \quad x = 2y - 3$$

or, solving for y,

$$f^{-1}: \quad y = \frac{x + 3}{2}$$

Thus,

$$f^{-1}(x) = \frac{x + 3}{2}$$

Now,

$$f[f^{-1}(x)] = 2[f^{-1}(x)] - 3$$
$$= 2\left(\frac{x + 3}{2}\right) - 3$$
$$= x + 3 - 3 = x$$

PROBLEM 7   For $g(x) = 3x + 2$, find $g^{-1}(x)$ and show that $g^{-1}[g(x)] = x$.

ANSWERS TO MATCHED PROBLEMS

5. (A) $R^{-1}: \quad |x| = y$ (B)

(C) R is not a function; R^{-1} is a function.

6. g 7. $g^{-1}(x) = \dfrac{x - 2}{3}$; $g^{-1}[g(x)] = \dfrac{g(x) - 2}{3} = \dfrac{(3x + 2) - 2}{3} = x$

EXERCISE B.2   *Find the inverse of each relation and indicate whether it is a function.*

1. $R = \{(-1, 2), (0, 1), (1, 2)\}$
2. $G = \{(-2, 4), (-1, 1), (0, 0), (1, 1), (2, 4)\}$
3. $H = \{(-1, 0.5), (0, 1), (1, 2), (2, 4)\}$
4. $F = \{(-5, 0), (-2, 1), (0, 2), (1, 4), (2, 7)\}$

Graph on the same coordinate system along with y = x.

5. R and R^{-1} in Problem 1

6. G and G^{-1} in Problem 2

7. H and H^{-1} in Problem 3

8. F and F^{-1} in Problem 4

Find the inverse for each relation in the form of an equation. Indicate whether it is a function.

9. M: $y = 2x - 7$

10. N: $y = \dfrac{x}{2} + 1$

11. P: $y = \dfrac{x + 3}{2}$

12. Q: $y = \dfrac{x - 2}{3}$

13. F: $y = 2x^2$

14. G: $y = x^2 - 4$

15. f: $y = 2x^2, \quad x \geq 0$

16. g: $y = x^2 - 4, \quad x \geq 0$

Graph the relation and its inverse on the same coordinate system, along with the graph of y = x.

17. M and M^{-1} in Problem 9

18. N and N^{-1} in Problem 10

19. F and F^{-1} in Problem 13

20. G and G^{-1} in Problem 14

21. f and f^{-1} in Problem 15

22. g and g^{-1} in Problem 16

Solve.

23. For $f(x) = 2x - 7$, find $f^{-1}(x)$ and $f^{-1}(3)$.

24. For $g(x) = (x/2) + 1$, find $g^{-1}(x)$ and $g^{-1}(-3)$.

25. For $h(x) = (x/3) + 1$, find $h^{-1}(x)$ and $h^{-1}(2)$.

26. For $m(x) = 3x + 2$, find $m^{-1}(x)$ and $m^{-1}(5)$.

27. Find $f[f^{-1}(4)]$ for Problem 23.

28. Find $g^{-1}[g(2)]$ for Problem 24.

29. Find $h^{-1}[h(x)]$ for Problem 25.

30. Find $m[m^{-1}(x)]$ for Problem 26.

APPENDIX: C
TABLES

TABLE I TRIGONOMETRIC FUNCTIONS—DEGREES AND MINUTES

→	sin	cos	tan	cot	sec	csc	
0°00′	0.0000	1.000	0.0000	—	1.000	—	90°00′
10′	0.0029	1.000	0.0029	343.8	1.000	343.8	89°50′
20′	0.0058	1.000	0.0058	171.9	1.000	171.9	40′
30′	0.0087	1.000	0.0087	114.6	1.000	114.6	30′
40′	0.0116	0.9999	0.0116	85.94	1.000	85.95	20′
0°50′	0.0145	0.9999	0.0145	68.75	1.000	68.76	10′
1°00′	0.0175	0.9998	0.0175	57.29	1.000	57.30	89°00′
10′	0.0204	0.9998	0.0204	49.10	1.000	49.11	88°50′
20′	0.0233	0.9997	0.0233	42.96	1.000	42.98	40′
30′	0.0262	0.9997	0.0262	38.19	1.000	38.20	30′
40′	0.0291	0.9996	0.0291	34.37	1.000	34.38	20′
1°50′	0.0320	0.9995	0.0320	31.24	1.001	31.26	10′
2°00′	0.0349	0.9994	0.0349	28.64	1.001	28.65	88°00′
10′	0.0378	0.9993	0.0378	26.43	1.001	26.45	87°50′
20′	0.0407	0.9992	0.0407	24.54	1.001	24.56	40′
30′	0.0436	0.9990	0.0437	22.90	1.001	22.93	30′
40′	0.0465	0.9989	0.0466	21.47	1.001	21.49	20′
2°50′	0.0494	0.9988	0.0495	20.21	1.001	20.23	10′
3°00′	0.0523	0.9986	0.0524	19.08	1.001	19.11	87°00′
10′	0.0552	0.9985	0.0553	18.07	1.002	18.10	86°50′
20′	0.0581	0.9983	0.0582	17.17	1.002	17.20	40′
30′	0.0610	0.9981	0.0612	16.35	1.002	16.38	30′
40′	0.0640	0.9980	0.0641	15.60	1.002	15.64	20′
3°50′	0.0669	0.9978	0.0670	14.92	1.002	14.96	10′
4°00′	0.0698	0.9976	0.0699	14.30	1.002	14.34	86°00′
10′	0.0727	0.9974	0.0729	13.73	1.003	13.76	85°50′
20′	0.0756	0.9971	0.0758	13.20	1.003	13.23	40′
30′	0.0785	0.9969	0.0787	12.71	1.003	12.75	30′
40′	0.0814	0.9967	0.0816	12.25	1.003	12.29	20′
4°50′	0.0843	0.9964	0.0846	11.83	1.004	11.87	10′
5°00′	0.0872	0.9962	0.0875	11.43	1.004	11.47	85°00′
10′	0.0901	0.9959	0.0904	11.06	1.004	11.10	84°50′
20′	0.0929	0.9957	0.0934	10.71	1.004	10.76	40′
30′	0.0958	0.9954	0.0963	10.39	1.005	10.43	30′
40′	0.0987	0.9951	0.0992	10.08	1.005	10.13	20′
5°50′	0.1016	0.9948	0.1022	9.788	1.005	9.839	10′
6°00′	0.1045	0.9945	0.1051	9.514	1.006	9.567	84°00′
10′	0.1074	0.9942	0.1080	9.255	1.006	9.309	83°50′
20′	0.1103	0.9939	0.1110	9.010	1.006	9.065	40′
30′	0.1132	0.9936	0.1139	8.777	1.006	8.834	30′
40′	0.1161	0.9932	0.1169	8.556	1.007	8.614	20′
6°50′	0.1190	0.9929	0.1198	8.345	1.007	8.405	10′
7°00′	0.1219	0.9925	0.1228	8.144	1.008	8.206	83°00′
	cos	sin	cot	tan	csc	sec	←

TABLE I (CONTINUED)

	sin	cos	tan	cot	sec	csc	
7°00′	0.1219	0.9925	0.1228	8.144	1.008	8.206	83°00′
10′	0.1248	0.9922	0.1257	7.953	1.008	8.016	82°50′
20′	0.1276	0.9918	0.1287	7.770	1.008	7.834	40′
30′	0.1305	0.9914	0.1317	7.596	1.009	7.661	30′
40′	0.1334	0.9911	0.1346	7.429	1.009	7.496	20′
7°50′	0.1363	0.9907	0.1376	7.269	1.009	7.337	10′
8°00′	0.1392	0.9903	0.1405	7.115	1.010	7.185	82°00′
10′	0.1421	0.9899	0.1435	6.968	1.010	7.040	81°50′
20′	0.1449	0.9894	0.1465	6.827	1.011	6.900	40′
30′	0.1478	0.9890	0.1495	6.691	1.011	6.765	30′
40′	0.1507	0.9886	0.1524	6.561	1.012	6.636	20′
8°50′	0.1536	0.9881	0.1554	6.435	1.012	6.512	10′
9°00′	0.1564	0.9877	0.1584	6.314	1.012	6.392	81°00′
10′	0.1593	0.9872	0.1614	6.197	1.013	6.277	80°50′
20′	0.1622	0.9868	0.1644	6.084	1.013	6.166	40′
30′	0.1650	0.9863	0.1673	5.976	1.014	6.059	30′
40′	0.1679	0.9858	0.1703	5.871	1.014	5.955	20′
9°50′	0.1708	0.9853	0.1733	5.769	1.015	5.855	10′
10°00′	0.1736	0.9848	0.1763	5.671	1.015	5.759	80°00′
10′	0.1765	0.9843	0.1793	5.576	1.016	5.665	79°50′
20′	0.1794	0.9838	0.1823	5.485	1.016	5.575	40′
30′	0.1822	0.9833	0.1853	5.396	1.017	5.487	30′
40′	0.1851	0.9827	0.1883	5.309	1.018	5.403	20′
10°50′	0.1880	0.9822	0.1914	5.226	1.018	5.320	10′
11°00′	0.1908	0.9816	0.1944	5.145	1.019	5.241	79°00′
10′	0.1937	0.9811	0.1974	5.066	1.019	5.164	78°50′
20′	0.1965	0.9805	0.2004	4.989	1.020	5.089	40′
30′	0.1994	0.9799	0.2035	4.915	1.020	5.016	30′
40′	0.2022	0.9793	0.2065	4.843	1.021	4.945	20′
11°50′	0.2051	0.9787	0.2095	4.773	1.022	4.876	10′
12°00′	0.2079	0.9781	0.2126	4.705	1.022	4.810	78°00′
10′	0.2108	0.9775	0.2156	4.638	1.023	4.745	77°50′
20′	0.2136	0.9769	0.2186	4.574	1.024	4.682	40′
30′	0.2164	0.9763	0.2217	4.511	1.024	4.620	30′
40′	0.2193	0.9757	0.2247	4.449	1.025	4.560	20′
12°50′	0.2221	0.9750	0.2278	4.390	1.026	4.502	10′
13°00′	0.2250	0.9744	0.2309	4.331	1.026	4.445	77°00′
10′	0.2278	0.9737	0.2339	4.275	1.027	4.390	76°50′
20′	0.2306	0.9730	0.2370	4.219	1.028	4.336	40′
30′	0.2334	0.9724	0.2401	4.165	1.028	4.284	30′
40′	0.2363	0.9717	0.2432	4.113	1.029	4.232	20′
13°50′	0.2391	0.9710	0.2462	4.061	1.030	4.182	10′
14°00′	0.2419	0.9703	0.2493	4.011	1.031	4.134	76°00′
	cos	sin	cot	tan	csc	sec	

TABLE I TRIGONOMETRIC FUNCTIONS—DEGREES AND MINUTES (CONTINUED)

↳→	sin	cos	tan	cot	sec	csc	
14°00′	0.2419	0.9703	0.2493	4.011	1.031	4.134	76°00′
10′	0.2447	0.9696	0.2524	3.962	1.031	4.086	75°50′
20′	0.2476	0.9689	0.2555	3.914	1.032	4.039	40′
30′	0.2504	0.9681	0.2586	3.867	1.033	3.994	30′
40′	0.2532	0.9674	0.2617	3.821	1.034	3.950	20′
14°50′	0.2560	0.9667	0.2648	3.776	1.034	3.906	10′
15°00′	0.2588	0.9659	0.2679	3.732	1.035	3.864	75°00′
10′	0.2616	0.9652	0.2711	3.689	1.036	3.822	74°50′
20′	0.2644	0.9644	0.2742	3.647	1.037	3.782	40′
30′	0.2672	0.9636	0.2773	3.606	1.038	3.742	30′
40′	0.2700	0.9628	0.2805	3.566	1.039	3.703	20′
15°50′	0.2728	0.9621	0.2836	3.526	1.039	3.665	10′
16°00′	0.2756	0.9613	0.2867	3.487	1.040	3.628	74°00′
10′	0.2784	0.9605	0.2899	3.450	1.041	3.592	73°50′
20′	0.2812	0.9596	0.2931	3.412	1.042	3.556	40′
30′	0.2840	0.9588	0.2962	3.376	1.043	3.521	30′
40′	0.2868	0.9580	0.2994	3.340	1.044	3.487	20′
16°50′	0.2896	0.9572	0.3026	3.305	1.045	3.453	10′
17°00′	0.2924	0.9563	0.3057	3.271	1.046	3.420	73°00′
10′	0.2952	0.9555	0.3089	3.237	1.047	3.388	72°50′
20′	0.2979	0.9546	0.3121	3.204	1.048	3.356	40′
30′	0.3007	0.9537	0.3153	3.172	1.049	3.326	30′
40′	0.3035	0.9528	0.3185	3.140	1.049	3.295	20′
17°50′	0.3062	0.9520	0.3217	3.108	1.050	3.265	10′
18°00′	0.3090	0.9511	0.3249	3.078	1.051	3.236	72°00′
10′	0.3118	0.9502	0.3281	3.047	1.052	3.207	71°50′
20′	0.3145	0.9492	0.3314	3.018	1.053	3.179	40′
30′	0.3173	0.9483	0.3346	2.989	1.054	3.152	30′
40′	0.3201	0.9474	0.3378	2.960	1.056	3.124	20′
18°50′	0.3228	0.9465	0.3411	2.932	1.057	3.098	10′
19°00′	0.3256	0.9455	0.3443	2.904	1.058	3.072	71°00′
10′	0.3283	0.9446	0.3476	2.877	1.059	3.046	70°50′
20′	0.3311	0.9436	0.3508	2.850	1.060	3.021	40′
30′	0.3338	9.9426	0.3541	2.824	1.061	2.996	30′
40′	0.3365	0.9417	0.3574	2.798	1.062	2.971	20′
19°50′	0.3393	0.9407	0.3607	2.773	1.063	2.947	10′
20°00′	0.3420	0.9397	0.3640	2.747	1.064	2.924	70°00′
10′	0.3448	0.9387	0.3673	2.723	1.065	2.901	69°50′
20′	0.3475	0.9377	0.3706	2.699	1.066	2.878	40′
30′	0.3502	0.9367	0.3739	2.675	1.068	2.855	30′
40′	0.3529	0.9356	0.3772	2.651	1.069	2.833	20′
20°50′	0.3557	0.9346	0.3805	2.628	1.070	2.812	10′
21°00′	0.3584	0.9336	0.3839	2.605	1.071	2.790	69°00′
	cos	sin	cot	tan	csc	sec	←↑

TABLE I (CONTINUED)

	sin	cos	tan	cot	sec	csc	
21°00′	0.3584	0.9336	0.3839	2.605	1.071	2.790	69°00′
10′	0.3611	0.9325	0.3872	2.583	1.072	2.769	68°50′
20′	0.3638	0.9315	0.3906	2.560	1.074	2.749	40′
30′	0.3665	0.9304	0.3939	2.539	1.075	2.729	30′
40′	0.3692	0.9293	0.3973	2.517	1.076	2.709	20′
21°50′	0.3719	0.9283	0.4006	2.496	1.077	2.689	10′
22°00′	0.3746	0.9272	0.4040	2.475	1.079	2.669	68°00′
10′	0.3773	0.9261	0.4074	2.455	1.080	2.650	67°50′
20′	0.3800	0.9250	0.4108	2.434	1.081	2.632	40′
30′	0.3827	0.9239	0.4142	2.414	1.082	2.613	30′
40′	0.3854	0.9228	0.4176	2.394	1.084	2.595	20′
22°50′	0.3881	0.9216	0.4210	2.375	1.085	2.577	10′
23°00′	0.3907	0.9205	0.4245	2.356	1.086	2.559	67°00′
10′	0.3934	0.9194	0.4279	2.337	1.088	2.542	66°50′
20′	0.3961	0.9182	0.4314	2.318	1.089	2.525	40′
30′	0.3987	0.9171	0.4348	2.300	1.090	2.508	30′
40′	0.4014	0.9159	0.4383	2.282	1.092	2.491	20′
23°50′	0.4041	0.9147	0.4417	2.264	1.093	2.475	10′
24°00′	0.4067	0.9135	0.4452	2.246	1.095	2.459	66°00′
10′	0.4094	0.9124	0.4487	2.229	1.096	2.443	65°50′
20′	0.4120	0.9112	0.4522	2.211	1.097	2.427	40′
30′	0.4147	0.9100	0.4557	2.194	1.099	2.411	30′
40′	0.4173	0.9088	0.4592	2.177	1.100	2.396	20′
24°50′	0.4200	0.9075	0.4628	2.161	1.102	2.381	10′
25°00′	0.4226	0.9063	0.4663	2.145	1.103	2.366	65°00′
10′	0.4253	0.9051	0.4699	2.128	1.105	2.352	64°50′
20′	0.4279	0.9038	0.4734	2.112	1.106	2.337	40′
30′	0.4305	0.9026	0.4770	2.097	1.108	2.323	30′
40′	0.4331	0.9013	0.4806	2.081	1.109	2.309	20′
25°50′	0.4358	0.9001	0.4841	2.066	1.111	2.295	10′
26°00′	0.4384	0.8988	0.4877	2.050	1.113	2.281	64°00′
10′	0.4410	0.8975	0.4913	2.035	1.114	2.268	63°50′
20′	0.4436	0.8962	0.4950	2.020	1.116	2.254	40′
30′	0.4462	0.8949	0.4986	2.006	1.117	2.241	30′
40′	0.4488	0.8936	0.5022	1.991	1.119	2.228	20′
26°50′	0.4514	0.8923	0.5059	1.977	1.121	2.215	10′
27°00′	0.4540	0.8910	0.5095	1.963	1.122	2.203	63°00′
10′	0.4566	0.8897	0.5132	1.949	1.124	2.190	62°50′
20′	0.4592	0.8884	0.5169	1.935	1.126	2.178	40′
30′	0.4617	0.8870	0.5206	1.921	1.127	2.166	30′
40′	0.4643	0.8857	0.5243	1.907	1.129	2.154	20′
27°50′	0.4669	0.8843	0.5280	1.894	1.131	2.142	10′
28°00′	0.4695	0.8829	0.5317	1.881	1.133	2.130	62°00′
	cos	sin	cot	tan	csc	sec	

I

TABLE I TRIGONOMETRIC FUNCTIONS—DEGREES AND MINUTES (CONTINUED)

	sin	cos	tan	cot	sec	csc	
28°00′	0.4695	0.8829	0.5317	1.881	1.133	2.130	62°00′
10′	0.4720	0.8816	0.5354	1.868	1.134	2.118	61°50′
20′	0.4746	0.8802	0.5392	1.855	1.136	2.107	40′
30′	0.4772	0.8788	0.5430	1.842	1.138	2.096	30′
40′	0.4797	0.8774	0.5467	1.829	1.140	2.085	20′
28°50′	0.4823	0.8760	0.5505	1.816	1.142	2.074	10′
29°00′	0.4848	0.8746	0.5543	1.804	1.143	2.063	61°00′
10′	0.4874	0.8732	0.5581	1.792	1.145	2.052	60°50′
20′	0.4899	0.8718	0.5619	1.780	1.147	2.041	40′
30′	0.4924	0.8704	0.5658	1.767	1.149	2.031	30′
40′	0.4950	0.8689	0.5696	1.756	1.151	2.020	20′
29°50′	0.4975	0.8675	0.5735	1.744	1.153	2.010	10′
30°00′	0.5000	0.8660	0.5774	1.732	1.155	2.000	60°00′
10′	0.5025	0.8646	0.5812	1.720	1.157	1.990	59°50′
20′	0.5050	0.8631	0.5851	1.709	1.159	1.980	40′
30′	0.5075	0.8616	0.5890	1.698	1.161	1.970	30′
40′	0.5100	0.8601	0.5930	1.686	1.163	1.961	20′
30°50′	0.5125	0.8587	0.5969	1.675	1.165	1.951	10′
31°00′	0.5150	0.8572	0.6009	1.664	1.167	1.942	59°00′
10′	0.5175	0.8557	0.6048	1.653	1.169	1.932	58°50′
20′	0.5200	0.8542	0.6088	1.643	1.171	1.923	40′
30′	0.5225	0.8526	0.6128	1.632	1.173	1.914	30′
40′	0.5250	0.8511	0.6168	1.621	1.175	1.905	20′
31°50′	0.5275	0.8496	0.6208	1.611	1.177	1.896	10′
32°00′	0.5299	0.8480	0.6249	1.600	1.179	1.887	58°00′
10′	0.5324	0.8465	0.6289	1.590	1.181	1.878	57°50′
20′	0.5348	0.8450	0.6330	1.580	1.184	1.870	40′
30′	0.5373	0.8434	0.6371	1.570	1.186	1.861	30′
40′	0.5398	0.8418	0.6412	1.560	1.188	1.853	20′
32°50′	0.5422	0.8403	0.6453	1.550	1.190	1.844	10′
33°00′	0.5446	0.8387	0.6494	1.540	1.192	1.836	57°00′
10′	0.5471	0.8371	0.6536	1.530	1.195	1.828	56°50′
20′	0.5495	0.8355	0.6577	1.520	1.197	1.820	40′
30′	0.5519	0.8339	0.6619	1.511	1.199	1.812	30′
40′	0.5544	0.8323	0.6661	1.501	1.202	1.804	20′
33°50′	0.5568	0.8307	0.6703	1.492	1.204	1.796	10′
34°00′	0.5592	0.8290	0.6745	1.483	1.206	1.788	56°00′
10′	0.5616	0.8274	0.6787	1.473	1.209	1.781	55°50′
20′	0.5640	0.8258	0.6830	1.464	1.211	1.773	40′
30′	0.5664	0.8241	0.6873	1.455	1.213	1.766	30′
40′	0.5688	0.8225	0.6916	1.446	1.216	1.758	20′
34°50′	0.5712	0.8208	0.6959	1.437	1.218	1.751	10′
35°00′	0.5736	0.8192	0.7002	1.428	1.221	1.743	55°00′
	cos	sin	cot	tan	csc	sec	

TABLE I (CONTINUED)

→	sin	cos	tan	cot	sec	csc	
35°00′	0.5736	0.8192	0.7002	1.428	1.221	1.743	55°00′
10′	0.5760	0.8175	0.7046	1.419	1.223	1.736	54°50′
20′	0.5783	0.8158	0.7089	1.411	1.226	1.729	40′
30′	0.5807	0.8141	0.7133	1.402	1.228	1.722	30′
40′	0.5831	0.8124	0.7177	1.393	1.231	1.715	20′
35°50′	0.5854	0.8107	0.7221	1.385	1.233	1.708	10′
36°00′	0.5878	0.8090	0.7265	1.376	1.236	1.701	54°00′
10′	0.5901	0.8073	0.7310	1.368	1.239	1.695	53°50′
20′	0.5925	0.8056	0.7355	1.360	1.241	1.688	40′
30′	0.5948	0.8039	0.7400	1.351	1.244	1.681	30′
40′	0.5972	0.8021	0.7445	1.343	1.247	1.675	20′
36°50′	0.5995	0.8004	0.7490	1.335	1.249	1.668	10′
37°00′	0.6018	0.7986	0.7536	1.327	1.252	1.662	53°00′
10′	0.6041	0.7969	0.7581	1.319	1.255	1.655	52°50′
20′	0.6065	0.7951	0.7627	1.311	1.258	1.649	40′
30′	0.6088	0.7934	0.7673	1.303	1.260	1.643	30′
40′	0.6111	0.7916	0.7720	1.295	1.263	1.636	20′
37°50′	0.6134	0.7898	0.7766	1.288	1.266	1.630	10′
38°00′	0.6157	0.7880	0.7813	1.280	1.269	1.624	52°00′
10′	0.6180	0.7862	0.7860	1.272	1.272	1.618	51°50′
20′	0.6202	0.7844	0.7907	1.265	1.275	1.612	40′
30′	0.6225	0.7826	0.7954	1.257	1.278	1.606	30′
40′	0.6248	0.7808	0.8002	1.250	1.281	1.601	20′
38°50′	0.6271	0.7790	0.8050	1.242	1.284	1.595	10′
39°00′	0.6293	0.7771	0.8098	1.235	1.287	1.589	51°00′
10′	0.6316	0.7753	0.8146	1.228	1.290	1.583	50°50′
20′	0.6338	0.7735	0.8195	1.220	1.293	1.578	40′
30′	0.6361	0.7716	0.8243	1.213	1.296	1.572	30′
40′	0.6383	0.7698	0.8292	1.206	1.299	1.567	20′
39°50′	0.6406	0.7679	0.8342	1.199	1.302	1.561	10′
40°00′	0.6428	0.7660	0.8391	1.192	1.305	1.556	50°00′
10′	0.6450	0.7642	0.8441	1.185	1.309	1.550	49°50′
20′	0.6472	0.7623	0.8491	1.178	1.312	1.545	40′
30′	0.6494	0.7604	0.8541	1.171	1.315	1.540	30′
40′	0.6517	0.7585	0.8591	1.164	1.318	1.535	20′
40°50′	0.6539	0.7566	0.8642	1.157	1.322	1.529	10′
41°00′	0.6561	0.7547	0.8693	1.150	1.325	1.524	49°00′
10′	0.6583	0.7528	0.8744	1.144	1.328	1.519	48°50′
20′	0.6604	0.7509	0.8796	1.137	1.332	1.514	40′
30′	0.6626	0.7490	0.8847	1.130	1.335	1.509	30′
40′	0.6648	0.7470	0.8899	1.124	1.339	1.504	20′
41°50′	0.6670	0.7451	0.8952	1.117	1.342	1.499	10′
42°00′	0.6691	0.7431	0.9004	1.111	1.346	1.494	48°00′
	cos	sin	cot	tan	csc	sec	←

TABLE I TRIGONOMETRIC FUNCTIONS—DEGREES AND MINUTES (CONTINUED)

	sin	cos	tan	cot	sec	csc	
42°00′	0.6691	0.7431	0.9004	1.111	1.346	1.494	48°00′
10′	0.6713	0.7412	0.9057	1.104	1.349	1.490	47°50′
20′	0.6734	0.7392	0.9110	1.098	1.353	1.485	40′
30′	0.6756	0.7373	0.9163	1.091	1.356	1.480	30′
40′	0.6777	0.7353	0.9217	1.085	1.360	1.476	20′
42°50′	0.6799	0.7333	0.9271	1.079	1.364	1.471	10′
43°00′	0.6820	0.7314	0.9325	1.072	1.367	1.466	47°00′
10′	0.6841	0.7294	0.9380	1.066	1.371	1.462	46°50′
20′	0.6862	0.7274	0.9435	1.060	1.375	1.457	40′
30′	0.6884	0.7254	0.9490	1.054	1.379	1.453	30′
40′	0.6905	0.7234	0.9545	1.048	1.382	1.448	20′
43°50′	0.6926	0.7214	0.9601	1.042	1.386	1.444	10′
44°00′	0.6947	0.7193	0.9657	1.036	1.390	1.440	46°00′
10′	0.6967	0.7173	0.9713	1.030	1.394	1.435	45°50′
20′	0.6988	0.7153	0.9770	1.024	1.398	1.431	40′
30′	0.7009	0.7133	0.9827	1.018	1.402	1.427	30′
40′	0.7030	0.7112	0.9884	1.012	1.406	1.423	20′
44°50′	0.7050	0.7092	0.9942	1.006	1.410	1.418	10′
45°00′	0.7071	0.7071	1.000	1.000	1.414	1.414	45°00′
	cos	sin	cot	tan	csc	sec	

TABLE II TRIGONOMETRIC FUNCTIONS—DECIMAL DEGREES
(csc $\theta = 1/\sin \theta$; sec $\theta = 1/\cos \theta$)

↳→	sin	cos	tan	cot	
0.0	0.00000	1.0000	0.00000	∞	90.0
.1	.00175	1.0000	.00175	573.0	89.9
.2	.00349	1.0000	.00349	286.5	.8
.3	.00524	1.0000	.00524	191.0	.7
.4	.00698	1.0000	.00698	143.24	.6
.5	.00873	1.0000	.00873	114.59	.5
.6	.01047	0.9999	.01047	95.49	.4
.7	.01222	.9999	.01222	81.85	.3
.8	.01396	.9999	.01396	71.62	.2
.9	.01571	.9999	.01571	63.66	89.1
1.0	0.01745	0.9998	0.01746	57.29	89.0
.1	.01920	.9998	.01920	52.08	88.9
.2	.02094	.9998	.02095	47.74	.8
.3	.02269	.9997	.02269	44.07	.7
.4	.02443	.9997	.02444	40.92	.6
.5	.02618	.9997	.02619	38.19	.5
.6	.02792	.9996	.02793	35.80	.4
.7	.02967	.9996	.02968	33.69	.3
.8	.03141	.9995	.03143	31.82	.2
.9	.03316	.9995	.03317	30.14	88.1
2.0	0.03490	0.9994	0.03492	28.64	88.0
.1	.03664	.9993	.03667	27.27	87.9
.2	.03839	.9993	.03842	26.03	.8
.3	.04013	.9992	.04016	24.90	.7
.4	.04188	.9991	.04191	23.86	.6
.5	.04362	.9990	.04366	22.90	.5
.6	.04536	.9990	.04541	22.02	.4
.7	.04711	.9989	.04716	21.20	.3
.8	.04885	.9988	.04891	20.45	.2
.9	.05059	.9987	.05066	19.74	87.1
3.0	0.05234	0.9986	0.05241	19.081	87.0
.1	.05408	.9985	.05416	18.464	86.9
.2	.05582	.9984	.05591	17.886	.8
.3	.05756	.9983	.05766	17.343	.7
.4	.05931	.9982	.05941	16.832	.6
.5	.06105	.9981	.06116	16.350	.5
.6	.06279	.9980	.06291	15.895	.4
.7	.06453	.9979	.06467	15.464	.3
.8	.06627	.9978	.06642	15.056	.2
.9	.06802	.9977	.06817	14.669	86.1
4.0	0.06976	0.9976	0.06993	14.301	86.0
	cos	sin	cot	tan	←↰

TABLE II TRIGONOMETRIC FUNCTIONS—DECIMAL DEGREES
(csc $\theta = 1/\sin \theta$; sec $\theta = 1/\cos \theta$) (CONTINUED)

↓ →	sin	cos	tan	cot	
4.0	0.06976	0.9976	0.06993	14.301	86.0
.1	.07150	.9974	.07168	13.951	85.9
.2	.07324	.9973	.07344	13.617	.8
.3	.07498	.9972	.07519	13.300	.7
.4	.07672	.9971	.07695	12.996	.6
.5	.07846	.9969	.07870	12.706	.5
.6	.08020	.9968	.08046	12.429	.4
.7	.08194	.9966	.08221	12.163	.3
.8	.08368	.9965	.08397	11.909	.2
.9	.08542	.9963	.08573	11.664	85.1
5.0	0.08716	0.9962	0.08749	11.430	85.0
.1	.08889	.9960	.08925	11.205	84.9
.2	.09063	.9959	.09101	10.988	.8
.3	.09237	.9957	.09277	10.780	.7
.4	.09411	.9956	.09453	10.579	.6
.5	.09585	.9954	.09629	10.385	.5
.6	.09758	.9952	.09805	10.199	.4
.7	.09932	.9951	.09981	10.019	.3
.8	.10106	.9949	.10158	9.845	.2
.9	.10279	.9947	.10334	9.677	84.1
6.0	0.10453	0.9945	0.10510	9.514	84.0
.1	.10626	.9943	.10687	9.357	83.9
.2	.10800	.9942	.10863	9.205	.8
.3	.10973	.9940	.11040	9.058	.7
.4	.11147	.9938	.11217	8.915	.6
.5	.11320	.9936	.11394	8.777	.5
.6	.11494	.9934	.11570	8.643	.4
.7	.11667	.9932	.11747	8.513	.3
.8	.11840	.9930	.11924	8.386	.2
.9	.12014	.9928	.12101	8.264	83.1
7.0	0.12187	0.9925	0.12278	8.144	83.0
.1	.12360	.9923	.12456	8.028	82.9
.2	.12533	.9921	.12633	7.916	.8
.3	.12706	.9919	.12810	7.806	.7
.4	.12880	.9917	.12988	7.700	.6
.5	.13053	.9914	.13165	7.596	.5
.6	.13226	.9912	.13343	7.495	.4
.7	.13399	.9910	.13521	7.396	.3
.8	.13572	.9907	.13698	7.300	.2
.9	.13744	.9905	.13876	7.207	82.1
8.0	0.13917	0.9903	0.14054	7.115	82.0
	cos	sin	cot	tan	← ↑

TABLE II (CONTINUED)

	sin	cos	tan	cot	
8.0	0.13917	0.9903	0.14054	7.115	82.0
.1	.14090	.9900	.14232	7.026	81.9
.2	.14263	.9898	.14410	6.940	.8
.3	.14436	.9895	.14588	6.855	.7
.4	.14608	.9893	.14767	6.772	.6
.5	.14781	.9890	.14945	6.691	.5
.6	.14954	.9888	.15124	6.612	.4
.7	.15126	.9885	.15302	6.535	.3
.8	.15299	.9882	.15481	6.460	.2
.9	.15471	.9880	.15660	6.386	81.1
9.0	0.15643	0.9877	0.15838	6.314	81.0
.1	.15816	.9874	.16017	6.243	80.9
.2	.15988	.9871	.16196	6.174	.8
.3	.16160	.9869	.16376	6.107	.7
.4	.16333	.9866	.16555	6.041	.6
.5	.16505	.9863	.16734	5.976	.5
.6	.16677	.9860	.16914	5.912	.4
.7	.16849	.9857	.17093	5.850	.3
.8	.17021	.9854	.17273	5.789	.2
.9	.17193	.9851	.17453	5.730	80.1
10.0	0.1736	0.9848	0.1763	5.671	80.0
.1	.1754	.9845	.1781	5.614	79.9
.2	.1771	.9842	.1799	5.558	.8
.3	.1788	.9839	.1817	5.503	.7
.4	.1805	.9836	.1835	5.449	.6
.5	.1822	.9833	.1853	5.396	.5
.6	.1840	.9829	.1871	5.343	.4
.7	.1857	.9826	.1890	5.292	.3
.8	.1874	.9823	.1908	5.242	.2
.9	.1891	.9820	.1926	5.193	79.1
11.0	0.1908	0.9816	0.1944	5.145	79.0
.1	.1925	.9813	.1962	5.097	78.9
.2	.1942	.9810	.1980	5.050	.8
.3	.1959	.9806	.1998	5.005	.7
.4	.1977	.9803	.2016	4.959	.6
.5	.1994	.9799	.2035	4.915	.5
.6	.2011	.9796	.2053	4.872	.4
.7	.2028	.9792	.2071	4.829	.3
.8	.2045	.9789	.2089	4.787	.2
.9	.2062	.9785	.2107	4.745	78.1
12.0	0.2079	0.9781	0.2126	4.705	78.0
	cos	sin	cot	tan	

II

TABLE II TRIGONOMETRIC FUNCTIONS—DECIMAL DEGREES
(csc $\theta = 1/\sin\theta$; sec $\theta = 1/\cos\theta$) (CONTINUED)

↓⟶	sin	cos	tan	cot	
12.0	0.2079	0.9781	0.2126	4.705	78.0
.1	.2096	.9778	.2144	4.665	77.9
.2	.2113	.9774	.2162	4.625	.8
.3	.2130	.9770	.2180	4.586	.7
.4	.2147	.9767	.2199	4.548	.6
.5	.2164	.9763	.2217	4.511	.5
.6	.2181	.9759	.2235	4.474	.4
.7	.2198	.9755	.2254	4.437	.3
.8	.2215	.9751	.2272	4.402	.2
.9	.2233	.9748	.2290	4.366	77.1
13.0	0.2250	0.9744	0.2309	4.331	77.0
.1	.2267	.9740	.2327	4.297	76.9
.2	.2284	.9736	.2345	4.264	.8
.3	.2300	.9732	.2364	4.230	.7
.4	.2317	.9728	.2382	4.198	.6
.5	.2334	.9724	.2401	4.165	.5
.6	.2351	.9720	.2419	4.134	.4
.7	.2368	.9715	.2438	4.102	.3
.8	.2385	.9711	.2456	4.071	.2
.9	.2402	.9707	.2475	4.041	76.1
14.0	0.2419	0.9703	0.2493	4.011	76.0
.1	.2436	.9699	.2512	3.981	75.9
.2	.2453	.9694	.2530	3.952	.8
.3	.2470	.9690	.2549	3.923	.7
.4	.2487	.9686	.2568	3.895	.6
.5	.2504	.9681	.2586	3.867	.5
.6	.2521	.9677	.2605	3.839	.4
.7	.2538	.9673	.2623	3.812	.3
.8	.2554	.9668	.2642	3.785	.2
.9	.2571	.9664	.2661	3.758	75.1
15.0	0.2588	0.9659	0.2679	3.732	75.0
.1	.2605	.9655	.2698	3.706	74.9
.2	.2622	.9650	.2717	3.681	.8
.3	.2639	.9646	.2736	3.655	.7
.4	.2656	.9641	.2754	3.630	.6
.5	.2672	.9636	.2773	3.606	.5
.6	.2689	.9632	.2792	3.582	.4
.7	.2706	.9627	.2811	3.558	.3
.8	.2723	.9622	.2830	3.534	.2
.9	.2740	.9617	.2849	3.511	74.1
16.0	0.2756	0.9613	0.2867	3.487	74.0
	cos	sin	cot	tan	⟵↑

TABLE II (CONTINUED)

$\searrow$	sin	cos	tan	cot	
16.0	0.2756	0.9613	0.2867	3.487	74.0
.1	.2773	.9608	.2886	3.465	73.9
.2	.2790	.9603	.2905	3.442	.8
.3	.2807	.9598	.2924	3.420	.7
.4	.2823	.9593	.2943	3.398	.6
.5	.2840	.9588	.2962	3.376	.5
.6	.2857	.9583	.2981	3.354	.4
.7	.2874	.9578	.3000	3.333	.3
.8	.2890	.9573	.3019	3.312	.2
.9	.2907	.9568	.3038	3.291	73.1
17.0	0.2924	0.9563	0.3057	3.271	73.0
.1	.2940	.9558	.3076	3.251	72.9
.2	.2957	.9553	.3096	3.230	.8
.3	.2974	.9548	.3115	3.211	.7
.4	.2990	.9542	.3134	3.191	.6
.5	.3007	.9537	.3153	3.172	.5
.6	.3024	.9532	.3172	3.152	.4
.7	.3040	.9527	.3191	3.133	.3
.8	.3057	.9521	.3211	3.115	.2
.9	.3074	.9516	.3230	3.096	72.1
18.0	0.3090	0.9511	0.3249	3.078	72.0
.1	.3107	.9505	.3269	3.060	71.9
.2	.3123	.9500	.3288	3.042	.8
.3	.3140	.9494	.3307	3.024	.7
.4	.3156	.9489	.3327	3.006	.6
.5	.3173	.9483	.3346	2.989	.5
.6	.3190	.9478	.3365	2.971	.4
.7	.3206	.9472	.3385	2.954	.3
.8	.3223	.9466	.3404	2.937	.2
.9	.3239	.9461	.3424	2.921	71.1
19.0	0.3256	0.9455	0.3443	2.904	71.0
.1	.3272	.9449	.3463	2.888	70.9
.2	.3289	.9444	.3482	2.872	.8
.3	.3305	.9438	.3502	2.856	.7
.4	.3322	.9432	.3522	2.840	.6
.5	.3338	.9426	.3541	2.824	.5
.6	.3355	.9421	.3561	2.808	.4
.7	.3371	.9415	.3581	2.793	.3
.8	.3387	.9409	.3600	2.778	.2
.9	.3404	.9403	.3620	2.762	70.1
20.0	0.3420	0.9397	0.3640	2.747	70.0
	cos	sin	cot	tan	$\nwarrow$

II

TABLE II TRIGONOMETRIC FUNCTIONS—DECIMAL DEGREES
(csc θ = 1/sin θ; sec θ = 1/cos θ) (CONTINUED)

↓⟶	sin	cos	tan	cot	
20.0	0.3420	0.9397	0.3640	2.747	70.0
.1	.3437	.9391	.3659	2.733	69.9
.2	.3453	.9385	.3679	2.718	.8
.3	.3469	.9379	.3699	2.703	.7
.4	.3486	.9373	.3719	2.689	.6
.5	.3502	.9367	.3739	2.675	.5
.6	.3518	.9361	.3759	2.660	.4
.7	.3535	.9354	.3779	2.646	.3
.8	.3551	.9348	.3799	2.633	.2
.9	.3567	.9342	.3819	2.619	69.1
21.0	0.3584	0.9336	0.3839	2.605	69.0
.1	.3600	.9330	.3859	2.592	68.9
.2	.3616	.9323	.3879	2.578	.8
.3	.3633	.9317	.3899	2.565	.7
.4	.3649	.9311	.3919	2.552	.6
.5	.3665	.9304	.3939	2.539	.5
.6	.3681	.9298	.3959	2.526	.4
.7	.3697	.9291	.3979	2.513	.3
.8	.3714	.9285	.4000	2.500	.2
.9	.3730	.9278	.4020	2.488	68.1
22.0	0.3746	0.9272	0.4040	2.475	68.0
.1	.3762	.9265	.4061	2.463	67.9
.2	.3778	.9259	.4081	2.450	.8
.3	.3795	.9252	.4101	2.438	.7
.4	.3811	.9245	.4122	2.426	.6
.5	.3827	.9239	.4142	2.414	.5
.6	.3843	.9232	.4163	2.402	.4
.7	.3859	.9225	.4183	2.391	.3
.8	.3875	.9219	.4204	2.379	.2
.9	.3891	.9212	.4224	2.367	67.1
23.0	0.3907	0.9205	0.4245	2.356	67.0
.1	.3923	.9198	.4265	2.344	66.9
.2	.3939	.9191	.4286	2.333	.8
.3	.3955	.9184	.4307	2.322	.7
.4	.3971	.9178	.4327	2.311	.6
.5	.3987	.9171	.4348	2.300	.5
.6	.4003	.9164	.4369	2.289	.4
.7	.4019	.9157	.4390	2.278	.3
.8	.4035	.9150	.4411	2.267	.2
.9	.4051	.9143	.4431	2.257	66.1
24.0	0.4067	0.9135	0.4452	2.246	66.0
	cos	sin	cot	tan	⟵↑

TABLE II (CONTINUED)

⌐→	sin	cos	tan	cot	
24.0	0.4067	0.9135	0.4452	2.246	66.0
.1	.4083	.9128	.4473	2.236	65.9
.2	.4099	.9121	.4494	2.225	.8
.3	.4115	.9114	.4515	2.215	.7
.4	.4131	.9107	.4536	2.204	.6
.5	.4147	.9100	.4557	2.194	.5
.6	.4163	.9092	.4578	2.184	.4
.7	.4179	.9085	.4599	2.174	.3
.8	.4195	.9078	.4621	2.164	.2
.9	.4210	.9070	.4642	2.154	65.1
25.0	0.4226	0.9063	0.4663	2.145	65.0
.1	.4242	.9056	.4684	2.135	64.9
.2	.4258	.9048	.4706	2.125	.8
.3	.4274	.9041	.4727	2.116	.7
.4	.4289	.9033	.4748	2.106	.6
.5	.4305	.9026	.4770	2.097	.5
.6	.4321	.9018	.4791	2.087	.4
.7	.4337	.9011	.4813	2.078	.3
.8	.4352	.9003	.4834	2.069	.2
.9	.4368	.8996	.4856	2.059	64.1
26.0	0.4384	0.8988	0.4877	2.050	64.0
.1	.4399	.8980	.4899	2.041	63.9
.2	.4415	.8973	.4921	2.032	.8
.3	.4431	.8965	.4942	2.023	.7
.4	.4446	.8957	.4964	2.014	.6
.5	.4462	.8949	.4986	2.006	.5
.6	.4478	.8942	.5008	1.997	.4
.7	.4493	.8934	.5029	1.988	.3
.8	.4509	.8926	.5051	1.980	.2
.9	.4524	.8918	.5073	1.971	63.1
27.0	0.4540	0.8910	0.5095	1.963	63.0
.1	.4555	.8902	.5117	1.954	62.9
.2	.4571	.8894	.5139	1.946	.8
.3	.4586	.8886	.5161	1.937	.7
.4	.4602	.8878	.5184	1.929	.6
.5	.4617	.8870	.5206	1.921	.5
.6	.4633	.8862	.5228	1.913	.4
.7	.4648	.8854	.5250	1.905	.3
.8	.4664	.8846	.5272	1.897	.2
.9	.4679	.8838	.5295	1.889	62.1
28.0	0.4695	0.8829	0.5317	1.881	62.0
	cos	sin	cot	tan	←⌐

TABLE II TRIGONOMETRIC FUNCTIONS—DECIMAL DEGREES
(csc $\theta = 1/\sin \theta$; sec $\theta = 1/\cos \theta$) (CONTINUED)

↓⟶	sin	cos	tan	cot	
28.0	0.4695	0.8829	0.5317	1.881	62.0
.1	.4710	.8821	.5340	1.873	61.9
.2	.4726	.8813	.5362	1.865	.8
.3	.4741	.8805	.5384	1.857	.7
.4	.4756	.8796	.5407	1.849	.6
.5	.4772	.8788	.5430	1.842	.5
.6	.4787	.8780	.5452	1.834	.4
.7	.4802	.8771	.5475	1.827	.3
.8	.4818	.8763	.5498	1.819	.2
.9	.4833	.8755	.5520	1.811	61.1
29.0	0.4848	0.8746	0.5543	1.804	61.0
.1	.4863	.8738	.5566	1.797	60.9
.2	.4879	.8729	.5589	1.789	.8
.3	.4894	.8721	.5612	1.782	.7
.4	.4909	.8712	.5635	1.775	.6
.5	.4924	.8704	.5658	1.767	.5
.6	.4939	.8695	.5681	1.760	.4
.7	.4955	.8686	.5704	1.753	.3
.8	.4970	.8678	.5727	1.746	.2
.9	.4985	.8669	.5750	1.739	60.1
30.0	0.5000	0.8660	0.5774	1.7321	60.0
.1	.5015	.8652	.5797	1.7251	59.9
.2	.5030	.8643	.5820	1.7182	.8
.3	.5045	.8634	.5844	1.7113	.7
.4	.5060	.8625	.5867	1.7045	.6
.5	.5075	.8616	.5890	1.6977	.5
.6	.5090	.8607	.5914	1.6909	.4
.7	.5105	.8599	.5938	1.6842	.3
.8	.5120	.8590	.5961	1.6775	.2
.9	.5135	.8581	.5985	1.6709	59.1
31.0	0.5150	0.8572	0.6009	1.6643	59.0
.1	.5165	.8563	.6032	1.6577	58.9
.2	.5180	.8554	.6056	1.6512	.8
.3	.5195	.8545	.6080	1.6447	.7
.4	.5210	.8536	.6104	1.6383	.6
.5	.5225	.8526	.6128	1.6319	.5
.6	.5240	.8517	.6152	1.6255	.4
.7	.5255	.8508	.6176	1.6191	.3
.8	.5270	.8499	.6200	1.6128	.2
.9	.5284	.8490	.6224	1.6066	58.1
32.0	0.5299	0.8480	0.6249	1.6003	58.0
	cos	sin	cot	tan	⟵↑

TABLE II (CONTINUED)

	sin	cos	tan	cot	
32.0	0.5299	0.8480	0.6249	1.6003	58.0
.1	.5314	.8471	.6273	1.5941	57.9
.2	.5329	.8462	.6297	1.5880	.8
.3	.5344	.8453	.6322	1.5818	.7
.4	.5358	.8443	.6346	1.5757	.6
.5	.5373	.8434	.6371	1.5697	.5
.6	.5388	.8425	.6395	1.5637	.4
.7	.5402	.8415	.6420	1.5577	.3
.8	.5417	.8406	.6445	1.5517	.2
.9	.5432	.8396	.6469	1.5458	57.1
33.0	0.5446	0.8387	0.6494	1.5399	57.0
.1	.5461	.8377	.6519	1.5340	56.9
.2	.5476	.8368	.6544	1.5282	.8
.3	.5490	.8358	.6569	1.5224	.7
.4	.5505	.8348	.6594	1.5166	.6
.5	.5519	.8339	.6619	1.5108	.5
.6	.5534	.8329	.6644	1.5051	.4
.7	.5548	.8320	.6669	1.4994	.3
.8	.5563	.8310	.6694	1.4938	.2
.9	.5577	.8300	.6720	1.4882	56.1
34.0	0.5592	0.8290	0.6745	1.4826	56.0
.1	.5606	.8281	.6771	1.4770	55.9
.2	.5621	.8271	.6796	1.4715	.8
.3	.5635	.8261	.6822	1.4659	.7
.4	.5650	.8251	.6847	1.4605	.6
.5	.5664	.8241	.6873	1.4550	.5
.6	.5678	.8231	.6899	1.4496	.4
.7	.5693	.8221	.6924	1.4442	.3
.8	.5707	.8211	.6950	1.4388	.2
.9	.5721	.8202	.6976	1.4335	55.1
35.0	0.5736	0.8192	0.7002	1.4281	55.0
.1	.5750	.8181	.7028	1.4229	54.9
.2	.5764	.8171	.7054	1.4176	.8
.3	.5779	.8161	.7080	1.4124	.7
.4	.5793	.8151	.7107	1.4071	.6
.5	.5807	.8141	.7133	1.4019	.5
.6	.5821	.8131	.7159	1.3968	.4
.7	.5835	.8121	.7186	1.3916	.3
.8	.5850	.8111	.7212	1.3865	.2
.9	.5864	.8100	.7239	1.3814	54.1
36.0	0.5878	0.8090	0.7265	1.3764	54.0
	cos	sin	cot	tan	

TABLE II TRIGONOMETRIC FUNCTIONS—DECIMAL DEGREES
$(\csc \theta = 1/\sin \theta; \sec \theta = 1/\cos \theta)$ (CONTINUED)

	sin	cos	tan	cot	
36.0	0.5878	0.8090	0.7265	1.3764	54.0
.1	.5892	.8080	.7292	1.3713	53.9
.2	.5906	.8070	.7319	1.3663	.8
.3	.5920	.8059	.7346	1.3613	.7
.4	.5934	.8049	.7373	1.3564	.6
.5	.5948	.8039	.7400	1.3514	.5
.6	.5962	.8028	.7427	1.3465	.4
.7	.5976	.8018	.7454	1.3416	.3
.8	.5990	.8007	.7481	1.3367	.2
.9	.6004	.7997	.7508	1.3319	53.1
37.0	0.6018	0.7986	0.7536	1.3270	53.0
.1	.6032	.7976	.7563	1.3222	52.9
.2	.6046	.7965	.7590	1.3175	.8
.3	.6060	.7955	.7618	1.3127	.7
.4	.6074	.7944	.7646	1.3079	.6
.5	.6088	.7934	.7673	1.3032	.5
.6	.6101	.7923	.7701	1.2985	.4
.7	.6115	.7912	.7729	1.2938	.3
.8	.6129	.7902	.7757	1.2892	.2
.9	.6143	.7891	.7785	1.2846	52.1
38.0	0.6157	0.7880	0.7813	1.2799	52.0
.1	.6170	.7869	.7841	1.2753	51.9
.2	.6184	.7859	.7869	1.2708	.8
.3	.6198	.7848	.7898	1.2662	.7
.4	.6211	.7837	.7926	1.2617	.6
.5	.6225	.7826	.7954	1.2572	.5
.6	.6239	.7815	.7983	1.2527	.4
.7	.6252	.7804	.8012	1.2482	.3
.8	.6266	.7793	.8040	1.2437	.2
.9	.6280	.7782	.8069	1.2393	51.1
39.0	0.6293	0.7771	0.8098	1.2349	51.0
.1	.6307	.7760	.8127	1.2305	50.9
.2	.6320	.7749	.8156	1.2261	.8
.3	.6334	.7738	.8185	1.2218	.7
.4	.6347	.7727	.8214	1.2174	.6
.5	.6361	.7716	.8243	1.2131	.5
.6	.6374	.7705	.8273	1.2088	.4
.7	.6388	.7694	.8302	1.2045	.3
.8	.6401	.7683	.8332	1.2002	.2
.9	.6414	.7672	.8361	1.1960	50.1
40.0	0.6428	0.7660	0.8391	1.1918	50.0
	cos	sin	cot	tan	

TABLE II (CONTINUED)

↓⟶	sin	cos	tan	cot	
40.0	0.6428	0.7660	0.8391	1.1918	50.0
.1	.6441	.7649	.8421	1.1875	49.9
.2	.6455	.7638	.8451	1.1833	.8
.3	.6468	.7627	.8481	1.1792	.7
.4	.6481	.7615	.8511	1.1750	.6
.5	.6494	.7604	.8541	1.1708	.5
.6	.6508	.7593	.8571	1.1667	.4
.7	.6521	.7581	.8601	1.1626	.3
.8	.6534	.7570	.8632	1.1585	.2
.9	.6547	.7559	.8662	1.1544	49.1
41.0	0.6561	0.7547	0.8693	1.1504	49.0
.1	.6574	.7536	.8724	1.1463	48.9
.2	.6587	.7524	.8754	1.1423	.8
.3	.6600	.7513	.8785	1.1383	.7
.4	.6613	.7501	.8816	1.1343	.6
.5	.6626	.7490	.8847	1.1303	.5
.6	.6639	.7478	.8878	1.1263	.4
.7	.6652	.7466	.8910	1.1224	.3
.8	.6665	.7455	.8941	1.1184	.2
.9	.6678	.7443	.8972	1.1145	48.1
42.0	0.6691	0.7431	0.9004	1.1106	48.0
.1	.6704	.7420	.9036	1.1067	47.9
.2	.6717	.7408	.9067	1.1028	.8
.3	.6730	.7396	.9099	1.0990	.7
.4	.6743	.7385	.9131	1.0951	.6
.5	.6756	.7373	.9163	1.0913	.5
.6	.6769	.7361	.9195	1.0875	.4
.7	.6782	.7349	.9228	1.0837	.3
.8	.6794	.7337	.9260	1.0799	.2
.9	.6807	.7325	.9293	1.0761	47.1
43.0	0.6820	0.7314	0.9325	1.0724	47.0
.1	.6833	.7302	.9358	1.0686	46.9
.2	.6845	.7290	.9391	1.0649	.8
.3	.6858	.7278	.9424	1.0612	.7
.4	.6871	.7266	.9457	1.0575	.6
.5	.6884	.7254	.9490	1.0538	.5
.6	.6896	.7242	.9523	1.0501	.4
.7	.6909	.7230	.9556	1.0464	.3
.8	.6921	.7218	.9590	1.0428	.2
.9	.6934	.7206	.9623	1.0392	46.1
44.0	0.6947	0.7193	0.9657	1.0355	46.0
	cos	sin	cot	tan	⟵↑

TABLE II TRIGONOMETRIC FUNCTIONS—DECIMAL DEGREES
(csc θ = 1/sin θ; sec θ = 1/cos θ) (CONTINUED)

↓⟶	sin	cos	tan	cot	
44.0	0.6947	0.7193	0.9657	1.0355	46.0
.1	.6959	.7181	.9691	1.0319	45.9
.2	.6972	.7169	.9725	1.0283	.8
.3	.6984	.7157	.9759	1.0247	.7
.4	.6997	.7145	.9793	1.0212	.6
.5	.7009	.7133	.9827	1.0176	.5
.6	.7022	.7120	.9861	1.0141	.4
.7	.7034	.7108	.9896	1.0105	.3
.8	.7046	.7096	.9930	1.0070	.2
.9	.7059	.7083	.9965	1.0035	45.1
45.0	0.7071	0.7071	1.0000	1.0000	45.0
	cos	sin	cot	tan	⟵↑

TABLE III TRIGONOMETRIC FUNCTIONS—RADIANS OR REAL NUMBERS

↓ →	sin	cos	tan	cot	sec	csc
.00	.0000	1.0000	.0000	—	1.000	—
.01	.0100	1.0000	.0100	99.997	1.000	100.00
.02	.0200	.9998	.0200	49.993	1.000	50.00
.03	.0300	.9996	.0300	33.323	1.000	33.34
.04	.0400	.9992	.0400	24.987	1.001	25.01
.05	.0500	.9988	.0500	19.983	1.001	20.01
.06	.0600	.9982	.0601	16.647	1.002	16.68
.07	.0699	.9976	.0701	14.262	1.002	14.30
.08	.0799	.9968	.0802	12.473	1.003	12.51
.09	.0899	.9960	.0902	11.081	1.004	11.13
.10	.0998	.9950	.1003	9.967	1.005	10.02
.11	.1098	.9940	.1104	9.054	1.006	9.109
.12	.1197	.9928	.1206	8.293	1.007	8.353
.13	.1296	.9916	.1307	7.649	1.009	7.714
.14	.1395	.9902	.1409	7.096	1.010	7.166
.15	.1494	.9888	.1511	6.617	1.011	6.692
.16	.1593	.9872	.1614	6.197	1.013	6.277
.17	.1692	.9856	.1717	5.826	1.015	5.911
.18	.1790	.9838	.1820	5.495	1.016	5.586
.19	.1889	.9820	.1923	5.200	1.018	5.295
.20	.1987	.9801	.2027	4.933	1.020	5.033
.21	.2085	.9780	.2131	4.692	1.022	4.797
.22	.2182	.9759	.2236	4.472	1.025	4.582
.23	.2280	.9737	.2341	4.271	1.027	4.386
.24	.2377	.9713	.2447	4.086	1.030	4.207
.25	.2474	.9689	.2553	3.916	1.032	4.042
.26	.2571	.9664	.2660	3.759	1.035	3.890
.27	.2667	.9638	.2768	3.613	1.038	3.749
.28	.2764	.9611	.2876	3.478	1.041	3.619
.29	.2860	.9582	.2984	3.351	1.044	3.497
.30	.2955	.9553	.3093	3.233	1.047	3.384
.31	.3051	.9523	.3203	3.122	1.050	3.278
.32	.3146	.9492	.3314	3.018	1.053	3.179
.33	.3240	.9460	.3425	2.920	1.057	3.086
.34	.3335	.9428	.3537	2.827	1.061	2.999
.35	.3429	.9394	.3650	2.740	1.065	2.916
.36	.3523	.9359	.3764	2.657	1.068	2.839
.37	.3616	.9323	.3879	2.578	1.073	2.765
.38	.3709	.9287	.3994	2.504	1.077	2.696
.39	.3802	.9249	.4111	2.433	1.081	2.630
	sin	cos	tan	cot	sec	csc

III

TABLE III TRIGONOMETRIC FUNCTIONS—RADIANS OR REAL NUMBERS
(CONTINUED)

↓⟶	sin	cos	tan	cot	sec	csc
.40	.3894	.9211	.4228	2.365	1.086	2.568
.41	.3986	.9171	.4346	2.301	1.090	2.509
.42	.4078	.9131	.4466	2.239	1.095	2.452
.43	.4169	.9090	.4586	2.180	1.100	2.399
.44	.4259	.9048	.4708	2.124	1.105	2.348
.45	.4350	.9004	.4831	2.070	1.111	2.299
.46	.4439	.8961	.4954	2.018	1.116	2.253
.47	.4529	.8916	.5080	1.969	1.122	2.208
.48	.4618	.8870	.5206	1.921	1.127	2.166
.49	.4706	.8823	.5334	1.875	1.133	2.125
.50	.4794	.8776	.5463	1.830	1.139	2.086
.51	.4882	.8727	.5594	1.788	1.146	2.048
.52	.4969	.8678	.5726	1.747	1.152	2.013
.53	.5055	.8628	.5859	1.707	1.159	1.978
.54	.5141	.8577	.5994	1.668	1.166	1.945
.55	.5227	.8525	.6131	1.631	1.173	1.913
.56	.5312	.8473	.6269	1.595	1.180	1.883
.57	.5396	.8419	.6410	1.560	1.188	1.853
.58	.5480	.8365	.6552	1.526	1.196	1.825
.59	.5564	.8309	.6696	1.494	1.203	1.797
.60	.5646	.8253	.6841	1.462	1.212	1.771
.61	.5729	.8196	.6989	1.431	1.220	1.746
.62	.5810	.8139	.7139	1.401	1.229	1.721
.63	.5891	.8080	.7291	1.372	1.238	1.697
.64	.5972	.8021	.7445	1.343	1.247	1.674
.65	.6052	.7961	.7602	1.315	1.256	1.652
.66	.6131	.7900	.7761	1.288	1.266	1.631
.67	.6210	.7838	.7923	1.262	1.276	1.610
.68	.6288	.7776	.8087	1.237	1.286	1.590
.69	.6365	.7712	.8253	1.212	1.297	1.571
.70	.6442	.7648	.8423	1.187	1.307	1.552
.71	.6518	.7584	.8595	1.163	1.319	1.534
.72	.6594	.7518	.8771	1.140	1.330	1.517
.73	.6669	.7452	.8949	1.117	1.342	1.500
.74	.6743	.7385	.9131	1.095	1.354	1.483
.75	.6816	.7317	.9316	1.073	1.367	1.467
.76	.6889	.7248	.9505	1.052	1.380	1.452
.77	.6961	.7179	.9697	1.031	1.393	1.437
.78	.7033	.7109	.9893	1.011	1.407	1.422
.79	.7104	.7038	1.009	.9908	1.421	1.408
	sin	cos	tan	cot	sec	

TABLE III (CONTINUED)

↓ →	sin	cos	tan	cot	sec	csc
.80	.7174	.6967	1.030	.9712	1.435	1.394
.81	.7243	.6895	1.050	.9520	1.450	1.381
.82	.7311	.6822	1.072	.9331	1.466	1.368
.83	.7379	.6749	1.093	.9146	1.482	1.355
.84	.7446	.6675	1.116	.8964	1.498	1.343
.85	.7513	.6600	1.138	.8785	1.515	1.331
.86	.7578	.6524	1.162	.8609	1.533	1.320
.87	.7643	.6448	1.185	.8437	1.551	1.308
.88	.7707	.6372	1.210	.8267	1.569	1.297
.89	.7771	.6294	1.235	.8100	1.589	1.287
.90	.7833	.6216	1.260	.7936	1.609	1.277
.91	.7895	.6137	1.286	.7774	1.629	1.267
.92	.7956	.6058	1.313	.7615	1.651	1.257
.93	.8016	.5978	1.341	.7458	1.673	1.247
.94	.8076	.5898	1.369	.7303	1.696	1.238
.95	.8134	.5817	1.398	.7151	1.719	1.229
.96	.8192	.5735	1.428	.7001	1.744	1.221
.97	.8249	.5653	1.459	.6853	1.769	1.212
.98	.8305	.5570	1.491	.6707	1.795	1.204
.99	.8360	.5487	1.524	.6563	1.823	1.196
1.00	.8415	.5403	1.557	.6421	1.851	1.188
1.01	.8468	.5319	1.592	.6281	1.880	1.181
1.02	.8521	.5234	1.628	.6142	1.911	1.174
1.03	.8573	.5148	1.665	.6005	1.942	1.166
1.04	.8624	.5062	1.704	.5870	1.975	1.160
1.05	.8674	.4976	1.743	.5736	2.010	1.153
1.06	.8724	.4889	1.784	.5604	2.046	1.146
1.07	.8772	.4801	1.827	.5473	2.083	1.140
1.08	.8820	.4713	1.871	.5344	2.122	1.134
1.09	.8866	.4625	1.917	.5216	2.162	1.128
1.10	.8912	.4536	1.965	.5090	2.205	1.122
1.11	.8957	.4447	2.014	.4964	2.249	1.116
1.12	.9001	.4357	2.066	.4840	2.295	1.111
1.13	.9044	.4267	2.120	.4718	2.344	1.106
1.14	.9086	.4176	2.176	.4596	2.395	1.101
1.15	.9128	.4085	2.234	.4475	2.448	1.096
1.16	.9168	.3993	2.296	.4356	2.504	1.091
1.17	.9208	.3902	2.360	.4237	2.563	1.086
1.18	.9246	.3809	2.427	.4120	2.625	1.082
1.19	.9284	.3717	2.498	.4003	2.691	1.077
	sin	cos	tan	cot	sec	csc

III

TABLE III TRIGONOMETRIC FUNCTIONS—RADIANS OR REAL NUMBERS (CONTINUED)

↓⟶	sin	cos	tan	cot	sec	csc
1.20	.9320	.3624	2.572	.3888	2.760	1.073
1.21	.9356	.3530	2.650	.3773	2.833	1.069
1.22	.9391	.3436	2.733	.3659	2.910	1.065
1.23	.9425	.3342	2.820	.3546	2.992	1.061
1.24	.9458	.3248	2.912	.3434	3.079	1.057
1.25	.9490	.3153	3.010	.3323	3.171	1.054
1.26	.9521	.3058	3.113	.3212	3.270	1.050
1.27	.9551	.2963	3.224	.3102	3.375	1.047
1.28	.9580	.2867	3.341	.2993	3.488	1.044
1.29	.9608	.2771	3.467	.2884	3.609	1.041
1.30	.9636	.2675	3.602	.2776	3.738	1.038
1.31	.9662	.2579	3.747	.2669	3.878	1.035
1.32	.9687	.2482	3.903	.2562	4.029	1.032
1.33	.9711	.2385	4.072	.2456	4.193	1.030
1.34	.9735	.2288	4.256	.2350	4.372	1.027
1.35	.9757	.2190	4.455	.2245	4.566	1.025
1.36	.9779	.2092	4.673	.2140	4.779	1.023
1.37	.9799	.1994	4.913	.2035	5.014	1.021
1.38	.9819	.1896	5.177	.1931	5.273	1.018
1.39	.9837	.1798	5.471	.1828	5.561	1.017
1.40	.9854	.1700	5.798	.1725	5.883	1.015
1.41	.9871	.1601	6.165	.1622	6.246	1.013
1.42	.9887	.1502	6.581	.1519	6.657	1.011
1.43	.9901	.1403	7.055	.1417	7.126	1.010
1.44	.9915	.1304	7.602	.1315	7.667	1.009
1.45	.9927	.1205	8.238	.1214	8.299	1.007
1.46	.9939	.1106	8.989	.1113	9.044	1.006
1.47	.9949	.1006	9.887	.1011	9.938	1.005
1.48	.9959	.0907	10.983	.0910	11.029	1.004
1.49	.9967	.0807	12.350	.0810	12.390	1.003
1.50	.9975	.0707	14.101	.0709	14.137	1.003
1.51	.9982	.0608	16.428	.0609	16.458	1.002
1.52	.9987	.0508	19.670	.0508	19.695	1.001
1.53	.9992	.0408	24.498	.0408	24.519	1.001
1.54	.9995	.0308	32.461	.0308	32.476	1.000
1.55	.9998	.0208	48.078	.0208	48.089	1.000
1.56	.9999	.0108	92.620	.0108	92.626	1.000
1.57	1.0000	.0008	1,255.8	.0008	1,255.8	1.000
1.58	1.0000	−.0092	−108.65	−.0092	−108.65	1.000
1.59	.9998	−.0192	−52.067	−.0192	−52.08	1.000
1.60	.9996	−.0292	−34.233	−.0292	−34.25	1.000
	sin	cos	tan	cot	sec	csc

III

TABLE IV COMMON LOGARITHMS

↓→	0	1	2	3	4	5	6	7	8	9
1.0	.0000	.004321	.008600	.01284	.01703	.02119	.02531	.02938	.03342	.03743
1.1	.04139	.04532	.04922	.05308	.05690	.06070	.06446	.06819	.07188	.07555
1.2	.07918	.08279	.08636	.08991	.09342	.09691	.1004	.1038	.1072	.1106
1.3	.1139	.1173	.1206	.1239	.1271	.1303	.1335	.1367	.1399	.1430
1.4	.1461	.1492	.1523	.1553	.1584	.1614	.1644	.1673	.1703	.1732
1.5	.1761	.1790	.1818	.1847	.1875	.1903	.1931	.1959	.1987	.2014
1.6	.2041	.2068	.2095	.2122	.2148	.2175	.2201	.2227	.2253	.2279
1.7	.2304	.2330	.2355	.2380	.2405	.2430	.2455	.2480	.2504	.2529
1.8	.2553	.2577	.2601	.2625	.2648	.2673	.2695	.2718	.2742	.2765
1.9	.2788	.2810	.2833	.2856	.2878	.2900	.2923	.2945	.2967	.2989
2.0	.3010	.3032	.3054	.3075	.3096	.3118	.3139	.3160	.3181	.3201
2.1	.3222	.3243	.3263	.3284	.3304	.3324	.3345	.3365	.3385	.3404
2.2	.3424	.3444	.3464	.3483	.3502	.3522	.3541	.3560	.3579	.3598
2.3	.3617	.3636	.3655	.3674	.3692	.3711	.3729	.3747	.3766	.3784
2.4	.3802	.3820	.3838	.3856	.3874	.3892	.3909	.3927	.3945	.3962
2.5	.3979	.3997	.4014	.4031	.4048	.4065	.4082	.4099	.4116	4133
2.6	.4150	.4166	.4183	.4200	.4216	.4232	.4249	.4265	.4281	.4298
2.7	.4314	.4330	.4346	.4362	.4378	.4393	.4409	.4425	.4440	.4456
2.8	.4472	.4487	.4502	.4518	.4533	.4548	.4564	.4579	.4594	.4609
2.9	.4624	.4639	.4654	.4669	.4683	.4698	.4713	.4728	.4742	.4757
3.0	.4771	.4786	.4800	.4814	.4829	.4843	.4857	.4871	.4886	.4900
3.1	.4914	.4928	.4942	.4955	.4969	.4983	.4997	.5011	.5024	.5083
3.2	.5051	.5065	.5079	.5092	.5105	.5119	.5132	.5145	.5159	.5172
3.3	.5185	.5198	.5211	.5224	.5237	.5250	.5263	.5276	.5289	.5302
3.4	.5315	.5328	.5340	.5353	.5366	.5378	.5391	.5403	.5416	.5428
3.5	.5441	.5453	.5465	.5478	.5490	.5502	.5514	.5527	.5539	.5551
3.6	.5563	.5575	.5587	.5599	.5611	.5623	.5635	.5647	.5658	.5670
3.7	.5682	.5694	.5705	.5717	.5729	.5740	.5752	.5763	.5775	.5786
3.8	.5798	.5809	.5821	.5832	.5843	.5855	.5866	.5877	.5888	.5899
3.9	.5911	.5922	.5933	.5944	.5955	.5966	.5977	.5988	.5999	.6010
4.0	.6021	.6031	.6042	.6053	.6064	.6075	.6085	.6096	.6107	.6117
4.1	.6128	.6138	.6149	.6160	.6170	.6180	.6191	.6201	.6212	.6222
4.2	.6232	.6243	.6253	.6263	.6274	.6284	.6294	.6304	.6314	.6325
4.3	.6335	.6345	.6355	.6365	.6375	.6385	.6395	.6405	.6415	.6425
4.4	.6435	.6444	.6454	.6464	.6474	.6484	.6493	.6503	.6513	.6522
4.5	.6532	.6542	.6551	.6561	.6571	.6580	.6590	.6599	.6609	.6618
4.6	.6628	.6637	.6646	.6656	.6665	.6675	.6684	.6693	.6702	.6712
4.7	.6721	.6730	.6739	.6749	.6758	.6767	.6776	.6785	.6794	.6803
4.8	.6812	.6821	.6830	.6839	.6848	.6857	.6866	.6875	.6884	.6893
4.9	.6902	.6911	.6920	.6928	.6937	.6946	.6955	.6964	.6972	.6981

IV

TABLE IV COMMON LOGARITHMS (CONTINUED)

	0	1	2	3	4	5	6	7	8	9
5.0	.6990	.6998	.7007	.7016	.7024	.7033	.7042	.7050	.7059	.7067
5.1	.7076	.7084	.7093	.7101	.7110	.7118	.7126	.7135	.7143	.7152
5.2	.7160	.7168	.7177	.7185	.7193	.7202	.7210	.7218	.7226	.7235
5.3	.7243	.7251	.7259	.7267	.7275	.7284	.7292	.7300	.7308	.7316
5.4	.7324	.7332	.7340	.7348	.7356	.7364	.7372	.7380	.7388	.7396
5.5	.7404	.7412	.7419	.7427	.7435	.7443	.7451	.7459	.7466	.7474
5.6	.7482	.7490	.7497	.7505	.7513	.7520	.7528	.7536	.7543	.7551
5.7	.7559	.7566	.7574	.7582	.7589	.7597	.7604	.7612	.7619	.7627
5.8	.7634	.7642	.7649	.7657	.7664	.7672	.7679	.7686	.7694	.7701
5.9	.7709	.7716	.7723	.7731	.7738	.7745	.7752	.7760	.7767	.7774
6.0	.7782	.7789	.7796	.7803	.7810	.7818	.7825	.7832	.7839	.7846
6.1	.7853	.7860	.7868	.7875	.7882	.7889	.7896	.7903	.7910	.7917
6.2	.7924	.7931	.7938	.7945	.7952	.7959	.7966	.7973	.7980	.7987
6.3	.7993	.8000	.8007	.8014	.8021	.8028	.8035	.8041	.8048	.8055
6.4	.8062	.8069	.8075	.8082	.8089	.8096	.8102	.8109	.8116	.8122
6.5	.8129	.8136	.8142	.8149	.8156	.8162	.8169	.8176	.8182	.8189
6.6	.8195	.8202	.8209	.8215	.8222	.8228	.8235	.8241	.8248	.8254
6.7	.8261	.8267	.8274	.8280	.8287	.8293	.8299	.8306	.8312	.8319
6.8	.8325	.8331	.8338	.8344	.8351	.8357	.8363	.8370	.8376	.8382
6.9	.8388	.8395	.8401	.8407	.8414	.8420	.8426	.8432	.8439	.8445
7.0	.8451	.8457	.8463	.8470	.8476	.8482	.8488	.8494	.8500	.8506
7.1	.8513	.8519	.8525	.8531	.8537	.8543	.8549	.8555	.8561	.8567
7.2	.8573	.8579	.8585	.8591	.8597	.8603	.8609	.8615	.8621	.8627
7.3	.8633	.8639	.8645	.8651	.8657	.8663	.8669	.8675	.8681	.8686
7.4	.8692	.8698	.8704	.8710	.8716	.8722	.8727	.8733	.8739	.8745
7.5	.8751	.8756	.8762	.8768	.8774	.8779	.8785	.8791	.8797	.8802
7.6	.8808	.8814	.8820	.8825	.8831	.8837	.8842	.8848	.8854	.8859
7.7	.8865	.8871	.8876	.8882	.8887	.8893	.8899	.8904	.8910	.8915
7.8	.8921	.8927	.8932	.8938	.8943	.8949	.8954	.8960	.8965	.8971
7.9	.8976	.8982	.8987	.8993	.8998	.9004	.9009	.9015	.9020	.9025
8.0	.9031	.9036	.9042	.9047	.9053	.9058	.9063	.9069	.9074	.9079
8.1	.9085	.9090	.9096	.9101	.9106	.9112	.9117	.9122	.9128	.9133
8.2	.9138	.9143	.9149	.9154	.9159	.9165	.9170	.9175	.9180	.9186
8.3	.9191	.9196	.9201	.9206	.9212	.9217	.9222	.9227	.9232	.9238
8.4	.9243	.9248	.9253	.9258	.9263	.9269	.9274	.9279	.9284	.9289
8.5	.9294	.9299	.9304	.9309	.9315	.9320	.9325	.9330	.9335	.9340
8.6	.9345	.9350	.9355	.9360	.9365	.9370	.9375	.9380	.9385	.9390
8.7	.9395	.9400	.9405	.9410	.9415	.9420	.9425	.9430	.9435	.9440
8.8	.9445	.9450	.9455	.9460	.9465	.9469	.9474	.9479	.9484	.9489
8.9	.9494	.9499	.9504	.9509	.9513	.9518	.9523	.9528	.9533	.9538

TABLE IV (CONTINUED)

→↓	0	1	2	3	4	5	6	7	8	9
9.0	.9542	.9547	.9552	.9557	.9562	.9566	.9571	.9576	.9581	.9586
9.1	.9590	.9595	.9600	.9605	.9609	.9614	.9619	.9624	.9628	.9633
9.2	.9638	.9643	.9647	.9652	.9657	.9661	.9666	.9671	.9675	.9680
9.3	.9685	.9689	.9694	.9699	.9703	.9708	.9713	.9717	.9722	.9727
9.4	.9731	.9736	.9741	.9745	.9750	.9754	.9759	.9763	.9768	.9773
9.5	.9777	.9782	.9786	.9791	.9795	.9800	.9805	.9809	.9814	.9818
9.6	.9823	.9827	.9832	.9836	.9841	.9845	.9850	.9854	.9859	.9863
9.7	.9868	.9872	.9877	.9881	.9886	.9890	.9894	.9899	.9903	.9908
9.8	.9912	.9917	.9921	.9926	.9930	.9934	.9939	.9943	.9948	.9952
9.9	.9956	.9961	.9965	.9969	.9974	.9978	.9983	.9987	.9991	.9996

IV

SELECTED ANSWERS

CHAPTER 1   EXERCISE 1.2

1. 180°   **3.** 45°   **5.** 240°   **7.** 633′   **9.** 4,212″   **11.** 1°12′   **13.** 43.351°   **15.** 2.213°
17. 103.295°   **19.** 100 cm   **21.** 450 km   **23.** 2.4 × 10⁵ mi   **25.** 1,440 cm (to three significant figures)
27. 34.9 m   **29.** 865,000 mi   **31.** 1.151

EXERCISE 1.3

1. $b' = 6$ **3.** $c = 108$ **5.** $c' = 0.2$ **7.** $b = 21$ m, $c = 23$ m   **9.** $a = 24$ in., $c = 55$ in.
11. $a = 8.5 \times 10^4$ km, $b = 1.8 \times 10^5$ km   **13.** $b = 50$ km, $c = 55$ km   **15.** $a = 1.2 \times 10^9$ yd, $c = 2.7 \times 10^9$ yd
17. $a = 3.6 \times 10^{-5}$ mm, $b = 7.6 \times 10^{-5}$ mm   **19.** $c \approx 106$ ft   **21.** 60 ft   **23.** 1.1 km

EXERCISE 1.4

1. a/c **3.** b/a **5.** c/a **7.** $\sin \theta$ **9.** $\tan \theta$ **11.** $\csc \theta$ **13.** 0.432   **15.** 0.709   **17.** 1.41
19. 0.294   **21.** 0.703   **23.** 1.08   **25.** $90° - \theta = 31°20'$, $a = 7.80$ mm, $b = 12.8$ mm
27. $90° - \theta = 6.3°$, $a = 0.354, c = 3.23$ **29.** $90° - \theta = 18.5°$, $a = 4.28$ in., $c = 13.5$ in.
31. $\theta = 56°20'$, $b = 33.6$ cm, $c = 40.4$ cm   **33.** $\theta = 28°30', 90° - \theta = 61°30'$, $a = 118$ ft
35. $\theta = 50.7°, 90° - \theta = 39.3°$, $c = 171$ mi   **37.** $90° - \theta = 52.54°$, $a = 6.939$ cm, $c = 8.742$ cm
39. $90° - \theta = 6°48'$, $b = 199.8$ mi, $c = 201.2$ mi   **41.** $\theta = 37°6', 90° - \theta = 52°54'$, $c = 70.28$ cm

43. $\theta = 44.11°, 90° - \theta = 45.89°$, $a = 36.17$ cm   **45.** $(\sin \theta)^2 + (\cos \theta)^2 = \left(\dfrac{b}{c}\right)^2 + \left(\dfrac{a}{c}\right)^2 = \dfrac{a^2 + b^2}{c^2} = \dfrac{c^2}{c^2} = 1$

EXERCISE 1.5

1. 6.9 m   **3.** 211 m   **5.** 1.1 km   **7.** 29,400 m, or 29.4 km   **9.** 3.5 m   **11.** 5.69 cm
13. 420 ft (to two significant figures)   **15.** 870 m   **19.** $g = 32.0$ ft/sec²

EXERCISE 1.6   CHAPTER REVIEW

1. 7,280″   **2.** 60°   **3.** 8,000   **4.** 36.38°   **5.** 560 ft (to two significant digits)
6. (A) b/c (B) c/a (C) b/a (D) c/b (E) a/c (F) a/b
7. $90° - \theta = 54.8°$, $a = 16.5$ cm, $b = 11.6$ cm   **8.** 144°
9. 4.19 in.   **10.** (A) $\cos \theta$ (B) $\tan \theta$ (C) $\sin \theta$ (D) $\sec \theta$ (E) $\csc \theta$ (F) $\cot \theta$
11. $90° - \theta = 27°40'$, $b = 7.63 \times 10^{-8}$ m, $c = 8.61 \times 10^{-8}$ m   **12.** $\theta = 40.3°, 90° - \theta = 49.7°$, $c = 20.6$ mm
13. $\theta = 40°20', 90° - \theta = 49°40'$ **14.** 8.7 ft   **15.** 940 ft   **16.** $\theta = 66°17'$, $a = 93.56$ km, $b = 213.0$ km
17. $\theta = 60.28°, 90° - \theta = 29.72°$, $b = 4,241$ m   **18.** 1.0496

CHAPTER 2   EXERCISE 2.2

1. $\pi/2$ rad   **3.** $\pi/3$ rad   **5.** $2\pi/3$ rad   **7.** 45°   **9.** 30°   **11.** 150°
13. $\pi/6$ rad, $2\pi/6$ rad or $\pi/3$ rad, $3\pi/6$ rad or $\pi/2$ rad, $4\pi/6$ rad or $2\pi/3$ rad, $5\pi/6$ rad, $6\pi/6$ rad or π rad
15. **17.** **19.**

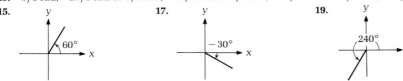

21. (A) 2 rad   (B) 1.5 rad   **23.** (A) 10 mm   (B) 1.5 mm   **25.** $\pi/10$ rad   **27.** $3\pi/20$ rad

29. $13\pi/18$ rad   **31.** $(288/\pi)°$ **33.** 15°   **35.** 3°

37.

39.

41.

43. I   **45.** III   **47.** II   **49.** 1.0008 rad   **51.** 18.2430°   **53.** 0.4605 rad   **55.** $7\pi/12$ rad, or 1.83 rad

57. $14\pi/365 = \pi/26$ rad, or 0.12 rad   **59.** 100 rad; $100/2\pi \approx 15.9$ revolutions

EXERCISE 2.3

1. $\sin\theta = \frac{4}{5}$, $\csc\theta = \frac{5}{4}$, $\cos\theta = \frac{3}{5}$, $\sec\theta = \frac{5}{3}$, $\tan\theta = \frac{4}{3}$, $\cot\theta = \frac{3}{4}$; same values for Q(6, 8) (Why?)

3. $\sin\theta = -\frac{3}{5}$, $\csc\theta = -\frac{5}{3}$, $\cos\theta = \frac{4}{5}$, $\sec\theta = \frac{5}{4}$, $\tan\theta = -\frac{3}{4}$, $\cot\theta = -\frac{4}{3}$; same values for Q(12, −9) (Why?)

5. $\sin\theta = \frac{4}{5}$, $\tan\theta = \frac{4}{3}$, $\csc\theta = \frac{5}{4}$, $\sec\theta = \frac{5}{3}$, $\cot\theta = \frac{3}{4}$

7. $\sin\theta = -\frac{4}{5}$, $\tan\theta = -\frac{4}{3}$, $\csc\theta = -\frac{5}{4}$, $\sec\theta = \frac{5}{3}$, $\cot\theta = -\frac{3}{4}$

9. $\sin\theta = -\frac{4}{5}$, $\cos\theta = -\frac{3}{5}$, $\tan\theta = \frac{4}{3}$, $\sec\theta = -\frac{5}{3}$, $\cot\theta = \frac{3}{4}$

11. $\sin\theta = -\frac{4}{5}$, $\cos\theta = \frac{3}{5}$, $\tan\theta = -\frac{4}{3}$, $\sec\theta = \frac{5}{3}$, $\cot\theta = -\frac{3}{4}$

13. 57.29   **15.** −0.9900   **17.** 3.236   **19.** −2.904   **21.** 0.4202   **23.** −0.4577

25. $\sin\theta = \frac{1}{2}$, $\cos\theta = \sqrt{3}/2$, $\tan\theta = 1/\sqrt{3}$, $\csc\theta = 2$, $\sec\theta = 2/\sqrt{3}$, $\cot\theta = \sqrt{3}$

27. $\sin\theta = -\sqrt{3}/2$, $\cos\theta = \frac{1}{2}$, $\tan\theta = -\sqrt{3}$, $\csc\theta = -2/\sqrt{3}$, $\sec\theta = 2$, $\cot\theta = -1/\sqrt{3}$

29. $\sin\theta = -\sqrt{2}/2$, $\cos\theta = \sqrt{2}/2$, $\tan\theta = -1$, $\csc\theta = -\sqrt{2}$, $\sec\theta = -\sqrt{2}$, $\cot\theta = -1$

31. I, IV   **33.** I, III   **35.** I, IV   **37.** III, IV   **39.** II, IV   **41.** III, IV

43. $\cos\theta = -\sqrt{5}/3$, $\tan\theta = 2/\sqrt{5}$, $\csc\theta = -\frac{3}{2}$, $\sec\theta = -3/\sqrt{5}$, $\cot\theta = \sqrt{5}/2$

45. $\cos\theta = \sqrt{5}/3$, $\tan\theta = -2/\sqrt{5}$, $\csc\theta = -\frac{3}{2}$, $\sec\theta = 3/\sqrt{5}$, $\cot\theta = -2/\sqrt{5}$

47. $\sin\theta = -\sqrt{2}/3$, $\cos\theta = 1/\sqrt{3}$, $\tan\theta = -\sqrt{2}$, $\csc\theta = -\sqrt{3}/\sqrt{2}$, $\cot\theta = -1/\sqrt{2}$

49. 0.6191   **51.** 4.938   **53.** −0.08169   **55.** −1.553   **57.** −0.8812   **59.** 1.079   **61.** tangent and secant

65. $I = -21$ amperes   **67.** $L = 12\csc\theta + 8\sec\theta$

EXERCISE 2.4

1. 1   **3.** $\frac{1}{2}$ **5.** 1   **7.** 1   **9.** $1/\sqrt{3}$, or $\sqrt{3}/3$ **11.** 0   **13.** $-\frac{1}{2}$ **15.** 0   **17.** $-\sqrt{3}$

19. $\sqrt{3}/2$ **21.** Not defined   **23.** $-\sqrt{3}$ **25.** $-\frac{1}{2}$ **27.** −1   **29.** $-1/\sqrt{2}$, or $-\sqrt{2}/2$ **31.** −1

33. $-\sqrt{3}/2$ **35.** $-\sqrt{3}$ **37.** $1/\sqrt{2}$, or $\sqrt{2}/2$ **39.** $2/\sqrt{3}$ **41.** 0

43. (A) 90°, 270°   (B) $\pi/2$, $3\pi/2$ **45.** (A) 0°, 180°   (B) 0, π **47.** (A) 30°   (B) $\pi/6$

49. (A) 120°   (B) $2\pi/3$ **51.** (A) 120°   (B) $2\pi/3$ **53.** 240°, 300°   **55.** $3\sqrt{3}$ cm²   **57.** $150\sqrt{3}$ in.²

EXERCISE 2.5

1. 10°   **3.** 15°   **5.** 20°   **7.** 35°   **9.** 20°   **11.** 20°   **13.** 14°20′   **15.** 15°10′   **17.** 14.63°

19. 36.86°   **21.** 0.14 rad   **23.** 0.72 rad   **25.** 0.94 rad   **27.** 0.25 rad   **29.** −   **31.** −   **33.** −

35. +   **37.** −   **39.** −   **41.** −0.766   **43.** −2.75   **45.** 0.819   **47.** 0.248   **49.** −0.262

51. Table: 0.909; calculator: 0.909   **53.** Table: 0.288; calculator: 0.291   **55.** Table: 4.80; calculator: 4.76

57. Table: −0.150; calculator: −0.146

EXERCISE 2.6

1. sin x increases from 0 to 1 **3.** sin x decreases from 1 to 0 **5.** sin x decreases from 0 to -1
7. sin x increases from -1 to 0 **9.** $\pi/2, 5\pi/2$ **11.** $0, \pi, 2\pi, 3\pi, 4\pi$ **13.** $0, \pi, 2\pi, 3\pi, 4\pi$ **15.** $-2\pi, 0, 2\pi$
17. $-3\pi/2, -\pi/2, \pi/2, 3\pi/2$ **19.** $\pi/2, 3\pi/2, 5\pi/2, 7\pi/2$ **21.** $0, \pi, 2\pi, 3\pi, 4\pi$ **23.** -0.2088 **25.** 7.228
27. 119.2 **29.** $-\sqrt{2}/2$ **31.** $-\sqrt{2}$ **33.** -1
35. (A) Identity (4) (B) Identity (9) (C) Identity (2) **37.** 2π **39.** All real y such that $-1 \le y \le 1$

EXERCISE 2.7

1.

θ	.01	.05	.10	.20	.25	.30	.35	.40
$\sin \theta$	.01	.05	.10	.20	.25	.30	.34	.39

3. (A) 262 m (B) 264 m **5.** (A) 15.2 m (B) 15.1 m

7.

x	.05	.15	.25	.35	.45
$x - \left(\dfrac{x^3}{6}\right)$	.04998	.1494	.2474	.3429	.4348
$\sin x$	.04998	.1494	.2474	.3429	.4350

9.

x	.05	.15	.25	.35	.45
$1 - \left(\dfrac{x^2}{2}\right)$	.9988	.9888	.9688	.9388	.8988
$\cos x$	.9988	.9888	.9689	.9394	.9004

EXERCISE 2.8 CHAPTER REVIEW

1. (A) $\pi/3$ (B) $\pi/4$ (C) $\pi/2$ **2.** (A) $30°$ (B) $90°$ (C) $45°$ **3.** $\sin \theta = \frac{3}{5}$; $\tan \theta = -\frac{3}{4}$
4. (A) 0.7355 (B) 1.085 **5.** (A) 0.9171 (B) 0.9099 **6.** (A) 0.9394 (B) 5.177
7. (A) $\sqrt{3}/2$ (B) $1/\sqrt{2}$, or $\sqrt{2}/2$ (C) 0 **8.** $42°$ **9.** 3 rad **10.** 6 cm **11.** $53\pi/45$ **12.** $15°$
13. 0; not defined; 0; not defined **14.** (A) I (B) IV **15.** (A) $7.4°$ (B) $76°40'$ (C) $37°40'$
16. (A) 0.75 rad (B) 1.28 rad (C) 0.86 rad **17.** -0.992 **18.** 0.130 **19.** 0.973 **20.** -4.34
21. -1.30 **22.** 1.26 **23.** Calculator: 0.683 (Table: 0.682) **24.** Calculator: -1.07 (Table: -1.07)
25. Calculator: 0.284 (Table: 0.287) **26.** Calculator: -3.38 (Table: -3.34)
27. Calculator: 0.757 (Table: 0.758) **28.** Calculator: -0.864 (Table: -0.861)
29. (A) $\pi/6$ (B) $\pi/4$ (C) $\pi/3$ **30.** $-\sqrt{3}/2$ **31.** $-1/\sqrt{3}$ **32.** $-1/\sqrt{2}$ **33.** -1 **34.** -1
35. 0 **36.** $\sqrt{3}/2$ **37.** -2 **38.** -1 **39.** Not defined **40.** $\sqrt{3}/2$ **41.** $\frac{1}{2}$ **42.** 0.40724
43. -0.33884 **44.** 0.64692 **45.** 0.49639 **46.** $\cos \theta = \frac{3}{5}$, $\tan \theta = -\frac{4}{3}$ **47.** $7\pi/6$ **48.** Both 2π
49. D
50. (A) Increases from 0 to 1 (B) Decreases from 1 to 0
      (C) Decreases from 0 to -1 (D) Increases from -1 to 0
51. (A) Decreases from 1 to 0 (B) Decreases from 0 to -1
      (C) Increases from -1 to 0 (D) Increases from 0 to 1
52. 200 rad, $100/\pi \approx 31.8$ revolutions

CHAPTER 3 **EXERCISE 3.2**

1.
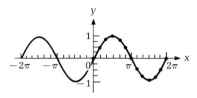

3.
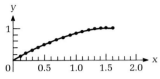

5. $-2\pi, -\pi, 0, \pi, 2\pi$

7. One unit

9.
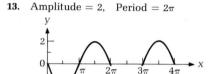

11. Amplitude $= 3$,   Period $= 2\pi$
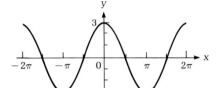

13. Amplitude $= 2$,   Period $= 2\pi$
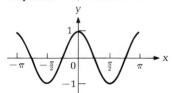

15. Amplitude $= \frac{1}{2}$,   Period $= 2\pi$
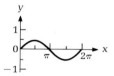

17. Amplitude $= 1$,   Period $= \pi$
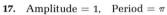

19. Amplitude $= 1$,   Period $= 1$
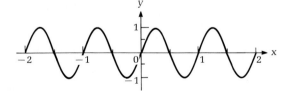

21. Amplitude $= 1$,   Period $= 8\pi$
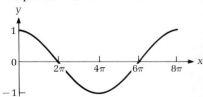

23. Amplitude = 2,   Period = $\pi/2$

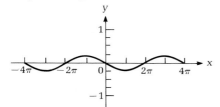

25. Amplitude = $\frac{1}{3}$,   Period = 1

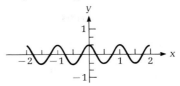

27. Amplitude = $\frac{1}{4}$,   Period = 4π

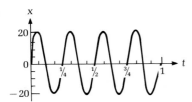

29. Amplitude = 110,   Period = $\frac{1}{60}$ sec,
Frequency = 60 Hz

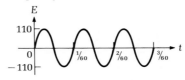

31.

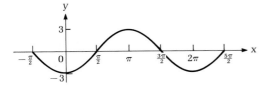

EXERCISE 3.3

1. Phase shift is $\pi/2$ to the left

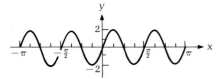

3. Phase shift is $\pi/4$ to the right

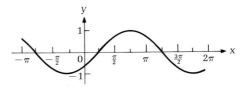

5. Phase shift is $\pi/2$ to the right

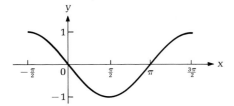

7. Amplitude = 1,   Period = 1,   Phase shift = $\frac{1}{2}$ right

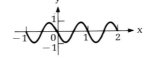

9. Amplitude = 4, Period = 2, Phase shift = ¼ left

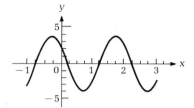

11. Amplitude = 2, Period = π, Phase shift = $\pi/2$ left

13. Amplitude = 2, Period = $2\pi/3$, Phase shift = $\pi/6$ right

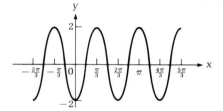

15. Both have same graph; thus, $\cos(x - \pi/2) = \sin x$ for all x

17. Amplitude = 5 m, Period = 12 sec, Phase shift = 3 m left

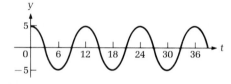

19. Amplitude = 30, Period = $\frac{1}{60}$, Frequency = 60 Hz, Phase shift = $\frac{1}{120}$ right

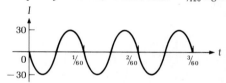

21. $x = 20 \sin(8\pi t + \pi/2)$; Amplitude = 20, Period = ¼, Phase shift = $\frac{1}{16}$ left

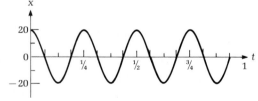

EXERCISE 3.4

1.

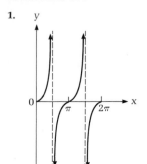

3.

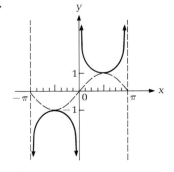

5. Period = $\pi/2$

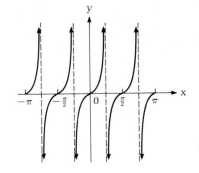

7. Period = ½

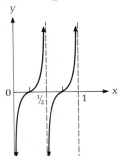

9. Period = 2

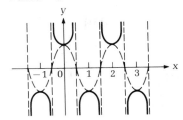

11. Period = 2π

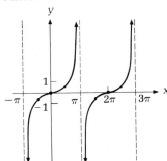

13. Period = 4π

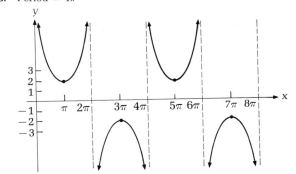

15. Period = π, Phase shift = $\pi/2$ right

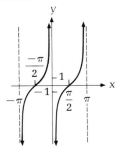

17. Period = $\pi/2$, Phase shift = $\pi/2$ left

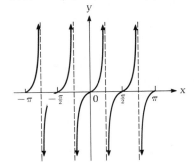

EXERCISE 3.5

1.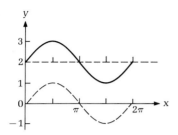

3.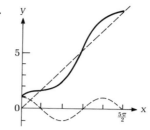

5.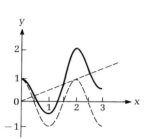

7.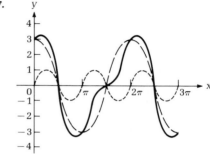

9.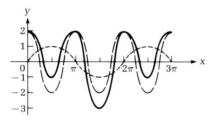

11.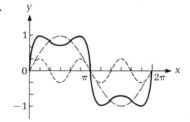

13.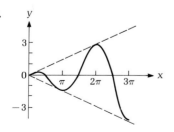

15.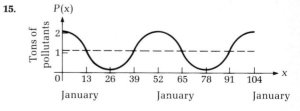

EXERCISE 3.6   CHAPTER REVIEW

1.

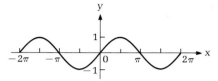

2.

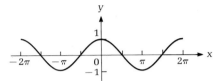

3.

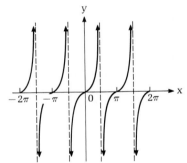

4.

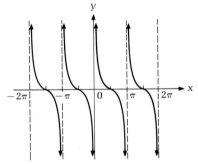

5.

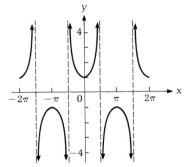

6.

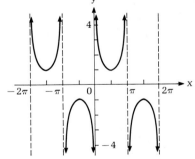

7.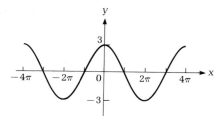

8.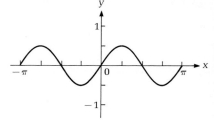

9.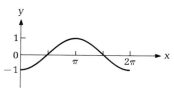

10.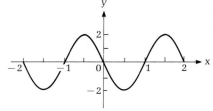

11.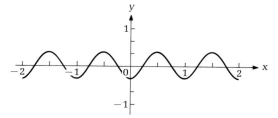

12.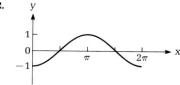

13.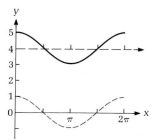

14.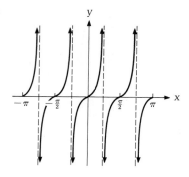

15.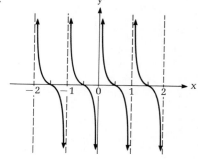

16.
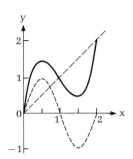

17.
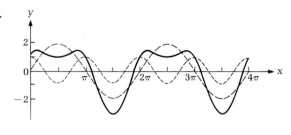

18. Period = 1, Amplitude = ⅓ **19.** 1 **20.** $\pi/2$ to the right

21. Amplitude = 3, Period = 2, Phase shift = 1 left

22.
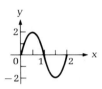

23.
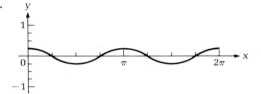

24.
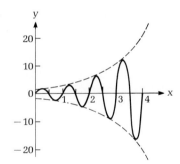

CHAPTER 4 EXERCISE 4.1

1. A **3.** E **5.** H **7.** J **9.** L **11.** K **63.** I, II **65.** II, III **67.** All **69.** I, IV

EXERCISE 4.2

13. $\dfrac{\sqrt{2}}{2}(\sin x - \cos x)$ **15.** $-\cos x$ **17.** $\dfrac{1 - \tan x}{1 + \tan x}$ **19.** $\dfrac{\sqrt{3} + 1}{2\sqrt{2}}$ **21.** $\dfrac{\sqrt{3} + 1}{2\sqrt{2}}$ **23.** $\sqrt{3}/2$

25. $\sqrt{3}$ **27.** $\sin(x - y) = \dfrac{-2 - \sqrt{75}}{12}$; $\tan(x + y) = \dfrac{\sqrt{75} - 2}{\sqrt{5} + 2\sqrt{15}}$

29. $\sin(x - y) = \dfrac{-2\sqrt{8} - 1}{3\sqrt{5}}$; $\tan(x + y) = \dfrac{1 - 4\sqrt{2}}{2 + 2\sqrt{2}}$

47. $0.6115 = 0.6115$; $-1.155 = -1.155$ **49.** $0.9756 = 0.9756$; $-0.4895 = -0.4895$

EXERCISE 4.3

1. $\dfrac{\sqrt{3}}{2} = \dfrac{\sqrt{3}}{2}$ **3.** $-\sqrt{3} = -\sqrt{3}$ **5.** $\dfrac{\sqrt{2+\sqrt{3}}}{2}$ **7.** $\dfrac{\sqrt{2-\sqrt{3}}}{\sqrt{2+\sqrt{3}}}$

23. $\sin 2x = {}^{24}\!/_{25}$, $\cos 2x = {}^{7}\!/_{25}$, $\tan 2x = {}^{24}\!/_{7}$ **25.** $\sin 2x = -{}^{24}\!/_{25}$, $\cos 2x = {}^{7}\!/_{25}$, $\tan 2x = -{}^{24}\!/_{7}$

27. $\sin 2x = -{}^{120}\!/_{169}$, $\cos 2x = -{}^{119}\!/_{169}$, $\tan 2x = {}^{120}\!/_{119}$ **29.** $\sin(x/2) = \sqrt{3}/3$, $\cos(x/2) = \sqrt{6}/3$

31. $\sin(x/2) = \sqrt{\dfrac{3+2\sqrt{2}}{6}}$, $\cos(x/2) = -\sqrt{\dfrac{3-2\sqrt{2}}{6}}$ **33.** $\sin(x/2) = -2\sqrt{5}/5$, $\cos(x/2) = \sqrt{5}/5$

41. (A) $-0.72335 = -0.72335$ (B) $-0.58821 = -0.58821$

43. (A) $-3.2518 = -3.2518$ (B) $0.89280 = 0.89280$ **45.** $d = \dfrac{v_0^2 \sin 2\theta}{32 \text{ ft/sec}^2}$

EXERCISE 4.4

1. $\frac{1}{2}\cos 12A + \frac{1}{2}\cos 2A$ **3.** $\frac{1}{2}\sin 5\theta + \frac{1}{2}\sin\theta$ **5.** $2\cos 6\theta\cos\theta$ **7.** $-2\cos 3u\sin 2u$ **9.** $(2-\sqrt{3})/4$

11. $^3\!/_4$ **13.** $\sqrt{2}/2$ **15.** $\sqrt{2}/2$

31. (A) $0.19853 = 0.19853$ (B) $1.5918 = 1.5918$ **33.** (A) $0.57285 = 0.57285$ (B) $1.8186 = 1.8186$

EXERCISE 4.5

1. $y = \sqrt{2}\sin(t + \pi/4)$; Amplitude $= \sqrt{2}$, Period $= 2\pi$, Frequency $= 1/2\pi$, Phase shift $= \pi/4$ (left)

3. $y = 2\sin(t + 5\pi/6)$; Amplitude $= 2$, Period $= 2\pi$, Frequency $= 1/2\pi$, Phase shift $= 5\pi/6$ (left)

5. $y = 2\sin(2t - \pi/6)$; Amplitude $= 2$, Period $= \pi$, Frequency $= 1/\pi$, Phase shift $= \pi/12$ (right)

7. $y = \sqrt{2}\sin\left(\dfrac{t}{2} - \dfrac{3\pi}{4}\right)$; Amplitude $= \sqrt{2}$, Period $= 4\pi$, Frequency $= 1/4\pi$, Phase shift $= 3\pi/2$ (right)

9. $y = 5\sin(\pi t - 0.64)$; Amplitude $= 5$, Period $= 2$, Frequency $= \frac{1}{2}$, Phase shift $\approx 0.64/\pi \approx 0.20$ (right)

11. $y = \sqrt{34}\sin(3t + 2.60)$; Amplitude $= \sqrt{34}$, Period $= 2\pi/3$, Frequency $= 3/2\pi$, Phase shift $= 0.87$ (left)

13. $y = 5\sin(8t + 4.07)$; Amplitude $= 5$ cm, Period $= \pi/4$ sec, Frequency $= 4/\pi$ Hz, Phase shift $= 0.51$ sec (left)

EXERCISE 4.6 CHAPTER REVIEW

1. $1/\csc x$ **2.** $\cos x/\sin x$ **3.** $1 - \cos^2 x$ **4.** $\cos x$ **14.** $1/2 = 1/2$ **15.** $1/\sqrt{2} = 1/\sqrt{2}$

16. $3/2 = 3/2$ **24.** $2\cos 2x\sin x$ **25.** $\cos 7x + \cos 3x$

26. $y = 2\sin(6t - 5\pi/6)$, Amplitude $= 2$, Period $= \pi/3$, Phase shift $= 5\pi/36$ right

CHAPTER 5 EXERCISE 5.1

1. $\pi/2$ **3.** 0 **5.** $\pi/6$ **7.** $\pi/4$ **9.** $\pi/4$ **11.** π **13.** $-\pi/6$ **15.** $3\pi/4$ **17.** -0.7 **19.** 0.37

21. 0 **23.** $\sqrt{3}/2$ **25.** $^4\!/_5$ **27.** $\sqrt{0.91}$ **29.** 0.6 **31.** $\pi/6 + 2k\pi$, $5\pi/6 + 2k\pi$, k any integer

33. $\pi/4 + 2k\pi$, $3\pi/4 + 2k\pi$, k any integer **35.** $2\pi/3 + 2k\pi$, $4\pi/3 + 2k\pi$, k any integer

37. $5\pi/4 + 2k\pi$, $7\pi/4 + 2k\pi$, k any integer **39.** 0.911 **41.** -0.103 **43.** 0.972 **45.** -0.062

47. $^{24}\!/_{25}$ **49.** $-\frac{1}{2}$ **51.** $-{}^{24}\!/_{25}$

EXERCISE 5.2

1. 0 **3.** $\pi/6$ **5.** $\pi/4$ **7.** $1/\sqrt{2}$ **9.** $^4\!/_3$ **11.** $3\pi/4$ **13.** $2\pi/3$ **15.** $-\pi/3$ **17.** Not defined

19. $-\sqrt{3}$ **21.** $^4\!/_5$ **23.** $-\sqrt{13}/2$ **25.** $-\frac{1}{2}$ **27.** 33.4 **29.** -4 **31.** $\pi/6 + k\pi$, k any integer

33. $\pi/6 + 2k\pi$, $11\pi/6 + 2k\pi$, k any integer **35.** $2\pi/3 + k\pi$, k any integer

37. $7\pi/6 + 2k\pi$, $11\pi/6 + 2k\pi$, k any integer **39.** $\frac{1}{2}$ **41.** 1.114 **43.** 0.315 **45.** 2.105 **47.** 1.022

49. $-{}^{8}\!/_{19}$ **51.** $^{24}\!/_{7}$ **61.** $\mathrm{Cot}^{-1}(-1) = 3\pi/4$, $\mathrm{Tan}^{-1} 1/-1 = -\pi/4$

EXERCISE 5.3

1. $0°$ **3.** $60°$ **5.** $60°$ **7.** $\pi/6$ **9.** $\pi/4$ **11.** $\pi/6$ **13.** $30°$, $150°$ **15.** $30°$, $90°$, $150°$, $270°$
17. $0°$, $90°$, $180°$, $360°$ **19.** $60°$, $120°$, $240°$, $300°$ **21.** 0, $\pi/3$, π, $5\pi/3$, 2π
23. $\pi/6$, $5\pi/6$, $7\pi/6$, $11\pi/6$ **25.** $\pi/6$, $\pi/3$, $7\pi/6$, $4\pi/3$ **27.** $\pi/2$ **29.** $3\pi/2$ **31.** $\pi/6$, $\pi/2$, $5\pi/6$
33. π **35.** $\pi/2, 3\pi/2$ **37.** $\pi/6, 5\pi/6, 3\pi/2$ **39.** $\pi/4, 5\pi/4$ **41.** $\pi/3, \pi, 5\pi/3$
43. $30° + k360°$, $150° + k360°$, k any integer   **45.** $\pi/4 + k\pi$, k any integer
47. $\pi/3 + 2k\pi, \pi + 2k\pi, 5\pi/3 + 2k\pi, 2k\pi$, k any integer   **49.** $3\pi/2 + 2k\pi$, k any integer
51. 0.72; $41.41°$ **53.** 0.92; $52.46°$ **55.** $0.37, 2.77$ **57.** $0.37, 2.77$ **59.** $0, \pi/2, 2\pi$ **61.** $0, 2\pi$
63. $(r, \theta) = (1, 30°), (1, 150°)$ **65.** $0.002\ 613\ \text{sec}$ **67.** $64.1°$

EXERCISE 5.4   CHAPTER REVIEW

1. $\pi/6$ **2.** $\pi/6$ **3.** $\pi/4$ **4.** $\pi/3$ **5.** $\pi/6$ **6.** $\pi/6$ **7.** $-\pi/6$ **8.** $2\pi/3$ **9.** $3\pi/4$ **10.** $-\pi/4$
11. $2\pi/3$ **12.** $-\pi/6$ **13.** $\pi/6$, $11\pi/6$ **14.** 0, π, 2π, $\pi/6$, $5\pi/6$ **15.** $\pi/6$, $5\pi/6$, $7\pi/6$, $11\pi/6$
16. $2\pi/3$, π, $4\pi/3$ **17.** 0.315 **18** $-\tfrac{3}{5}$ **19.** $-\sqrt{5}$ **20.** $-\tfrac{1}{2}$ **21.** $180°$ **22.** $210°$, $330°$
23. $90°$, $270°$ **24.** $15°$, $75°$, $195°$, $255°$ **25.** $60°$, $240°$
26. $\pi/6 + 2k\pi$, $5\pi/6 + 2k\pi$, k any integer; $30° + k360°$, $150° + k360°$, k any integer
27. $3\pi/4 + k\pi$, k any integer; $135° + k180°$, k any integer   **28.** $\pi/6 + 2k\pi$, $11\pi/6 + 2k\pi$, k any integer
29. $\pi/2 + 2k\pi$, $3\pi/2 + 2k\pi$, $7\pi/6 + 2k\pi$, $11\pi/6 + 2k\pi$, k any integer   **30.** 1.88 **31.** 1.34 **32.** 0.77
33. -0.54 **34.** $41.41°$ **35.** 0.89 **36.** $0.82, 2.32$ **37.** $\sqrt{1 - x^2}$, $-1 \le x \le 1$

CHAPTER 6   EXERCISE 6.1

1. $\alpha = 67°20'$, $a = 55.2\ \text{km}$, $c = 58.9\ \text{km}$ **3.** $\alpha = 98°$, $b = 4.32\ \text{mm}$, $c = 7.62\ \text{mm}$
5. $\alpha = 31°40'$, $\gamma = 96°20'$, $c = 15.1\ \text{cm}$ **7.** $\beta = 135°15'$, $\gamma = 25°50'$, $c = 65.00\ \text{yd}$
9. $\alpha = 141°$, $\alpha' = 39°$, $\beta = 9°$, $\beta' = 111°$, $b = 13\ \text{m}$, $b' = 75\ \text{m}$ **11.** No triangle   **13.** $27.06 \approx 27.04$
17. $8.37\ \text{mi from } A$; $4.50\ \text{mi from } B$ **19.** $4.43 \times 10^7\ \text{km}, 2.40 \times 10^8\ \text{km}$ **21.** $d = 2.8\ \text{nautical mi}$

EXERCISE 6.2

1. $a = 6.00\ \text{mm}$, $\beta = 65°$, $\gamma = 64°20'$ **3.** $c = 26.3\ \text{m}$, $\alpha = 32.3°$, $\beta = 12.7°$
5. $\alpha = 62.7°$, $\beta = 36.3°$, $\gamma = 81°$ **7.** $\alpha = 18°10'$, $\beta = 33°10'$, $\gamma = 128°40'$
9. $b^2 = a^2 + b^2 - 2ab \cos 90° = a^2 + b^2 - 0 = a^2 + b^2$ **11.** $-0.87 = -0.87$ **13.** $710\ \text{km}$ **15.** $121\ \text{cm}$
17. $\alpha = 49°30'$, $\beta = 32°10'$, $\gamma = 98°20'$

EXERCISE 6.3

1. $M_3 = 71\ \text{km/hr}, \theta = 29°$ **3.** $M_3 = 57\ \text{lb}, \theta = 33°$ **5.** $M_3 = 154\ \text{knots}, \theta = 21.9°$
7. $H = 33\ \text{lb}, V = 22\ \text{lb}$ **9.** $H = 11\ \text{knots}, V = 20\ \text{knots}$ **11.** $H = 178\ \text{km/hr}, V = 167\ \text{km/hr}$
13. $M_3 = 188\ \text{lb}, \theta = 16°$ **15.** $M_3 = 9\ \text{knots}, \theta = 12°$ **17.** $M_3 = 699\ \text{km/hr}, \theta = 7.4°$
19. Magnitude $= 5\ \text{km/hr}$, Direction $= 143°$ **21.** $351°$; $247\ \text{km/hr}$
23. Magnitude $= 2{,}500\ \text{lb}$, Direction $= 19°$ (relative to $\mathbf{F}_1$)   **25.** $5{,}100\ \text{ft-lb}$
27. $(2{,}500\ \text{lb}) \sin 15° = 650\ \text{lb}, (2{,}500\ \text{lb}) \cos 15° = 2{,}400\ \text{lb}$ **29.** To the left

EXERCISE 6.4

1. $102\ \text{m}^2$ **3.** $12\ \text{cm}^2$ **5.** $11.6\ \text{in.}^2$ **7.** $40{,}900\ \text{ft}^2$ **9.** $45.4\ \text{cm}^2$ **11.** $129\ \text{m}^2$
15. $A_1 = A_3 = \dfrac{ab}{2} \sin(180° - \theta) = \dfrac{ab}{2} \sin \theta = A_2 = A_4$

EXERCISE 6.5 CHAPTER REVIEW

1. $\gamma = 82°$, $a = 21$ m, $c = 22$m **2.** $a = 12$ in., $\beta = 51°$, $\gamma = 104°$ **3.** 9.4 at 32° **4.** $H = 9.8$, $V = 6.9$
5. 216 km **6.** 10.3 in.² **7.** $a = 51.2$ m, $\beta = 111.7°$, $\gamma = 40.8°$ **8.** $\beta = 43°30'$, $\gamma = 101°10'$, $c = 22.4$ in.
9. $\beta = 136°30'$, $\gamma = 8°10'$, $c = 3.24$ in. **10.** $|\mathbf{u} + \mathbf{v}| = 23.2$, $\theta = 16.1°$ **11.** 260 km/hr at 57° **12.** 75.9 m²
13. 252 m **14.** 5.9 cm, 17 cm **15.** 92.8 cm² (using Heron's formula)
16. (A) Magnitude = 443 lb, Direction = 13.4° (relative to $\mathbf{F}_1$) (B) 10,800 ft-lb

CHAPTER 7   EXERCISE 7.1

1.
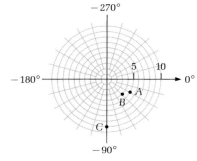

3.
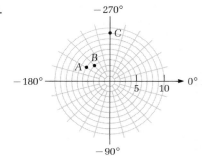

5.
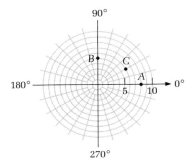

7.
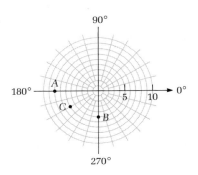

9.
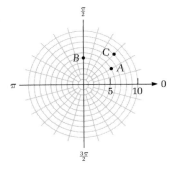

11.
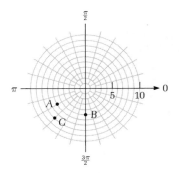

13.

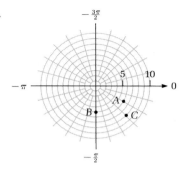

15.

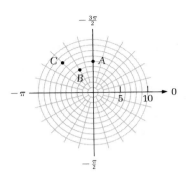

17. $(4, 4\sqrt{3})$ **19.** $(0, -9)$ **21.** $(-2\sqrt{2}, -2\sqrt{2})$ **23.** $(-5\sqrt{3}, 5)$ **25.** $(-3\sqrt{3}, 3)$ **27.** $(-2\sqrt{3}, 2)$
29. $(4, \pi/6)$ **31.** $(8, 3\pi/4)$ **33.** $(8, 4\pi/3)$ **35.** $(7, 3\pi/2)$

37.

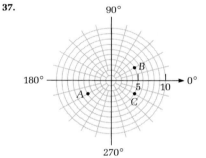

39.

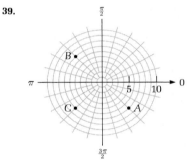

41. $r = 6 \cos \theta$ **43.** $r(2 \cos \theta + 3 \sin \theta) = 5$ **45.** $r^2 = 9$,   or $r = \pm 3$ **47.** $2x + y = 4$ **49.** $x^2 + y^2 = 8x$
51. $x^2 + y^2 = 16$ **53.** $y - 3 = \pm 2\sqrt{x^2 + y^2}$,   or $(y - 3)^2 = 4(x^2 + y^2)$

EXERCISE 7.2

1.

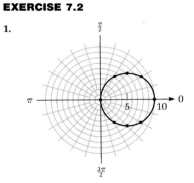

3.

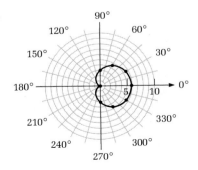

5.

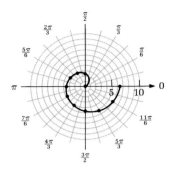

7.

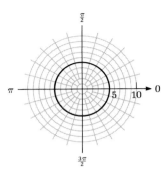

9.

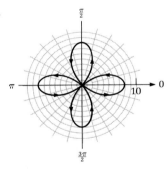

11.

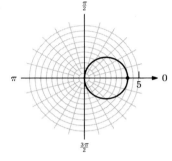

13.

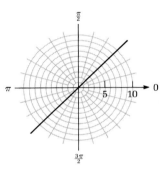

15.

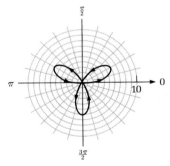

17.

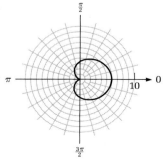

19.

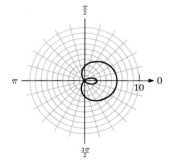

21.

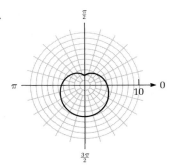

23.

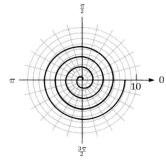

25.

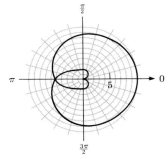

27.

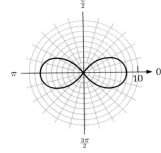

29.

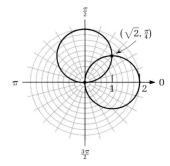

(B)   Faster at perihelion

31.   (A)

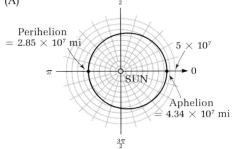

EXERCISE 7.3

1.

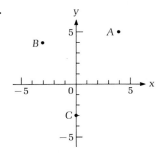

3.

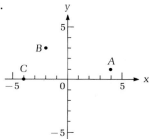

5.

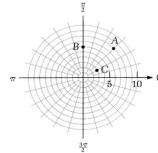

7.

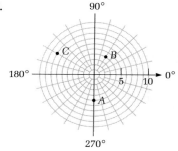

9. $z_1z_2 = 50$ cis $77°$, $z_1/z_2 = 2$ cis $13°$ **11.** $z_1z_2 = 21$ cis $265°$, $z_1/z_2 = \frac{7}{3}$ cis $61°$ **13.** $\sqrt{2}$ cis$(\pi/4)$
15. 2 cis$(5\pi/6)$ **17.** 4 cis$(\pi/2)$ **19.** 2 cis$(4\pi/3)$ **21.** 4 cis$(5\pi/3)$ **23.** 8 cis$(3\pi/2)$ **25.** $1 + i$
27. $-1 + i$ **29.** -8 **31.** $12i$ **33.** $-3 - 3\sqrt{3}i$ **35.** $2\sqrt{3} - 2i$ **37.** $2i$, 2 cis $90°$
39. $4i$, 4 cis$(\pi/2)$ **41.** $\sqrt{34}$ cis $59.0°$ **43.** $\sqrt{58}$ cis $156.8°$ **45.** $\sqrt{61}$ cis $320.2°$ **47.** $7.17 + 5.46i$
49. $-4.78 - 1.48i$ **51.** $8.59 - 6.87i$

EXERCISE 7.4

1. 27 cis $45°$ **3.** 32 cis $450° = 32$ cis $90°$ **5.** 64 cis $180°$ **7.** -4 **9.** $16\sqrt{3} + 16i$ **11.** 1
13. 2 cis $15°$, 2 cis $195°$ **15.** 2 cis $30°$, 2 cis $150°$, 2 cis $270°$
17. $2^{1/10}$ cis $27°$, $2^{1/10}$ cis $99°$, $2^{1/10}$ cis $171°$, $2^{1/10}$ cis $243°$, $2^{1/10}$ cis $315°$
19. 1 cis $0°$, 1 cis $60°$, 1 cis $120°$, 1 cis $180°$, 1 cis $240°$, 1 cis $300°$
23. 1, $0.309 + 0.951i$, $-0.809 + 0.588i$, $-0.809 - 0.588i$, $0.309 - 0.951i$
25. $0.855 + 1.481i$, -1.710, $0.855 - 1.481i$

EXERCISE 7.5 CHAPTER REVIEW

1.
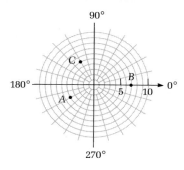

2.
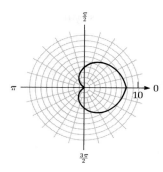

3.
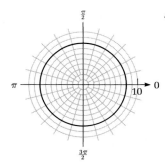

4. $(2, 2)$ **5.** $(2, 5\pi/6)$ **6.**
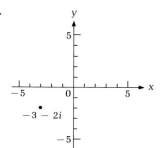

7.
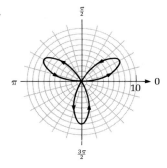

8. $z_1 z_2 = 27 \text{ cis } 79°, \quad z_1/z_2 = 3 \text{ cis } 5°$ **9.** $16 \text{ cis } 40°$

10.
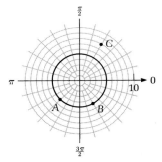

11.
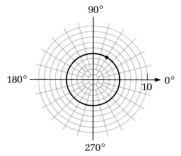

12.

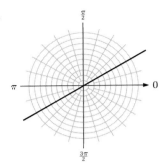

13. $r^2 = 8r \cos \theta$, or $r = 8 \cos \theta$ **14.** $3x - 2y = -2$ **15.** $x^2 + y^2 = -3x$ **16.** 2 cis 210°   **17.** $-3 + 3i$
18. 8 cis 195°   **19.** ½ cis 75°   **20.** -4 **21.** 2 cis 70°, 2 cis 190°, 2 cis 310°
22. $(2 \text{ cis } 30°)^2 = 4 \text{ cis } 60 = 4 + i4\sqrt{3}$

23.

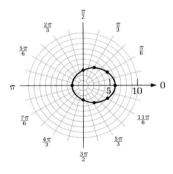

24.

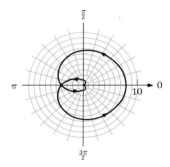

25. $y - 3 = \pm 2\sqrt{x^2 + y^2}$, or $(y - 3)^2 = 4(x^2 + y^2)$ **26.** 2.289, $-1.145 + 1.983i$, $-1.145 - 1.983i$

CHAPTER 8   EXERCISE 8.2

1. Amplitude $= 0.004$ m, Period $= \frac{1}{500}$ sec, Frequency $= 500$ Hz

3.

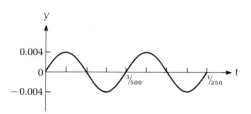

5.

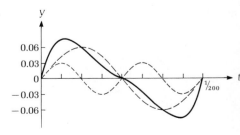

EXERCISE 8.3

1. Amplitude $= 10$,   Frequency $= 60$ Hz,   Phase shift is $\frac{1}{2}$ to the right

EXERCISE 8.4

1. Period $= 12$ months,   Amplitude ≈ 84 min

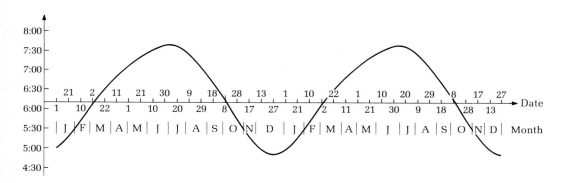

EXERCISE 8.5

1. 30 ft;   328 ft;   41 ft/sec

3.

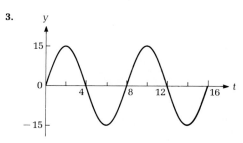

5. (A)

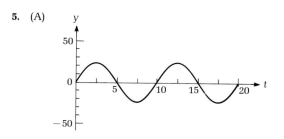

(B)

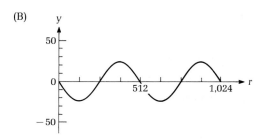

EXERCISE 8.6

1. 2,220 kg

3.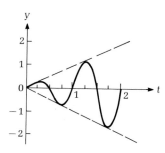

EXERCISE 8.7

1. Period $= 10^{-8}$ sec; $\lambda = 3$ m   **3.** $B = 2\pi \times 10^{18}$

5. Period $= 10^{-5}$ sec;   Frequency $= 10^5$ Hz;   No

EXERCISE 8.8

1. 40 km/hr   **3.** $\theta = 60°$ **5.** 7.7×10^{10} cm/sec

EXERCISE 8.9

1. 3 mm/sec   **3.** 17 rad/sec   **5.** 24π m/sec   **7.** 4.65 rad/hr   **9.** 223.5 rad/sec,   or 35.6 rps

11. $\pi/4{,}380$ rad/hr;   66,700 mph

EXERCISE 8.10

1. $\beta = 29.3°$ **3.** $\beta = 13.9°$ **5.** $\alpha = 24.4°$ **7.** Low

EXERCISE 8.11

1.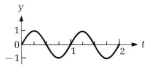

3.

EXERCISE 8.12

1. $M_1 = 676$ lb, $M_2 = 677$ lb

EXERCISE 8.13

1. $-6°$

APPENDIX A   EXERCISE A.1

1. 6.4×10^2 **3.** 5.46×10^9 **5.** 7.3×10^{-1} **7.** 3.2×10^{-7} **9.** 4.91×10^{-5} **11.** 6.7×10^{10}
13. 56,000   **15.** 0.0097   **17.** 0.108   **19.** 635,000   **21.** 86.8   **23.** 0.00465   **25.** 7.3×10^2
27. 4.0×10^{-2} **29.** 4.4×10^{-4} **31.** 3   **33.** 2   **35.** 1

EXERCISE A.2

1. $7 + 5i$ **3.** $-1 - 9i$ **5.** -18 **7.** $8 + 6i$ **9.** $-5 - 10i$ **11.** 34   **13.** $\frac{2}{5} - \frac{1}{5}i$ **15.** $\frac{4}{13} - \frac{7}{13}i$
17. $\frac{2}{25} + \frac{11}{25}i$ **19.** $5 - 2i$ **21.** $3 + 5i$ **23.** $13 - 19i$ **25.** $1 - \frac{3}{2}i$ **27.** 0   **29.** 1

APPENDIX B   EXERCISE B.1

1. 3   **3.** -5 **5.** -1 **7.** 0   **9.** -20 **11.** -6 **13.** 5   **15.** $-\frac{5}{3}$ **17.** 6   **19.** $-3 - 2h$
21. -2 **23.** $g[f(2)] = g(-3) = -5$ **25.** Not a function   **27.** Function   **29.** Not a function
31. Function   **33.** $Y = \{1, 3, 7\}$ **35.** F; $X = \{-2, -1, 0\}$, $Y = \{0, 1\}$ **37.** 0 m;   4.88 m;   19.52 m;   43.92 m
39. $19.52 + 4.88h$;   The ratio tends to 19.52 m/sec, the speed at $t = 2$

EXERCISE B.2

1. $R^{-1} = \{(2, -1), (1, 0), (2, 1)\}$;   Not a function   **3.** $H^{-1} = \{(0.5, -1), (1, 0), (2, 1), (4, 2)\}$;   Function
5.   **7.**

9. M^{-1}: $x = 2y - 4$,   or $y = \dfrac{x + 7}{2}$;   Function   **11.** p^{-1}: $x = \dfrac{y + 3}{2}$,   or $y = 2x - 3$;   Function
13. F^{-1}: $x = 2y^2$,   or $y = \pm\sqrt{x/2}$;   Not a function   **15.** f^{-1}: $y = \sqrt{x/2}$;   Function
17.   **19.**   **21.**

23. $f^{-1}(x) = \dfrac{x + 7}{2}$; $f^{-1}(3) = 5$ **25.** $h^{-1}(x) = 3x - 3$; $h^{-1}(2) = 3$ **27.** 4   **29.** x

INDEX

INVERSE TRIGONOMETRIC FUNCTIONS (5.1, 5.2)

$y = \text{Sin}^{-1}\, x$ means $x = \sin y$

   where $-\dfrac{\pi}{2} \le y \le \dfrac{\pi}{2}$ and $-1 \le x \le 1$

$y = \text{Cos}^{-1}\, x$ means $x = \cos y$

   where $0 \le y \le \pi$ and $-1 \le x \le 1$

$y = \text{Tan}^{-1}\, x$ means $x = \tan y$

   where $-\dfrac{\pi}{2} < y < \dfrac{\pi}{2}$ and x is any real number

$y = \text{Cot}^{-1}\, x$ means $x = \cot y$

   where $0 < y < \pi$ and x is any real number

$y = \text{Sec}^{-1}\, x$ means $x = \sec y$

   where $0 \le y \le \pi,\ y \ne \dfrac{\pi}{2}$, and $x \le -1$ or $x \ge 1$

$y = \text{Csc}^{-1}\, x$ means $x = \csc y$

   where $-\dfrac{\pi}{2} \le y \le \dfrac{\pi}{2},\ y \ne 0$, and $x \le -1$ or $x \ge 1$

(Principal values for $\text{Sec}^{-1}\, x$ and $\text{Csc}^{-1}\, x$ are not universally established.)

LAW OF SINES (6.1)

$$\frac{\sin \alpha}{a} = \frac{\sin \beta}{b} = \frac{\sin \gamma}{c}$$

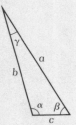

LAW OF COSINES (6.2)

$a^2 = b^2 + c^2 - 2bc \cos \alpha$
$b^2 = a^2 + c^2 - 2ac \cos \beta$
$c^2 = a^2 + b^2 - 2ab \cos \gamma$

POLAR COORDINATES (7.1)

$r^2 = x^2 + y^2$
$x = r \cos \theta$
$y = r \sin \theta$
$\tan \theta = y/x$

TRIGONOMETRIC FORM OF A COMPLEX NUMBER (7.3)

$x + iy = r[\cos(\theta + 2n\pi) + i \sin(\theta + 2n\pi)]$
$\qquad = r \operatorname{cis}(\theta + 2n\pi), \quad n \in I$

DE MOIVRE'S THEOREM (7.4)

nth Power of z

$z^n = (x + iy)^n = (r \operatorname{cis} \theta)^n = r^n \operatorname{cis} n\theta, \quad n \in N$

nth Roots of z

$r^{1/n} \operatorname{cis} \dfrac{\theta + 2k\pi}{n}, \quad k = 0, 1, \ldots, (n-1)$